The Freedom of Speech

The Freedom of Speech

Talk and Slavery in the Anglo-Caribbean World

MILES OGBORN

The University of Chicago Press
Chicago and London

The University of Chicago Press, Chicago 60637
The University of Chicago Press, Ltd., London
© 2019 by The University of Chicago
Published 2019
Printed in the United States of America

28 27 26 25 24 23 22 21 20 19   1 2 3 4 5

ISBN-13: 978-0-226-65592-5 (cloth)
ISBN-13: 978-0-226-65768-4 (paper)
ISBN-13: 978-0-226-65771-4 (e-book)
DOI: https://doi.org/10.7208/chicago/9780226657714.001.0001

Library of Congress Cataloging-in-Publication Data

Names: Ogborn, Miles, author.
Title: The freedom of speech : talk and slavery in the Anglo-Caribbean world /
    Miles Ogborn.
Description: Chicago ; London : The University of Chicago Press, 2019. |
    Includes bibliographical references and index.
Identifiers: LCCN 2019008630 | ISBN 9780226655925 (cloth : alk. paper) |
    ISBN 9780226657684 (pbk. : alk. paper) | ISBN 9780226657714 (e-book)
Subjects: LCSH: Slaves—Jamaica—Social conditions. | Slaves—Barbados—
    Social conditions. | Oral communication—Jamaica. | Oral
    communication—Barbados. | Slavery—Jamaica—History. |
    Slavery—Barbados—History.
Classification: LCC HT1096 .O34 2019 | DDC 306.3/62097292—dc23
LC record available at https://lccn.loc.gov/2019008630

To Jane, Catherine, and Eve

CONTENTS

ABBREVIATIONS

BA	Barbados Archives
BL	British Library
BL Add. MS	British Library Additional Manuscripts
BLOU	Bodleian Library, Oxford University
BLYU	Beinecke Library, Yale University
BMHS	Barbados Museum and Historical Society
HCC	*The History Civil and Commercial of the British Colonies of the West Indies*, 3rd ed., vol. 2 (London, 1801), Bryan Edwards
HJ	*The History of Jamaica* (London, 1774), [Edward Long]
JAJ	*Journals of the Assembly of Jamaica*
JPTPR	"Judicial Proceedings relative to the Trial and Punishment of Rebels, or alleged Rebels, in the Island of Jamaica, since the 1st of January 1823," in *Papers Relating to the Manumission, Government and Population of the Slaves in the West-Indies, 1822–1824*, House of Commons, *British Parliamentary Papers*, 1825 (66) xxv: 37–132
LCP	Library Company of Philadelphia
LDS	*London Debating Societies, 1776–1799*, vol. 30, ed. Donna Andrew (London: London Record Society, 1994)
LPL	Lambeth Palace Library, London
LSF	Library of the Society of Friends, London
LT	Lucas Transcripts, Barbados Public Library
LTST (1789)	*Report of the Lords of the Committee of Council Appointed for the Consideration of all Matters Relating to Trade and Foreign Plantations . . . Concerning the Present State of the Trade to Africa, and Particularly the Trade in Slaves* (London, 1789)
NLJ	National Library of Jamaica

ODNB	*Oxford Dictionary of National Biography*
SEAST	Society for Effecting the Abolition of the Slave Trade
SLNSW Banks Papers	State Library of New South Wales, Papers of Sir Joseph Banks
SPG	Society for the Propagation of the Gospel
TNA	The National Archives, Kew
USPG	United Society for the Propagation of the Gospel
WMMS	Wesleyan Methodist Missionary Society Archive, School of Oriental and African Studies, London

With One Little Blast of Their Mouths:
Speech, Humanity, and Slavery

This is a book about speech and slavery. It is about oaths, orations, orders, mutterings, rumors, incantations, debates, whispers, conversations, prayers, and proclamations. It argues for the need to hear the many forms that speech took, and the proliferation of speech among many speakers as well as their silencing, to understand the power relations of slavery in the islands of the Anglo-Caribbean world between their colonization by the English in the seventeenth century and the end of slavery there in the 1830s. It does so through the always partial evidence of what was said, where and how it was said, and what that was taken to mean. Indeed, given the nature of the archive's silences, it is remarkable what evidence of speech does remain and what might be heard in it.[1] The extraordinary power of speech in these places, its intimate relationship with the violence of slavery, and the complexities of oral exchanges are evident in two brief accounts of what was said between the enslaved and those who enslaved them in these islands. The first, from Barbados in 1683, reports an "enquiry and Examination" by the authorities into a suspected conspiracy among the enslaved to revolt. It concluded that "nothing is or can be made out against any Negroes only some insolent bold Negroes four or five for Example who were well whip[p]ed for terror to others." The bloody sight of this use of state-sanctioned physical violence to terrify the enslaved back into silence had, instead, prompted an "old Negro Man" to say "that the Negroes ere long would serve the Christians so." Hearing this statement "struck so much terror into his mistress," Madam Sharp, that the man was burned alive for "uttering" such "insolent words."[2] He suffered and died just for what he had said.

The second account is from the mid-eighteenth-century Jamaican plantation overseer Thomas Thistlewood, whose diaries of his life in the Carib-

bean from 1750 to 1786 are drawn on across the chapters that follow.[3]
He noted in December 1752 that Quashie—the driver on the Egypt plantation in western Jamaica that Thistlewood was new to managing—had
told him, in the field "before all ye Negroes," that Thistlewood would not
last long there. Thistlewood asked if Quashie meant to poison or murder
him, and "after a Pause, he reply'd neither, but he intended to invent Some
good lye, and go tell his Master, Mr. Dorrill, to get me turned away &c. &c."
Indeed, Quashie went on to say that "his Master would believe a Negroe
before a White man, and gave an Instance, off ye same at Bowens, Where
Mr Dorrill Said he Would take a Negroe's Word before Roberts tho' he had
a White face, &c. about some Peas!"[4] What prompted Thistlewood to recall
this threat, and its challenge to his word as a white man and the power
he thought that conferred, was a more violent recent confrontation where
what he said had often come to naught.

Several days before, Thistlewood had gone for an evening walk. At sun-
set, by a small morass, he encountered "Congo Sam," who had escaped
from the plantation almost a month previously. They fought—Sam with
a blunt bill for cutting canes and Thistlewood with a pimento stick—and
they spoke. As Thistlewood recorded it:

> What ye most shew'd his intention when I kept him off from me with my
> Stick Saying, you Villain runaway, away with you, &c. he answered in the
> Negro manner, I will kill you, I will kill you now, and Came upon me with
> greater Vigour. I Call'd out, Murder, and help ffor God's Sake, very loud, but
> no assistance came, so I had no prospect but to loose my Liffe, till I threw
> myself at him and fortunately seized hold of ye blade of ye Bill.

Holding either end of the weapon, they went as far as the watch-hut, but
Sam would go no further. They both appealed to others for help. Thistlewood recorded that Bella and Abigail, two enslaved women from Egypt
plantation, were there "but would not assist me," and that Sam "spoke to
them in his language and I was much afraid of them." After a tense stand-
off, Sam suddenly released the bill and threw himself into the river. Thistlewood followed, attacking Sam with the bill. Finally, trapped in Thistlewood's embrace, Sam was forced to stand waist deep in the water, while, as
his captor recorded, "5 Negro men, and 3 Women, Strangers, went over y^e
bridge and would by no means assist me, neither for Threats nor Promises:
One Saying he was Sick, the others y^t they were in a hurry." After ten minutes, London, also from Egypt, did do as Thistlewood said. However, Sam
escaped once more into the bush, having slipped the knots London had

tied with Thistlewood's handkerchief, although he was eventually captured with the help of two passing white men and tied up.[5]

This incident led Thistlewood not only to remember Quashie's threat out in the field but also to record that he had "reason to believe y^t many off ye Negroes, as Quashe, Ambo, Phibbah, &c. knew y^t Sam had an intent to Murder me, when we should meet, by what I heard them Speak one day in ye Cook room, when I was in ye back Piaza reading." He also had his doubts about London, suspecting that he "had no good intent, when in ye Bush with Sam, iff he had not heard Company Coming with me." In turn, Thistlewood reported everything to Mr. Dorrill. Sam was taken before the magistrates and sent to jail to await trial, and Abigail and Bella had to suffer the torment of a hundred lashes each, later running away to protest their treatment. Yet, when the case came to court at Savanna-la-mar, Thistlewood recorded that London "told me he would not go" and give evidence. Sam was not, it seems, convicted of attempted murder.[6]

Neither of these accounts speaks more to the truth of speech and slavery than the other, and both are from the archive of the enslavers, where the speech of the enslaved is recorded only under particular circumstances.[7] Moreover, it is not that one is ultimately about "power" and the other "resistance," since both demonstrate the pervasive use of violence to enforce slavery and the deployment of what have come to be seen as the weapons of the weak. Instead, these accounts show in various ways how speech, freedom, and bondage were intimately connected in this Anglo-Caribbean world of slaveholders and the enslaved, where, as Natalie Zacek puts it, the answer to Gayatri Spivak's question—"Can the subaltern speak?"—"is a cautious 'yes, but . . .'"[8] Speaking out by the enslaved might be punished by death, but such words were still spoken, although we cannot know how many were silenced by threats and the whip. Yet it is also clear that in different circumstances, the same words seem to have been said, heard, or acted on differently. Most significantly, both accounts show how the relationships among the enslaved, and between slaveholders and those they claimed to own, were shaped by different forms of speech, and the expectations that those different forms—either adhered to or not—held for speakers and listeners. This was true of conspiratorial conversations and judicial enquiries; orders, threats, and promises, heeded or ignored; whisperings among confidants (perhaps even those which were meant to be overheard to deliver a veiled message) or direct confrontations; and evidence given in court, or withheld in silence. Running through all these ways in which speech and silence took on specific meanings are attempts to make forms of speech align with racialized notions of freedom and slavery, so that "freedom" was a result of the policing of modes of speech. One very clear example of this is that in 1696 in Jamaica, it was made a capital crime for the enslaved to "imagine the Death of any white Person."[9] In a society of enforced illiteracy, the proof of that imagining would lie in what was said in court to have been said elsewhere. In this way, the burning alive of the "old Negro Man" established the freedom of speech, or modes of freedom defined in terms of speech, in that it violently established what free people could say, and what words the unfree could not be permitted to utter.

That, then, is the first sense of "the freedom of speech" in this book's title: the use and enforcing of ways of speaking to mark the boundaries between the free and the unfree. Forms of freedom and bondage made through speech: an interpretation that does not accord with the usual definition of the freedom of speech, since it emphasizes the disciplining of modes of speech in the making of identities, subjects, and social relations, which is ultimately, and sometimes immediately, underpinned by violence. However, as the brief accounts I have provided show, this interpretation is not sufficient on its own. Ideas of the freedom of speech also need to acknowledge speaking as a complex set of communicative practices that (almost) everyone routinely engages in as part of their everyday lives, even under such oppressive conditions as plantation slavery. Although multiple modes of silencing—structured by race, gender, status, and age—were certainly at work, it was still the case that the unutterable might be uttered, and that it would matter when those prevented from speaking spoke out (or, indeed, chose not to speak when expected to). Moreover, these utterances could be made because this communicative action did not depend on access to pen, paper, or the printing press—which, perhaps, accounts for the "terror" of Madam Sharp at hearing the old man utter those "insolent words," and Thistlewood's accounting of all those who failed to heed him, spoke against him, or questioned whether his word would have any currency at all. It is also evident in the work of the seventeenth-century clergyman Morgan Godwyn, who, questioning the legitimacy of racialized slavery, revealed in print the language of white supremacy among planters in Barbados—"*Opinions, which they are ashamed to own amongst better People*"—and argued that slaveholders could not deny the humanity of those they enslaved as a justification for slavery because they themselves could "with one little blast of their Mouths, even but a word or two, convert [them] into Men; and be at the same time the happy *Authors* of life to Souls, as well as freedom to Bodies."[10] For Godwyn, the slaveholders only needed to speak to give the enslaved their freedom, and he judged it but a "little" act. The price for the enslaved was instruction in Christianity, but that was always also shadowed by the sense that slavery was made with words and might be undone when others could speak (as we will see in chapter 4). There is, then, a freedom of speech in that talk is notoriously cheap, ephemeral, and ubiquitous, making it dangerously transgressive. This is a relationship between speech and mobility across boundaries evident in the name—"Talkee talkee"—that it was said the enslaved gave to the passes that allowed them to temporarily leave the confines of the plantation. Talk's freedom opens up another geography of slavery.[11]

The violent policing of freedom and bondage and the challenges posed by forms of everyday practice which exceeded those modes of regulation are often addressed separately within histories of Caribbean slavery. However, using speech to address them together acknowledges the importance of what Robin Blackburn presents as the central contradiction of racialized slavery, that "racist ideologies" meant that "slaves were conceived of as an inferior species, and treated as beasts of burden to be driven and inventoried like cattle," while, at the same time, "the slaves were useful to the planters precisely because they were men and women capable of understanding and executing complex orders, and of intricate co-operative techniques." As Blackburn puts it, "the most disturbing thing about slaves from the slaveholder's point of view was not cultural difference but the basic similarity between himself and his property."[12] More recently, Walter Johnson has put this idea even more succinctly, arguing that instead of dehumanizing them, "slavery depended upon the human capacities of enslaved people," deploying what Saidiya Hartman identifies as "the forms of violence and domination enabled by the recognition of [the] humanity" of the enslaved.[13]

While speech is only one way in which these characteristics of racialized slavery might be addressed, and not necessarily one prioritized by either Johnson or Hartman writing on the nineteenth-century United States, or in work on the terrors of the slave trade, it was of fundamental importance to eighteenth-century British slaveholders in the Caribbean, and to those they enslaved.[14] This was because of the way in which speech and humanity were aligned in various forms of thought and practice which slaveholders and their supporters sought to counter as the opposition to slavery and the slave trade gathered force in Europe and North America from the 1770s and, entwined with the resistance of the enslaved across the Atlantic world, sought the abolition of the trade and the emancipation of the enslaved with increasing force in the 1780s and 1790s. These connections between speech and the question of humanity within the politics of slavery are evident in the work of Edward Long: a London lawyer and periodical essayist who became—on the death of his father—first, a Jamaican planter, with land and enslaved workers on the Lucky Valley plantation in Clarendon, and then, after returning to England, the author of a three-volume history of Jamaica as an English colony and a slave society.

Race, Speech, and Slavery

In the second volume of his *History of Jamaica* (1774), Edward Long provided a lengthy description of the various peoples on the island, defined in terms of both race and legal status, from the white inhabitants to the "mulattoes," Indians, Maroons, and "Negroes." Here, following a political history of the island's constitutional freedoms and geographical descriptions of its productive and picturesque plantation landscapes and the future potential for commodity capitalism, Long elaborated his ideas on race to provide a justification for slavery. In particular, after insisting that blackness and whiteness are physiologically fixed rather than varying with climate, and having presented "some remarks upon the Negroes in general found on that part of the African continent, called Guiney, or Negro-land," where he describes them fixed in a permanent state of savagery, "almost incapable of making any progress in civility or science," he offers a theory of human difference:

> When we reflect on the nature of these men, and their dissimilarity to the rest of mankind, must we not conclude, that they are a different species of the same *genus*? Of other animals, it is well known, there are many kinds, each having its proper species subordinate thereto; and why shall we insist, that man alone, of all other animals, is undiversified in the same manner, when we find so many irresistible proofs which denote his conformity to the general system of the world?[15]

These thoughts on humans and other animals, and the nature of Nature, brought Long to "the monkey-kind" and, in particular, "the *oran-outang* species, who have some trivial resemblance to the ape-kind, but the strongest similitude to mankind, in countenance, figure, stature, organs, erect posture, actions or movements, food, temper, and manner of living." This "similitude" is then explored at length through anecdotes, examples, and arguments from an array of sources—especially natural historians and anatomists such as the Comte de Buffon, Carolus Linnaeus, and Edward Tyson, and European travelers such as Louis le Compte, Gemelli Carari, and Francois Pyrard—which all served to demonstrate that "the Indians are therefore excusable for associating him with the human race, under the appellation of oran-outang, or *wild man*, since he resembles man much more than he does the ape, or any other animal" (*HJ*, 2:358, 363). Long was also keen to provide evidence that humans and apes might interbreed.

In examining these highly charged arenas of similarity and difference, one question was of particular significance to Long: could the "oran-outang" speak? It was, he argued in a satirical vein, as yet unproven:

> How far an oran-outang might be brought to give utterance to those Euro-pean words (the signification of whose sounds, it is plain from Buffon, and others, he has the capacity to understand . . .), remains for experiment. If the trial were to be *impartially* made, he ought to pass regularly from his horn-book, through the regular steps of pupilage, to the school, and university, till the usual modes of culture are exhausted upon him. If he should be trained up in this manner from childhood . . . , to the age of 20 or 25, under fit pre-ceptors, it might then with certainty be determined, whether his tongue is incapable of articulating human language. (*HJ*, 2:370)

It would be "truly astonishing," Long thought, if the "oran-outang" could not do as well as "a natural idiot" or a "parrot." He thought it likely that these primates could talk to each other, and "that probability favours the opinion, that human organs were not given him for nothing: that this race have some language by which their meaning is communicated; whether it resembles the gabbling of turkies like that of the Hottentots, or the hissing of serpents, is of very little consequence, so long as it is intelligible among themselves" (*HJ*, 2:370).

Long's primary target was Buffon. The French natural historian had, in his *Histoire Naturelle*, compared the "Orang-outang" and the southern Af-rican "Hottentot" to establish a strict boundary between humans and ani-mals. Buffon argued, as Silvia Sebastiani puts it, that while there might be no absolute anatomical distinctions between them, "the faculties of the most savage man were separated from those of the most perfect animal by an infinite distance, articulated by thought and word." Moreover, that this distance could be accounted for by the fact "that man was endowed with a superior faculty, the soul, that allowed him to speak and think."[16] As Long described Buffon's position, "According to Mr. Buffon, he [the 'oran-outang'] has eyes, but sees not; ears has he, but hears not, he has a tongue, and the human organs of speech, but *speaks not*; he has the human brain, but does not *think*, forms no comparisons, draws no conclusions, makes no reflections, and is determined, like brute animals by a positive limited instinct" (*HJ*, 2:369). Yet Long would not accept this argument, writing,

"But how can we be sure of this Fact?" in the margin of his own copy of *The History of Jamaica*.[17] Indeed, he had already argued at length in that book that "when we compare the accounts of this race [the 'oran-outang'], so far as they appear credible, and to be relied on, we must, to form a candid judgement, be of opinion that Mr. Buffon has been rather too precipitate in some of his conclusions?" (*HJ*, 2:364). It was, Long thought, necessary to seek more evidence. His own later manuscript notes on this point, probably when preparing a second edition, show that he gathered additional material that would blur the sharp distinction between the human and the animal that Buffon aimed to inscribe. As in the published work, this was from travelers' accounts, including selections from Samuel Purchas, Herman Lopes de Castenenda, and Athanasius Kircher, and more from anatomists such as Tyson, and from Buffon himself. In addition, there were discussions of the characteristics of the animal and the human from Thomas Hobbes, Voltaire, and, especially, Jean-Jacques Rousseau, and more prosaic snippets on the linguistic abilities of a child in Lithuania raised by bears, a dog in Bristol taught "to speak certain words as articulately as a man," and Alexander Selkirk (the inspiration for Robinson Crusoe), who, alone on his island, became unable to speak whole words. A lengthy section was copied from Charles Burney's *History of Music*, presenting Cornelius de Pauw's argument "that the Orang-outang has been the prototype of all the fauns, satyrs, Pans, and *Sileni* described by the ancient poets," along with discussion of other classical authors, including Herodotus, Aristotle, and Philostratus, who "mix the negroes & these Wild men in the same class of Brutality, allowing them the Semblance of Men, without the intellectual qualifications essential to the perfect dignity & rank of the human Character."[18] Indeed, locating the "oran-outang" required work in several registers at once. As Long noted, quoting Rousseau, "Our Voyagers make Beasts under the name of Pongos, Mandrills, and oran outangs, of the very beings which the Antients exalted into Divinities under the name of Satyrs[,] Fauns and Sylvans—perhaps more exact Enquiries will show them to be men." Yet, what sort of men would they be? Long's various attempts at diagrams setting out particular sorts of humans, apes, and monkeys show that the "oran-outang" existed for him simultaneously inside and outside the category of the human.[19]

There was the danger for Long that while this plethora of material drawn from incommensurate sorts of sources could effectively blur the human–animal boundary, it would not decide the case for or against Buffon or deny speech a central role in determining the nature of humanity. On this key point Long found especially fruitful the ideas of James Burnett, Lord Monboddo, a fellow judge and follower of Rousseau.[20] The first volume of Monboddo's *Of the Origin and Progress of Language* had been published in 1773 and was a direct response to this fundamental Enlightenment question of speech, and its connection to reason, as "that which chiefly distinguishes us from the brute creation." Here Monboddo outlined a theory that tied language to human history as "civil society" gradually emerged from a state of nature. In particular, he argued that "it was impossible that language could have been invented without society, yet society, and even *civil* society, may have subsisted, perhaps for ages, before language was invented." This radical notion of the "progress" of the human from brute origins blurred the boundaries of "man" and "beast." Monboddo was prepared to countenance seventeenth-century Dutch stories of people with tails in the Nicobar Islands, arguing that "we have not yet discovered all the varieties of nature, not even in our own species," and he is best known for asserting that "Ouran Outangs" were members "of our species," but at a much lower level of development of the "arts," such that "they have not yet come to the length of language." Indeed, like wild children unable to speak, and the deaf and their education in language, the "Ouran Outang" was crucial to Monboddo. His theory depended on the possibility of nonspeaking humans and, in the "Ouran Outang"—or at least in the curious mixture of stories, pictures, live creatures, and dead specimens signified by that name—he had found "a whole nation" of them.[21]

This is what Long seized on. In an extensive footnote, he stated that "an ingenious modern author has suggested many strong reasons to prove, that the faculty of speech is not the gift of nature to man; that articulation is the work of art, or at least of a habit acquired by custom and exercise; and that mankind are truly in their natural state a *mutum pecus* [dumb cattle]." He took Monboddo's example of Peter, the wild youth of Hanover, "to prove that the want of articulation, or expressing ideas by speech, does not afford a positive indication of want of intellect" because speech "may to some organs be insurmountable." Yet while singular instances of nonspeaking humans were telling, it was the "oran-outang" that was definitive because, as Long put it, "To find a whole society of people labouring under the same impediment, would be really wonderful."[22]

Yet Long was only interested in admitting these "dumb animals" into the expanded ranks of humanity because it allowed him to question Buffon's delimitation of the human and to racialize this philosophical discourse:

We must then infer the strongest conclusion to establish our belief of a natural diversity of the human intellect . . . An oran-outang, in this case, is a human being, *quoad* his form and organs; but of an inferior species, *quoad* his intellect; has in form a much nearer resemblance to the Negroe race, than the latter bear to white men; the supposition then is well founded, that the brain, and intellectual organs, so far as they are dependent upon meer matter, though similar in texture and modification to those of other men, may in some of the Negroe race be so constituted as *not to result to the same effects;* for we cannot but allow, that the Deity might, if it was his pleasure, diversify his works in this manner, and either withhold the *superior principle* entirely, or in part only, or infuse it into the different classes and races of human creatures, in such portions, as to form the same gradual climax towards perfection in the human system, which is so evidently designed in every other. (*HJ*, 2:371)

In other words, if there could be humans who cannot (yet) speak, there could also be speakers who were not fully human. In Monboddo's history of speech, Long had found a basis for his notions of racial difference within the category of the human and with it a justification for slavery.[23]

In part, Long's intention was simply to deny that speech was the defining element of what Sylvia Wynter calls the "descriptive statement" of the human, since the connections between speech, reason, and humanity had long been used by European writers to question the cruel treatment of the enslaved, if not the legitimacy of slavery itself.[24] For example, in 1680, Morgan Godwyn had opposed those who aligned "Negro's" with "*Irrational* Creatures, such as the Ape and Drill, that do carry with them some resemblances of Men,*" and argued that "the consideration of the shape and figure of our *Negro's* Bodies, their Limbs and Members; their Voice and Countenance, in all things according with other Mens; together with their *Risibility* and *Discourse* (Man's *peculiar* Faculties) should be a sufficient Conviction" that they were human.[25] And in an anonymous pamphlet published in London in 1709 during parliamentary discussions over the Royal African Company's monopoly over the slave trade, and designed to demonstrate the trade's cruelty regardless of who ran it, the purported funeral speech of "a Black of Gardaloupe" included his questions: "What, are we not Men? Have we not the common Facultys and Passions with others? Why else has Nature given us human Shape and Speech?"[26] These arguments meant that demonstrations of verbal eloquence in print or in person were, in the seventeenth and eighteenth centuries, as much a statement of the humanity of the supposed speakers as were direct commentaries on the relationships between speech, humanity, and slavery. Indeed, late eighteenth-century abolitionists, including black Atlantic activists such as Olaudah Equiano, drew on these shared meanings to represent and perform, through speech, and against alternative representations and performances, the humanity of the enslaved and the "African" for British audiences. As we will see in chapter 5, speech became—in its form as well as its content—a crucial resource in the battle against slavery in both Europe and the Caribbean, and against proslavery writers such as Edward Long.[27]

In questioning the inseparability of humanity and speech, Long aligned himself with the Scottish Enlightenment philosopher and historian David Hume, who had notoriously argued that "negroes" were "naturally inferior to the whites," a claim he based on the case of the early eighteenth-century black Jamaican slaveholder, lawyer, poet, and mathematician Francis Williams, whose importance to the island's laws on speech—and the trouble he caused for Long's ideas of race—are considered in chapter 1. Hume stated that "in JAMAICA . . . they talk of one negroe as a man of parts and learning; but 'tis likely he is admired for very slender accomplishments, like a parrot, who speaks a few words plainly."[28] Long also sought to establish grounds of difference other than speech, arguing in his notes that "it is evident that the monkey does not belong to the human Species, not only because he wants the faculty of Speech, but above all, because his Species has not the *faculty of improving*, which is the *specific characteristic* of the human Species." Again, he argued that it remained to be seen whether this characteristic applied to the "oran-outang" and, given his discussion of Africa, the inhabitants of "Negro-land."[29]

Long's arguments also need to be understood within the immediate context of the early 1770s.[30] As a defense of Caribbean plantation slavery, *The History of Jamaica* was partly a direct response to the 1772 case of James Somerset, an enslaved man who had been brought to London from North America and who had sought the help of Granville Sharp to challenge his master's plan to transport him to Jamaica.[31] Sharp, an evangelical Anglican and government employee who played a key role in the early history of the abolition movement in Britain, had used such cases from the 1760s onward—accompanied by his own lengthy legal pamphlets—to try to establish in law that there could be no slavery in England.[32] While Lord Mansfield's judgment that Somerset could not be compelled to leave the country stopped short of abolishing slavery in England, it did, for many observers, suggest an ongoing reshaping of the geography of freedom and bondage in the British Empire. Those who saw their property rights challenged by it responded vociferously.

Like Samuel Estwick, the assistant agent for Barbados, Edward Long immediately penned an anonymous pamphlet questioning Mansfield's decision. Both works concerned themselves almost exclusively with questions of property and law, with Long arguing that the slaveholders held the right to property in people, and to move them in and out of England, via "the consent of the nation in Parliament" that gave sanction to slavery and the slave trade.[33] However, the arguments in *The History of Jamaica* were quite different and developed those presented in the second, and much expanded, edition of Estwick's *Considerations on the Negroe Cause* (1773). There Estwick raised the possibility that the enslavement of Africans was based on a "physical motive"—bodily differences that might question the "opinion *universally* received, that human nature is *universally* the same"— rather than a "political motive," the force of violence and law. Citing John Locke, Francis Hutcheson, and Hume, Estwick moved toward differences between "distinct and separate species of men" in terms of their "*moral sense* or moral powers," judging, as would Long, Africans as "altogether incapable of making any progress in civility or science" and without "any sort of plan or system of morality."[34] Since, for Long and Estwick, Lord Mansfield's judgment "that Negroes in this country are free," had turned law into opinion, "for opinion it must be, if there is no positive law to ground your opinion upon," then bodily difference seemed to offer a surer basis for the continued enslavement of Africans.[35]

Long's Enlightenment ideas of humanity and race combined notions of absolute difference and gradation, both between the human and the animal and between different sorts of humans. This is evident in the contradictory ways in which he sought not just to deny speech a central role in the definition of the human but also to rework the relationship between speech and humanity via the notion of perfectibility. As Long noted in relation to Buffon, "why may not there be a gradation of souls as well as bodies?—the chain admits of many intermediate links between imperfection & perfection."[36] Here he drew not only on Monboddo but also on Noël Antoine Pluche's *Spectacle de la nature* (in English from 1733), which provided a definition and description of the human based on the gendered body and on domination.[37] Long used Pluche's work to argue that man's body is designed to make him "*master of all*," including legs "adapted to the several exigencies of his government, but useless and denied to his *slaves, the inferior animals*"; arms and hands designed for "authority"; and a stomach and teeth, that gave him the advantage of omnivorousness. When it came to speech, it was not "the human voice, merely as a voice" that distinguished humanity—an idea from Aristotle's *Politics*—"since other animals have a voice as well as man." Instead, where "*speech* puts an immense distance between man and the animals" was, for Long, "in the variety of its inflections and universality of signification." While "Man" might express himself without speaking, via other parts of his body, "speech was superadded to all these signs, that man should not want any means of explaining himself clearly." What this meant was that "in every thing, man alone unites the prerogatives that have been granted but singly to any particular species, and his dignity arises from the *right use* to which his *reason* enables him to apply his corporeal powers and senses" (*HJ*, 2:365–66, 369). There is, then, a sense of absolute difference that emerges out of perfectibility and the combination of more specific differences.

As a result, far from simply denying the importance of speech, Long's *History of Jamaica* is full of discussions of it in practice on the island, its centrality for discerning the differences between types of people, and its importance for governing (or undermining) the relationships between them. Language appears both as a set of anxieties in making a settled and stable society built on slavery and as the means to that end. So when Long came to discuss "Creole Negroes," those born in Jamaica and whom he saw as the foundation of a stable system of slavery, he described them as "in general irascible, conceited, proud, indolent, lascivious, credulous, and very artful." This characterization was exhibited in their speech, with Long seeing them as "excellent dissemblers, and skilful flatterers," particularly when it came to "mulatto mistresses." Long describes the "language of the Creoles" as "bad English, larded with the Guiney dialect, owing to their adopting the African words, in order to make themselves understood by the imported slaves; which they find much easier than teaching these strangers to learn English." And he lampooned them for their modes of speech: "reduplications" for emphasis such as "*walky-walky, talky-talky,* [and] *washy-washy*"; "confound[ing] all the moods, tenses, cases and conjugations, without mercy"; and "catching at any hard word that the Whites happen to let fall in their hearing" and "misapply[ing] it in a strange manner" (*HJ*, 2:407, 426–27).[38] They are defined as different, and inferior, by their speech.

These linguistic tricks and inventive creolizations were, for Long, dangers that the colonial situation and its distinctive demography posed for social distinction through language, particularly for white women and children, especially girls.[39] Long lamented that "this sort of gibberish likewise infects many of the white Creoles, who learn it from their nurses in infancy, and meet with much difficulty, as they advance in years, to shake it entirely off, and express themselves with correctness." The regulation of women's speech mattered for the social, and familial, reproduction of white planter society, as evidenced by the father who, mindful of both class and race, ensured that his daughters never spoke such "drawling, dissonant gibberish" by using "all his vigilance to preserve their language and manners from this infection." He hired a tutor from England and ensured that "after her arrival, they were never suffered to converse with the Blacks." Speech mattered because it might prevent white women from becoming suitable wives. While the "ladies . . . who live in and about the towns" had sufficient "company with Europeans" to "become better qualified to fill the honourable station of a wife, and to head their table with grace and propriety," those on the plantations were "truly to be pitied." Their bodies and manners were affected, "lolling almost the whole day upon beds or settees," and "when she rouses from slumber, her speech is whining, languid, and childish. When arrived at maturer years, the consciousness of her ignorance makes her abscond from the sight or conversation of every rational creature. Her ideas are narrowed to the ordinary subjects that pass before her, the business of the plantation, the tittle-tattle of the parish; the tricks, superstitions, diversions, and profligate discourses, of black servants, equally illiterate and unpolished" (HJ, 2:427, 278–79).[40] Long's solution was "Education!"—the provision of appropriate schools, which "by making the women more desirable partners in marriage, would render the island more populous, and residence in it more eligible" (HJ, 2:280). Once again he drew on Hume's racialized and gendered notions of speech, since the ideal was the philosopher's advocacy of English heterosocial conversation, where the presence of women would reform white creole masculinity away from drinking, gambling, dueling, and, although Long does not say so directly, "mulatto mistresses."[41]

So Jamaica's historian was pleased to report that the postprandial practice of proposing indecent toasts to drive the women from the room had lost ground to "polite intercourse between the two sexes," and that "in the genteeler families, conversation between the two parties is kept up for a considerable time after dinner." While "scandal and gossiping are [still] in vogue here as well as in other countries" and "the warmth of this climate must co-operate with natural instinct to rouze the passions," Long presented white creole women as free of artifice, inclined to gaiety, and only guilty of "trivial follies," even if some of them could, in his view, be vain, proud, and overbearing. Indeed, he was pleased to note that the best among them would "endeavour, by diligent reading and observation, to enlarge her notions, banish her prejudices, and stock her intellect with such improvements, as may enable her to bear her part in a sensible conversation . . . [such that] she will entertain her company in a rational manner, and with correct language, and not expose her husband to be hooted at, for his folly in tying himself for life to a pretty idiot" (*HJ*, 2:281, 284).

Indeed, if the reformation and reproduction of a white settler society might come through Humean conversation, Long also presented forms of speech as evidence of relationships between the enslaved and their enslavers that simultaneously denied the criticisms of slavery that had emerged around the Somerset trial and promised a secure basis for racial slavery in the future—one free of the violent revolts that marked the Caribbean colonies and were dramatically described in his *History*. In response to Sharp's accusation that planters treated "their Negroes" "no better than dogs or horses," Long noted that any slaveholder "would the defamer might be present to observe with what freedom and confidence they [the enslaved] address him; not with the abject prostration of real slaves, but as their common friend and father." Such paternalism, Long insisted elsewhere, was in no need of the whip.[42] "I have known," he argued of "our Creole Blacks," "a very considerable number of them on a plantation kept in decorum for several years, with no other discipline than keen and well-timed rebukes." They could be governed, he wrote, via "encomiums on their praise-worthy conduct, and by stinging reproaches for their misdemeanours." This was an image of plantation slavery later reproduced in William Clark's early nineteenth-century ameliorationist depictions of Antigua, where the word replaced the whip, albeit still in evidence in the hand of the slave driver (figure 1). Yet this form of discipline required each overseer to "study well the temper of every Creole Black under his particular command" and to know how to govern his own temper so as to "never betray any sign of heat or passion in his admonitions." Once again, just as speech was used to differentiate people—in terms of who could say what, where, and to whom—it also necessarily connected them because slavery depended on spoken communication of many kinds. Most important, the combination of Long's attempt to dismantle the relationship between speech and humanity, and his demonstration via the practicalities of speech on plantations that it required a close attention to the "private history" of every one of the enslaved and a careful modulation of speech on the part of overseers and slaveholders, showed that these contradictory ways of speaking were a matter of concern to everyone on these islands (*HJ*, 2:411–12).

Figure 1. *Cutting the Sugar-Cane*, from William Clark, *Ten Views in the Island of Antigua* (London, 1823). Courtesy of the John Carter Brown Library at Brown University.

Speech Acts

Edward Long's concern with speech demonstrates what was at stake in relation to the spoken word in the slave societies of Barbados and Jamaica. While not all of these meanings of speech were brought to bear whenever anyone spoke, and while most speech was barked orders, the relationships between speech, humanity, and freedom were a constant and changing refrain across the period in the realms of law, politics, natural history, and religion. As with the Somerset case, this also made speech central to opposition to slavery and the slave trade, both on the islands and in Britain, where it emerged as a strong political force in the 1780s, leading eventually to the abolition of the slave trade in 1807 and the end of slavery in the 1830s. Once again, Long's concern to police the boundaries of speech in order to produce a particular relationship between speech and freedom, and his simultaneous, if inadvertent, demonstration of slavery's dependence on multiple forms of spoken communication whose "profligate discourses" would always exceed the boundaries put on them, show the two versions of "the freedom of speech" with which this book is concerned.

Indeed, while Long's is a virulent statement of the enslaver's view of the world, it demonstrates the value of putting speech, and its contradictions, at the heart of an analysis of slavery. Doing so focuses attention on speech as the asymmetrical common ground on and through which slavery worked, and where the enslaved and the slaveholders, as well as many others, all had a stake. As a result, the interpretation of the social relations of these societies can begin with a set of practices—much like Vincent Brown's concern with the meanings and ways of managing death in Jamaica—with which everyone was engaged, albeit in highly differentiated and unequal ways.[43] This starting point offers the promise of presenting an account of slavery that can avoid the tendency toward writing separate accounts of "power" or "resistance" which emphasize either the extraordinary apparatus of domination brought to bear on the enslaved population or the manifold forms of resistance that those same populations deployed. As the accounts of speech in action presented at the beginning of this chapter show, examining speech in this context does not necessarily lead to emphasizing either the imposition of power or challenges to it in direct or indirect forms, but to demonstrating the relationships between them where words are spoken, heard, and acted on. Realizing this promise, however, depends on understanding speech in particular ways that lead toward the sort of historical geography of speech practices undertaken in the rest of this book.

There is, first of all, the question of what counts as speech. In large part, this question can be treated pragmatically, and relatively unproblematically, by attending to when the archive shows that communication took oral forms, or when there was discussion of a particular form of speech—such as sermons, meetings among the enslaved, or testimony given in court. The chapters that follow present evidence that can help recover in some way the forms of speech that existed in the slave societies of Barbados and Jamaica, and how their meanings and uses shaped the relationships between people. As the discussion of Long's *History of Jamaica* shows, there was plenty of writing about speech that provides evidence of how primarily white, literate men thought it should work. Moreover, as the archival material presented at the beginning of the chapter indicates, there is also evidence that can be used to try to examine the complexities of speech in practice, by examining who said what, where, and when, and listening for the speech of women as well as men, and the enslaved as well as the enslavers. Yet, as Long's attempts to police the boundaries of speech also demonstrate, the alignment of speaking, reasoning, and humanity made the definition of what counted as speech, and of who could speak, a distinctly political act. This is obviously the case in relation to how and where the boundaries between speech and silence were drawn and violently enforced because the existence of those boundaries could not simply be assumed by the powerful. Thus, one eighteenth-century translation into English of words from "Papiamento Creol," the language spoken on the island of Curaçao, has "Ca-boca. Hold your tongue," and its range of "Curses of the Creols . . . to their negroes" then includes the chilling "Si bo no ta caya boca, mi Saca bo coraçon y mi bati na bocara" (if you don't hold your tongue, I will pull your heart out and shove it in your mouth).[44] This enforcing of silence worked out across the multiple dimensions of social difference, and feminist scholars in particular have drawn attention to what Toni Morrison called "unspeakable things unspoken" and Zora Neale Hurston "the muteness of slavery," and to the many histories and geographies of silencing and violence which have to be part of any account of slavery—and particularly of speech, gender, and slavery—and its legacies.[45]

When these silences are broken, the boundary between speech and noise is also a matter of contention. Indeed, Jacques Rancière argues that the very definition of the realm of the political is an aesthetic judgment based on the "distribution of the sensible," a contestation over what can be seen and heard. This judgment is evident, for him, in Aristotle's foundational political distinction—also drawn on by Long—between human "speech" and animal "voice" since, for Aristotle, only speech "serves to indicate what is useful and what is harmful, and so also what is just and what is unjust."[46] Thus, as Rancière puts it, "A speaking being, according to Aristotle, is a political being. If a slave understands the language of its rulers, however, he does not 'possess' it."[47] Rancière's intention is not to confirm these relationships between speech, noise, and politics, human and nonhuman animals, or the enslaved and their "rulers," but to suggest that they might be ordered very differently. This idea, with another form of slavery in mind, is also evident in Edouard Glissant's account of creole speech in his *Caribbean Discourse*. Here he argues that "for Caribbean man, the word is first and foremost sound. Noise is essential to speech. Din is discourse." In what he describes as "the unrelenting silence of the world of slavery" where "speech was forbidden, slaves camouflaged the word under the provocative intensity of the scream. No one could translate the meaning of what seemed to be nothing but a shout. It was taken to be nothing but the call of a wild animal. This is how the dispossessed man organized his speech by weaving it into the apparently meaningless texture of extreme noise."[48] While the essentialized and gendered terms of this statement need to be questioned, it does indicate an interpretation of the relationship between speech, sound, and politics which suggests that singing, muttering, groaning, yelling, or cries of pleasure and of pain cannot be simply ruled in or out of the category of speech because their inclusion or exclusion as meaningful communicative acts is always part of the politics of speech itself.[49] Bans on drumming among the enslaved, for example, have to be understood as a prohibition that renders such "talking drums" political as they were excluded from a role in organizing collectivities of the enslaved (as we shall see further in chapter 2).[50]

As a result, thinking about "speech" as a coherent and bounded category is less useful than considering the multiple forms that orality takes in practice and understanding them as more or less ordered speech practices for which particular attempts at definition and the establishment of boundaries reveal a great deal precisely because they can never be finally fixed in all that hot air. Doing so involves treating the forms of speech that feature in this book—oaths, testimony, orations, proclamations, debates, boasts, threats, sermons, catechizing, incantations, oracular prophecies, polite conversations, medical consultations, promises, declarations, and more—as social and cultural activities with their own complex, uneven, and contested histories and geographies. Speech practices need to be understood as embedded in their historical and geographical contexts and as creative responses to them. Interpreting these practices shifts the focus from exactly what was said, which can never be established from the written record, to the conditions of speaking: what Erving Goffman called "forms of talk."[51] These forms of talk were, and are, multiply differentiated, located, and embodied. Monboddo, for example, argued in 1792 that there was a distinction between *"Talking"* ("the tone of private conversation") and *"Speaking"* (in "Public"), and that as well as understanding "the management of his voice," a "public speaker" should learn "the motion and gesture of the body" to give "the look and appearance" of an orator. Along with this training in classical rhetoric, Monboddo argued that "our young student should learn to also make the distinction between *Talking* and Prating, and also between *Prating* and Pratling . . . , so rich is our language in words expressing the different tones and manners of utterance, richer than any language I know."[52] Indeed, it should be noted that, with a nod to the *Tatler*, Edward Long had, as a young London lawyer, published an essay periodical called the *Prater* in the 1750s which, through the persona of Nicholas Babble, took delight in satirizing modes of speech and behavior among London's middling sort.[53] Thus, as histories of particular speech practices in seventeenth- and eighteenth-century Britain and North America show— from literary and political coffeehouse or tea-table conversation and parliamentary or meetinghouse oratory to street-corner or milk-yard gossip and insult—the meanings of these forms of talk are socially differentiated and contested, associated with particular embodied performances, located in some spaces rather than others, shaped by the relationship between speakers and listeners (who are themselves often speakers), and the subject of much discussion in manuscript and in print over conventions that need to be established or conformed to, or have been transgressed in some way.[54] It is also evident from Diana Paton's work on the gendered violence of spoken insults in early nineteenth-century Jamaica that speech practices were shaped by the particular power relations of plantation slavery.[55]

Considering speech practices in this way emphasizes their geography: the relationship between speech and space. This relationship is most developed in terms of the locally situated nature of speech, what might be seen as a "microgeography" of the sites of speaking and listening—in courtrooms, taverns, and parlors, as well as coffeehouses, parliaments, and doorways—which provides the venues for talk and shapes the meanings of what was said. Such microgeographies of speech form a large part of what follows as speech is traced, as far as possible, across the variegated geography of the sugar islands and located in their particular spaces, in order to understand its role in underpinning or undermining the system of slavery. This geography is evident in the words spoken in hidden places or in public in the accounts that began this chapter: with words spoken in confidence, overheard through windows, or stated out in the fields in front of everyone. Indeed, the specific patterning of this geography, at least for the free, can be seen when the late eighteenth-century reforming Barbadian plantation owner Joshua Steele (of whom we will hear more in chapter 1) imagined his opponents arguing that "such visionary Notions [of changes in the rules on evidence giving] might be ornamental in the Oratory of the Pulpit; where, by Act of Parliament, no Man dares to contradict the Preacher; but in a Curing House [for boiling sugar], a Counting-House, or a Court-House, he would find the Statutes of the island were flatly against such dangerous Doctrines."[56] There was a strong sense of where such words could be said, and heard, and where they could not.

In turn, these microgeographies need to be situated within the broader historical geographies of Barbados and Jamaica, the islands on which this book focuses. Barbados (figure 2), an island of only 167 square miles, became an English colony in 1625 and passed through a rapid "sugar revolution" in the 1650s. Jamaica (figure 3), a much larger island (more than 4,200 square miles), was taken from the Spanish by the English in 1655 and became the most productive English sugar colony by 1700. Both islands share a history of imperial warfare, demographic and economic change, and slave revolt. The seventeenth- and early eighteenth-century Caribbean region was a scene of endemic plunder, piracy, and economic and imperial rivalry as the Spanish, French, Dutch, and English vied for control over colonies, shipping routes, bullion, enslaved people, and the products of their labor. After the transimperial suppression of piracy, the long eighteenth century was one of recurrent military conflict among Britain, France, and Spain, which means that Barbados and Jamaica need to be understood in relation to their near neighbors—St. Domingue and Cuba, Martinique and Trinidad—and to the continental Americas and the Atlantic world (figure 4). The war of the Spanish Succession (1701–1713), the

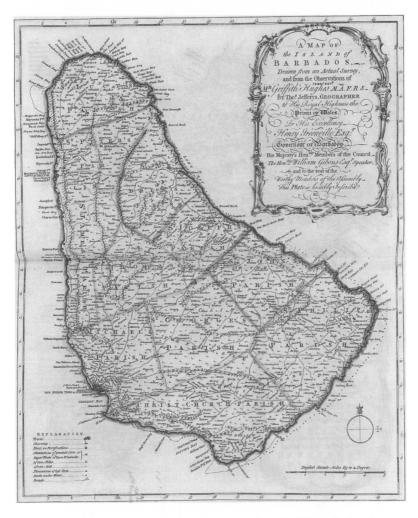

Figure 2. Barbados, 1750. Griffith Hughes, *A Map of the Island of Barbados* (1750). Courtesy of the John Carter Brown Library at Brown University.

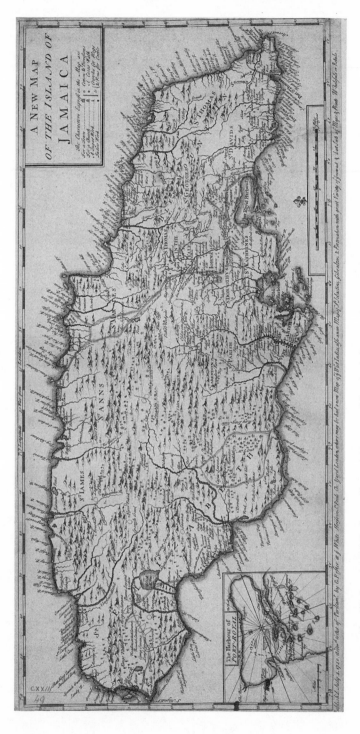

Figure 3. Jamaica, 1740. George Foster, *A New Map of the Island of Jamaica* (1740). © The British Library Board. K.Top.123.49.

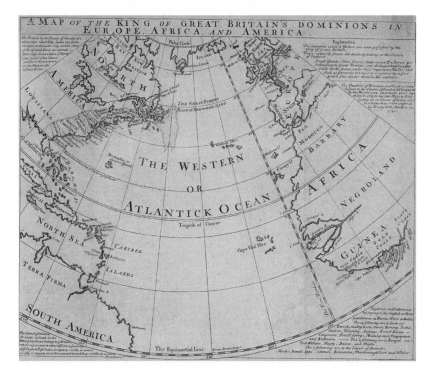

Figure 4. The British Atlantic. Emmanuel Bowen, *A Map of the King of Britain's Dominions in Europe, Africa and America* (1752). Courtesy of the John Carter Brown Library at Brown University.

war of Jenkins' Ear (1739–1748)—triggered by the punishment for smuggling of a British merchant captain by a Spanish garda costa off the coast of Florida—the Seven Years' War (1756–1763), and the American (1776–1783) and French (1792–1815) revolutionary wars were all played out in the Caribbean as well as elsewhere. Throughout the period, Barbados and Jamaica were kept in British hands, but other islands were won and lost, and all were subject to increasing imperial control, both during the wars and in relation to the taxes levied to pay for them.

What made these islands so valuable, and worth European powers fighting over them, was sugar. Barbados and Jamaica were the most profitable parts of the British Empire, and the focus of its Atlantic economy, and they were transformed as a result. Barbados was quickly deforested in the seventeenth century, divided into large plantations, particularly in the south and west of the island, and populated with enslaved African labor. In 1645, there were 5,580 enslaved Africans on the island; by 1698, there were

42,000; by 1786, there were more than 62,000 enslaved people, around 16,000 whites, and 800 free(d) people of color. Jamaica's demographic transformation was even more dramatic. Its enslaved population doubled to 150,000 between 1700 and 1750 and had reached 354,000 by 1808. The island's white population was about 25,000 by the early nineteenth century, and there was, by that time, a substantial population (about 40,000) of free(d) people of color. On both islands, however, this account of demographic growth obscures the extent to which the sugar economy was based on working enslaved men and women to death. Eight hundred thousand enslaved Africans had been sold to Jamaican planters between 1655 and 1807 (of more than five million brought to the Caribbean on the slave ships), yet the population in 1808 was less than half that.[57]

On both islands, most of the enslaved toiled on plantations, landscapes of modernity that combined large-scale agricultural land and labor with mechanized processing to produce commodities for export. The islands' economies, and their connections to other islands and into the Atlantic world, as well as their politics, were organized through increasingly substantial urban areas. In 1680, Bridgetown, Barbados, and Port Royal, Jamaica, with 2,900 people each, were the third and fourth largest towns in British America, after Boston and New York (figures 5 and 6). Kingston, established on a new grid plan when Port Royal was tipped into the sea by a massive earthquake in 1692, had grown to 11,000 people by the 1770s, matching Bridgetown, which had around 17,000 inhabitants by 1800. Both were smaller than the largest North American cities, but since many of Jamaica's political and administrative functions were served by the inland capital of Spanish Town (with 5,000 inhabitants by 1807), and much of Boston, New York, and Philadelphia's wealth was based on trade to the Caribbean, the sugar islands had a significant effect on urbanization across the American colonies. Yet these Caribbean cities were different from those in Europe and North America. They were more militarized, with prominent fortifications, dockyards, and parade grounds. They were less the site of significant residential investments for the wealthy, as sugar and slave money was poured into plantation houses, English country houses, and the urban renaissances of Bristol, London, and Liverpool. And they had distinctive demographies, with both much larger black and enslaved populations than cities in the northern colonies and Europe, and much smaller ratios of black to white, enslaved to free, and men to women than other parts of the sugar islands.[58]

Beyond the cities and the plantations, which, during the eighteenth century, were extended in periodic waves to the north and west as the sugar

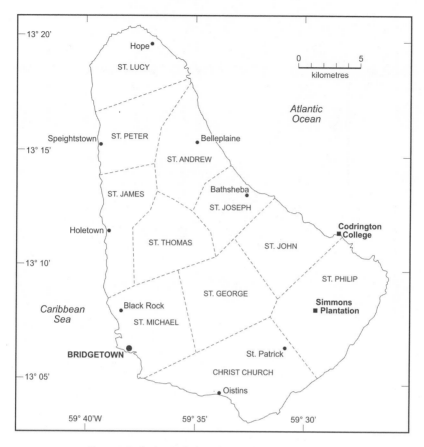

Figure 5. Barbados, including places mentioned in the text.

economy boomed, Jamaica's mountainous interior also housed communities of Maroons. These people were the descendants of slave communities who had freed themselves during the seventeenth-century conquest of the island and those who had subsequently escaped from slavery. They sought autonomy from the British through open warfare throughout the 1730s and in the 1790s, and via treaties in 1739 that accommodated them to the imperial state in return for partial sovereignty. This free black population, which formed separate "leeward" (including Trelawny Town and Accompong Town) and "windward" (including Scott's Hall and Nanny Town) communities, provided a constant reminder for all that to be black and to be enslaved were not synonymous. In addition, periodic revolts and conspiracies among the enslaved—and the threats and rumors of them that circulated within and between the Caribbean islands, especially in times

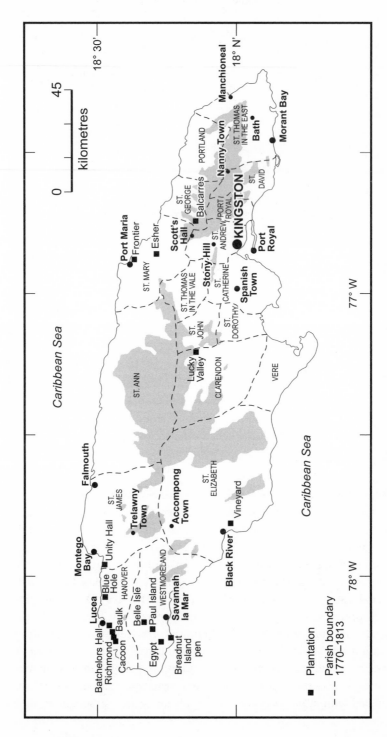

Figure 6. Jamaica, including places mentioned in the text.

of war—also offered the promise of freedom for those in bondage. There were large-scale plots to overthrow the plantation system on Barbados in the late seventeenth century, and an armed revolt broke out in 1816 as the enslaved sought to press home their desire to emancipate themselves. Extensive plots and revolts are also reported for Jamaica in the late 1600s and the 1740s, 1760s, 1770s, and 1790s (particularly in response to the successful slave revolt that started on St. Domingue in 1791 and led to the creation of Haiti as an independent state in 1804), as well as the early nineteenth century. The most significant armed insurrections were in Clarendon in 1690; the widespread revolt—first in St. Mary's and then across western Jamaica—in 1760 (often known as Tacky's Revolt) and the Baptist War of 1831–32. Although this book does not present systematic comparisons between Jamaica and Barbados, the similarities and differences between the islands do come to matter at various times in terms of the experience of slavery and freedom and the ways that shaped what was and could be said on these islands.

Trade (including that in people), warfare, and struggles for freedom meant that speech and other forms of communication were part of interisland patterns of circulation—both within and between empires—that were simultaneously knitted into a larger, Atlantic historical geography.[59] These circulations and connections have most often been understood as part of a black Atlantic diaspora made through the forced transportation of millions of African men and women across the ocean, with the creolized oral forms of the Caribbean taking their place alongside other cultural practices involving music, food, dress, dance, and belief.[60] Careful attention, and much debate, is given to the processes whereby new languages were adopted or formed, in Africa and in the Americas, through processes of adaptation, pidginization, and creolization in situations of asymmetry, and where what resulted was a complex linguistic geography as spoken forms were inventively shaped to and by the multiple demands of power-laden speech situations.[61] These processes were a matter for all those who found themselves (by force or by choice) on these islands, and for attempts—such as the story Long tells of the planter, his daughters, and their tutor, or of naming practices among the enslaved—to preserve language as it traveled as well as those speaking to change it or hearing it change.[62] This attention to the sonic Atlantic also asks important questions about the shifting and contested boundaries between speech and other sounds and embodied practices—particularly song, music, and dance—as a vital part of what Mary Caton Lingold calls "the sound-based knowledge systems cradled within the African diaspora."[63]

This book does not map a historical geography of the languages of the Caribbean, or even of differentiated linguistic usages in just Barbados and Jamaica. What follows also attends more to speaking than to either singing or silence, although each has its place. Indeed, this book's focus on forms of talk attempts to work beyond the two, quite different and somewhat polarized, ways in which speech in the Caribbean has generally been understood: as either the silence of slavery or the astonishing and inventive proliferation of creolized sonic forms. Partly as a result of the attention to forms of speech, partly because of the available evidence, but primarily as a way of interrogating the assertion that empires are oral cultures too, in what follows, very close attention is given to the modes of speech that provided the basis for claims of identity, power, and historical and geographical continuity with the metropole for the white inhabitants of Barbados and Jamaica—those the enslaved called "buckra"—and those elsewhere who supported racialized plantation slavery. Understanding their forms of talk as a carefully reproduced, but inevitably transformed, oral culture of the Atlantic world directs attention to the investment of imperial and colonial power in speech practices rather than just to the way in which empire involved a triumph of literacy over orality, and writing over speaking. It is therefore possible to talk about an imperial Atlantic geography of speech as well as a diasporic one.

Thinking this way means using the term *oral culture* in a particular way, one that does not depend on a "great divide" between orality and literacy. In various ways, anthropologists, philosophers, literary critics, and historians of both the spoken and written word have worked to dismantle strict temporal, spatial, and categorical divisions between writing and speaking.[64] As a result, oral cultures or, perhaps better, cultures of orality do not simply disappear but become part of new ecologies of communication alongside gesture, script, print, images, music, and material culture. Recovering the uses of orality means examining the archive for when it was that speech was required or chosen, at least for those who could require it or had a choice. There is not, therefore, as Jane Kamensky notes, a hierarchy of forms of communication, with printing at the top; instead, their differentiation serves particular historically and geographically situated purposes.[65] For example, speech, script, and print all played significant but distinct roles in state formation, imperial power, the politics of the public sphere, or the making of economies, and for some purposes it is speech that is necessary and valued.[66] It is also the case that once speech and writing are both involved, they become entangled in particular communicative practices as speech is written down and texts are read out loud. Indeed, it is this ecology of communicative forms that brought the geographies of speech—from the face-to-face to the oceanic—together in new ways for all the peoples of the Atlantic world. Print and script provided new means of allowing speech to carry, albeit in altered forms, across space and time, and speech provided new audiences for many forms of writing.[67]

Indeed, it is writing that provides the evidence of past speech practices on which this book is based. What follows draws on material in archives and libraries in Britain, Barbados, Jamaica, and North America, including church records, missionaries' journals, correspondence, diaries, works of natural history, codes of laws, the journals of the houses of assembly, and contemporary newspapers and pamphlets. Considering forms of talk rather than hoping to hear what was really said in the past warrants the use of this material, but it remains a set of archives and a set of textual practices that are fundamentally structured to deny voice to the enslaved, even as their words were elicited and recorded. Attending to other forms of talk, particularly those of the enslaved, which each of the following chapters tries to do, must both work against the purposes of the imperial archive to gain some limited access to the contours of suppressed and unheard modes of speech, and draw upon the writings of others in anthropology, history, and literature who have tried to trace the speech practices of the black Atlantic. In this respect, once again "the freedom of speech" signals both the multiplicity of forms of talk and the processes, which often begin or end in violence, which determined what might be heard.

Examining speech, I argue, offers a new way into the study of slavery in the Caribbean, but it might do even more. One of the most influential ways of understanding the power relations of empire, including those of slavery, has been to conceive of them as and through texts within which identities, relationships, and possibilities are inscribed in particular ways.[68] These texts are then understood as elements of "discourses" that provide the broader structures of meaning within which processes of reading and writing make sense. This approach has not gone unchallenged. First, work on the cultures of empire needs to admit the fragmented, contradictory, anxious, performative, relational, and processual nature of any imperial "discourse" or be judged "monolithic." Second, there are other welcome moves—both theoretical and substantive—away from interpretations of discursive formations to matters of practice, materiality, political economy, the emotions, and violence, where words are judged less important than feelings, actions, and things.[69] These different challenges to the notion of discourse have, however, produced a polarization between increasingly complex interpretations of textual meaning and claims by others that such work is not what should be done at all. My suggestion is that attention to speech practices, understood in the way set out in this book, offers—as talk, not text—a better guiding idea for, and empirical point of entry into, the ongoing and incomplete making of meaning within imperial and colonial settings and networks. As practice, not discourse (or, perhaps, discoursing, not "discourse"), attention to speech offers an arena in which material-ity, embodiment, affect, and violence are not somehow simply separated from, and opposed to, the universe of words and meanings but part of them from the beginning.[70] Rather than simply dethroning questions of representation in favor of notions of practice, this book argues that under-standing speech as situated, dialogic social practices grounded in mean-ingful cultural routines—such as praying, debating, or conversing—always needs to be about both representation and practice.[71] Thus, speaking is si-multaneously a practice and a mode of representation in the moment that it happens, and speech is also both practice and representation in terms of how it works within ecologies of communication that involve texts (in script and print), images, and objects. In these ways, this book aims to ad-dress the specific context of Caribbean slavery within the British empire and much wider issues of representation, communication, and power that are debated across the humanities and social sciences.

Speech has already been recruited for multiplicity against the determin-ing structures of language by Michel de Certeau. In his studies of *The Prac-tice of Everyday Life*, he took speech (rather than cooking or walking) as the model of the notion of practice itself. In particular, it was speech that demonstrated how "everyday life invents itself by poaching in countless ways on the property of others." For de Certeau, this perspective offered a challenge to Michel Foucault's theorization of "discourse," because "the act of speaking (with all the enunciative strategies that implies) is not re-ducible to a knowledge of the language. By adopting the point of view of enunciation . . . we privilege the act of speaking; according to that point of view, speaking operates within the field of a linguistic system; it effects an appropriation, or reappropriation of language by its speakers."[72] There is here an idea of both the rules of language and the "countless" ways, both conscious and unconscious, in which enunciations (re)appropriate them. It is also clear that the complexities rapidly multiply where there are many languages and many more ways of speaking them. Yet, how might this notion of enunciation be understood without arguing for the necessarily resistant properties of speech or its categorical differentiation from other modes of communication? How can forms of order or regularity in speech be identified while still recognizing the potential multiplicity of modes of speech and of enunciations within them?

It is important to note that Goffman's elaboration of "forms of talk" is not simply the first set of entries in an endless catalog of ways of speaking. In common with his other work, it is part of an analysis of how face-to-face interaction works in practice to constitute social relations. Significantly, he argues that in such interactions the meaning of what is said depends on context, but rather than that being whatever is needed to explain deviations between what is said and what is meant, it should be understood through J. L. Austin's idea of the "speech act." This concept, famously, is an analysis of "how to do things with words." It considers speech as a form of action in and on the world where words have "illocutionary force": they do something—give a verdict, enact a marriage, bind a promise—as they are said.[73] Yet, crucially, for words to act with effect, it depends on the context of the utterance, what are called its "felicity conditions"—who speaks, where, when, and how—rather than simply the words that are said. Therefore, as Judith Butler argues, speech, considered as a form of action, becomes embodied and relational, its meanings and effects open-ended and beyond the control of the speaker since utterances have historical geographies: they both carry with them long histories of previous speech acts and are performed in social spaces that shape their meanings.[74] Yet, both Austin and Goffman, even with their insistence on the importance of context, offer little in the way of a history, geography, or anthropology of speech acts which might be of use in situating talk somewhere like the long eighteenth-century Caribbean.

In contrast, but without simply multiplying concrete historical and geographical contexts, Bruno Latour has also used speech act theory within what he calls an "Anthropology of the Moderns" that aims to discern the differences between various "Modes of Existence." His is an extended investigation of the idea that the forms of "veridiction"—the ways truth is produced—characteristic of science, which Latour has done so much to unravel, are not the only ones. As part of this investigation, he is keen to demonstrate that there are other modes, working with other forms of truth making, that produce what we call law, religion, and politics. Throughout his explication of these forms, their relationships, and the problems that arise from confusing them, they are understood in terms of speech: legal speech, political speech, religious speech. And, drawing on Austin, Latour argues that the notions of felicity and infelicity conditions—what it takes to make a speech act work—"make it possible to contrast very different types of veridiction without reducing them to a single model."[75] It is clear that, despite the terminology, it is not only the physical act of speaking that matters here. Latour is as concerned as ever with inscriptions, images, objects, texts, and files.[76] In part, this is a philosophical move toward the heart of things, to the simplicity of the spoken utterance. Yet, understanding the anthropology of "the Moderns" in terms of scientific, legal, political, and religious speech acts does attend to all those things that must be assembled around these utterances to make them different from each other and effective in themselves. In its stripped-down philosophical manner, it is a reiteration of the need to understand communicative action in terms of context, embodiment, performance, and materiality.

For Latour, these forms of veridiction—understood as networks or assemblages—produce the effect of the different domains (law, science, religion, and politics) which "the Moderns" think about as separate geographical "territories," even while they are not and cannot be kept separate.[77] This conceptual geography of speech acts provides the basis for the overall division of modes of speech in the first four chapters of this book, bringing them together at the end in the discussion of abolition and emancipation. It is notable that these are also the key institutions of Western literate professional masculinity since the late seventeenth century—the law, the church, parliament, and the academy. It is not Latour's aim to show how they have developed historically, or where they have been either conjoined or kept apart in the past, and he is certainly insufficiently attentive to the ways in which these speech acts and forms of veridiction are both made by and make differences between men and women.[78] However, he does want to emphasize the particularity and peculiarity of these domains, deploying the imagined figure of an ethnographer who is constantly bemused by these curious people who, in saying one thing while doing another, can never have been as "Modern" as they think they are.[79] While this means being attentive, in a general philosophical sense, to the implications for those who wish to draw strong contrasts between what Latour calls, at times, "the Whites" ("Western, European, Modern") and "the 'Others,'" particularly pointing up where they are not so different after all, there is also a need to recognize that those conscripted into plantation slavery in the black Atlantic were crucial actors in the making of "modernity" too.[80]

So, while the chapters of this book attend to the different forms of speech that operated within, and produced, for "the Whites," the law, politics, belief, and the knowledge of nature, and shaped the relationships between the free and the enslaved within them, there is also a repeated insistence that such separations are culturally and historically specific and therefore unstable. In particular, for Atlantic Africans, there was little sense that law, politics, religion, and natural knowledge should be understood as organized into separate, if intimately related, "territories" or "domains" of speech, veridiction, practice, and institutional action. In particular, across the chapters that follow, speech practices associated with *obeah*—the name given in the Anglophone Caribbean to practices which, like vodun, Santería, or candomblé, engaged spirit powers with action in the world—recur as ways in which speech (as binding oaths, prescriptions, divinations, incantations, and proclamations) actively combined what others thought of as law, medicine, politics, and religion, to create something apart from plantation slavery and, at times, something that challenged it.[81]

Considering obeah raises once again the fundamental question of who can speak. It does so because speaking in this way (or what was judged to be that way by white lawyers, politicians, doctors, and priests) could, after the mid-eighteenth century, be punishable by death. While denying that the words of the obeahmen or obeahwomen had the spiritual power others attributed to them, they still needed to be silenced.[82] As important, however, obeah—and the ontology it and other Atlantic African and Amerindian cultural practices suggest—also admits many more beings into the realms of speech.[83] It is, of course, impossible to set out these notions in full. They were complex, changing, locally differentiated, and continually (re)produced ideas and practices that come to us through the imperial archive, memory, and the interpretation of current practice. Obeah is a name that denotes an uncertain and shifting object, forged under and through oppression.[84] However, it is clear that here the natural world and the world of spirits were not separate. While the gods themselves might be distant from human concerns, localized or particular spirits, and certainly the spirits of the ancestors—from the immediate and the more distant past—inhabited the landscape, animals, trees, and rivers, and were part of the lives of the living, for good or ill. Obeah promised to engage those spirits for human ends and to speak with them. While the historical evidence is inevitably meager, it does accord with anthropological fieldwork to indicate forms of divination, possession, and ritual work that involved speaking to and for the spirits in their own language. As Kenneth Bilby has argued for the Jamaican Maroons, their "deep language" of spirit possession—the language of the ancestors which holds many secrets—is intimately related to both the Suriname creole that would have been spoken by enslaved Africans transported to the island in the late seventeenth century and to the Akan languages of the Gold Coast.[85] And, while they are folklore rather than myth, the Anansi stories—of the trickster spider—

which have traveled from West Africa to the Caribbean, also produce a world, through the repeated and ritualized work of memory, song, and storytelling, where animals speak like humans and mediate between the spirit and natural worlds.[86] This set of relationships and powers could explain why some people might want to talk with or sound like animals and why what was most valued about Anansi was the way he, she, or it uses the play of language to move between and trick the powerful: gods, spirits, animals, and humans alike.[87]

This book argues, therefore, that who can speak and what they might say is a crucial question for the history of slavery in the Caribbean, and it explores in detail the various ways that speech and silence worked, and what they meant, in these islands and within slavery's violent struggle over humanity and freedom. So, when Edward Long came in his *History of Jamaica* to the question of whether "Africans" were cannibals, he drew on his reading to argue that since they ate all sorts of things, even the body parts of apes, whom "they esteem . . . as scarcely their inferiors in humanity," it was very likely that they were. Pressing the case, and buoyed up by the seeming agreement between his own vision of the closeness of the links in the great chain of being and that held by those "Africans," he noted their supposition that apes could talk. Yet what he reported as their reasoning sets out the contradiction at the heart of racialized slavery, and of this book, and overturned the terms of his own racialized hierarchy. As he put it, apes, "Africans" suppose, "are very able to talk, but so cunning withal, that, to avoid working, they dissemble their talent, and pretend to be dumb" (*HJ*, 2:382).[88] These human and nonhuman animals certainly understood what "the freedom of speech" meant in a world of enslavement.

ONE

On Our Bare Word: Oath Taking, Evidence Giving, and the Law

In 1789, writing under a pseudonym, the Barbadian slaveholder Joshua Steele put into the mouth of a fictional clergyman a heretical argument about speech and the law in the sugar islands. The clergyman argued for "swear[ing] Negroes, as *legal Evidences?*"

> By any Thing, of which they appear to have a solemn and awful Idea: *By the Sun! By the Moon! By the God, who made and governs the Universe!* (For I never yet spoke with a Negro, who had not something like an innate Idea of such a Supreme Being) or by *Grave Dirt,* if you will; as it is certain, the Solemnity of that Oath, appears to be connected with *their Ideas of the Servivance of departed Souls,* and *of future Rewards and Punishments under the Decrees of Divine Power.*[1]

In doing so, he emphasized that oaths had to be sworn to secure evidence in court—one form of speech guaranteeing the truth of another—and that, within the legal systems of Britain's Caribbean slave societies, the question of who could swear such oaths and give such evidence went right to the heart of definitions of the law, identity, and humanity. Who on these islands could give their word, and how could they do it? What would it mean to have enslaved Africans swearing on the earth taken from graves in the courtrooms of Jamaica and Barbados? Would it mean that they would then tell the truth, even about those who held them in bondage?

This chapter examines speech practices and the law in these slave societies, and, more particularly, oath taking and evidence giving as forms of talk that underpinned systems of law and violence. Such an examination involves some of the basic ground of speech act theory, considering statements, like oaths, that require rigorous felicity conditions. It is only with the right words said in the right way by and to the right people in the right place, and with whatever other necessary gestures, texts, and objects, that they will have effects as actions. In this case, acting as a guarantee that the words spoken are to be taken as truth, with whatever implications that has for liars. This relationship between speech and truth, as Latour argues for practices of "speaking *legally*," is based on a mode of veridiction peculiar to law as a "regime of enunciation." As he notes, it is both clear and frustrating that law is a separate "mode of existence" within which strong claims to truth can be made that are, at the same time, only narrowly applicable. Law has "on the one hand astonishing force, objectivity; on the other remarkable weakness." Where things are done legally, according to all the correct procedures, the full weight of the law can be brought to bear, even though what is "legal" may not be the same as what is "just," "effective," or "true" in other senses. However, any deviation from the correct "forms"—the necessary legal wording, rituals, and procedures—and all of that strength disappears. Unless the legal conditions are satisfied, the words have no effect. What this means that "law" actually is, he argues, can only be understood by examination of what legal enunciations do as they seek to "find out whether some fact corresponds, or not, to a definition [or principle] that will allow a judgment to be made." He concludes that as a mode of speech law's fundamental purpose and effect is to make utterances attributable to speaking subjects: "[It is] a regime that makes it possible to connect enunciators and utterances by invisible threads. . . . With law, characters become assigned to their acts and to their goods. They find themselves responsible, guilty, owners, authors, insured, protected. And this authorizes us to say that 'without law,' utterances would be quite simply *unattributable*." Significantly, despite the many forms of writing that characterize the law in its modern modes, it is speaking that is crucial. As Latour puts it, "even among peoples said to be 'without writing,' the anthropology of law attests to hundreds of astonishing procedures for attaching promises to their authors by solemn oaths and imposing rituals. On this point, writing has only accentuated the habit of already well-established links."[2] Yet this sense of legal speech's fundamental character also needs to be given a history and geography that questions how law's speakers, utterances, and truths are made in particular circumstances. What happens in situations where subjecthood and the possibility of enunciation are denied to some people based on race, gender, or status? What happens when there are different ways of attaching utterances to their authors? What then happens to the processes of attribution and the question of legal truth? What happens when law has an Atlantic, imperial, or diasporic geography?

I want to pursue these questions by considering the forms of talk at work within the law in the Anglo-Caribbean world—particularly the taking of oaths and the giving of evidence—and to examine the ways in which they were instantiated and contested within Barbados and Jamaica. I will first consider the Atlantic geographies of law and legal speech and then attend more closely to the taking of oaths and the giving of evidence to show how they simultaneously worked to make up and unmake the radically unequal social relations of slavery.

Speech, Law, and Space

In early modern England, the law was a domain within which the spoken word voiced in particular ways by specific people in defined spaces was of fundamental importance. Oath-bound judges and juries were at the heart of the English legal system.[3] They were part of formal procedures in which some texts were only effective if they were read out loud.[4] In criminal trials, male jurors heard testimony on oath from suspects and witnesses who were increasingly bound by legislation governing subpoena procedures to appear before the court and for whom perjury was a statutory crime.[5] This was a very localized historical geography of speech. Yet the law was also an arena of practice that structured the social and political relationships of the British Atlantic world. English "liberties" were the foundation of both empire and identity, and they were guaranteed by law. In particular, liberty at home and overseas "rested on two institutions for determining and making law: juries and Parliament."[6] Both of these institutions, examined in this chapter and the next, depended on speech practices—hearing evidence and engaging in debate—to preserve the legal rights of "freeborn Englishmen" wherever they were. Yet these liberties and speech practices were exclusionary ones: there were unequal rights and rights to speak for men and women, and both were largely denied to indigenous and enslaved peoples.[7]

There was, for the English, a particular geography of law and empire. Eliga Gould argues that while the conventions of the European law of nations (governing questions of war and peace) and the more specific provisions of the English common law (governing colonial societies) were understood as overarching frameworks for Atlantic governance, there was the recognition of significant differences between "zones of law" (in Europe and the settler colonies) and "zones of violence" (in the terrestrial and oceanic imperial peripheries).[8] This geographical division made possible significant legal differences between places: the continuation of warfare in the Caribbean among nations at peace in Europe, or vice versa; the legal foundation of chattel slavery in American plantation societies but not in Britain; and the coexistence of multiple legal jurisdictions—including Maroon communities—which were accepted by English traders, settlers, and imperial administrators.[9] Indeed, Lauren Benton has argued that relationships of sovereignty and space in Atlantic empires were so complex as to exceed this simple division of "law" and "violence" through a shifting and uneven cartography of corridors, enclaves, and anomalous zones, presenting a geography more akin to a tangle of strings or a fabric full of holes than a map of the territory.[10]

Establishing rules, rights, powers, and obligations within this differentiated legal geography mattered to those on either side of the Atlantic who took up the task of colonial governance. Barbados and Jamaica, like the other American settler colonies, insisted on establishing representative assemblies with lawmaking powers. Planters were keen to establish their property rights in English law (including those over other people) and, at the same time, insisted on their right to make laws, producing tensions with metropolitan authorities determined to ensure that local autonomy did not go too far or introduce contradictions between the laws of different places (as we will see in chapter 2).[11] The complexities can be seen in the manuscript notebook of an unknown late eighteenth-century student of property law and lawmaking who was trying to understand the constitutional differences between Barbados and Jamaica, which laws governed these island colonies, and the extent to which they were the same laws as in England. His reading of John Ogilby's *America* (1671), John Oldmixon's *British Empire in America* (1708), John Davies's translation of Charles de Rochefort's *History of the Caribby Islands* (1666), and Montesquieu's *De l'esprit des lois* (1749), as well as collections of the laws of Barbados, brought to light the principles underpinning arguments for both colonial difference and metropolitan centralization that Mary Jane Bilder has called the "transatlantic constitution."[12] He first noted, from Sir Matthew Hale's *History of the Common Law* (1713), the principle of *lex loci* (law of the place), that "every Country has circumstances peculiar to itself with respect to its Soil, situation, Inhabitants and Commerce, to all which convenient Laws ought to be adapted, as 'tis from thence that particular Customs & Usages do arise." Laws had to be suited to the place being governed, and therefore colonial law could differ from English law. However, he also noted the maxim that colonial laws should not be "repugnant" to the laws of England or invasive of the royal prerogative. This was not a blanket proscription of difference. For example, Hale argued that colonial laws "are not at once to be overthrown merely because they happen to be various from those of England."[13] In addition, avoiding significant differences was not just a matter of the imperial center's restriction of the autonomy of the colonies. Colonial property owners depended on English law to securely hold and transmit their wealth, and they wanted it applied. For example, in May 1711, when the council of Jamaica thought that the law passed on the island under Henry Morgan "for putting the laws of England in force here" had been repealed in England without anyone informing them, they invoked their right as the freeborn subjects of England to be governed by "the known-laws of England" and suggested informing Queen Anne herself that "great confusion must be introduced in the said island, even to the ruin and destruction thereof," if English law was not enforced there.[14]

So, while difference was acceptable, some differences—those judged "repugnant" to English law or transgressing the rights of freeborn English subjects in the Caribbean—were not. Thus, the "transatlantic constitution" was an ongoing debate between the colonies and the imperial metropolis over legal geography that tried to establish the basis for colonial law in general and colonial laws in particular. It also meant that things worked out differently in different places. The note-taking law student was most concerned with Barbados. However, matters were even less clear when it came to Jamaica. Oliver Cromwell's forces had taken the island from the Spanish in 1655. However, rather than being "the possession of an un-inhabited land," Jamaica's circumstances meant that there could not be a simple application of English law:

> That it was impossible the Laws of this Nation, by mere Conquest, without more, should take Place in a conquered Country; because, for a Time, there must want Officers, without which our Laws can have no force: That if our Law did take place, Yet they in Jamaica having Power to make new Laws, our general Laws may be altered by theirs in particular; Also they held, That in the Case of an Infidel Country, their Laws by conquest do not intirely cease, but only such as are against the Laws of God; and that in such Cases where the Laws are rejected or silent; the conquered Country shall be governed according to the Rule of natural Equity.[15]

Most important, the Jamaican assembly established its right to make laws—although they had to be ratified by the English monarch—on the basis that "though it was a Conquered Country, yet they are not a Conquered People, But descended from the Conquerors, who were Englishmen."[16] Indeed, as Englishmen (albeit transplanted ones), they argued that they should "have a deliberative power in the making of laws."[17] As chapter 2 will show, lawmaking through deliberation and debate among white men in the assembly was central to both the practicalities of colonial legislative practice and to the forms of identity that bound the English empire of liberty in the Atlantic world together and differentiated it from others.[18]

In Barbados and Jamaica, questions of the repugnancy of colonial laws were particularly acute when it came to slavery. Planters wanted to ensure that their legal title over property in people was secure, even when it was not guaranteed, and many argued was actively denied, by English law.[19] Through the assemblies' development of comprehensive slave acts in the late seventeenth century, a significant differentiation of legal subjectivities was set into law, with very different rights being accorded to groups that came to be defined as "Negro" and "white" (rather than Christian); and slave, servant, and free.[20] On Barbados, the 1688 *Act for the Governing of Negroes* made the case for legal differentiation between people and between the Caribbean colonies and the metropolis, stating that "the said Negroes and other Slaves brought unto the People of this Island . . . are of barbarous, wild and savage Natures, and such as renders them wholly unqualified, to be governed by the Laws, Customs and Practices of our Nations." Assemblies judged they had to make other laws to "restrain the Disorders, Rapines and Inhumanities to which they are naturally prone and inclined," to encourage good behavior, and to offer the enslaved protection "from the Cruelties and Insolencies of themselves or other ill-tempered People or Owners."[21] In turn, Jamaican legislators adapted their slave laws from Barbados and were careful to argue that these laws were both necessary for their specific circumstances and compatible with English law, although "there being in all the body of that Law noe perfect track to guide us where to walk Nor any expert rule sett us how to governe such Slaves as wee have."[22]

While these laws, legal subjectivities, and judicial practices governed the islands' courts and discussions over law between metropole and colony, they were not the only forms of law on these islands. Steele's heresy and the law student's notion that conquest did not extinguish all other laws at least signals the existence of a qualified legal pluralism that included the Maroon communities in the mountains of the Jamaican interior. It also opens the possibility of understanding other practices with the force of law—binding subjects and their utterances—if not other legal systems, among enslaved Africans in the Caribbean whose personal and ancestral connections to societies with strong legal traditions may have shaped the way they understood and participated in "speaking *legally*."

To take just one example—the law of the Asante, one of the centralizing and expansionist states of eighteenth-century Akan West Africa, from which many men and women were transported to the Caribbean by slave traders—it is possible to show both similarities to the English conceptions of law that those of Asante origins would have encountered in the Caribbean and some of the significant differences. At the risk of essentializing what would have been a long historical process of legal development, the rule of law (*mmra*) was understood as absolutely fundamental to Asante state and society, with law available to all via a "hierarchical network of 'courts' extending from the hearths of the extended families, through the councils of lineage elders, to the court of the *Asantehene*," or ruler.[23] There was a clear injunction to use the due process of the law to seek redress for personal injury, because not to do so was seen as reducing people to the status of "brute beasts" and threatened the security of the state. These courts operated according to rigorous rules of evidence that assessed the credibility of witnesses and the coherence of their testimony and ensured that those sitting in judgment would give an impartial and unbiased hearing to all, making judgments based on rules of justice and legal equity. They were to hold to account the rulers as well as the ruled.

Speech was central in Asante law, and it followed particular forms of words without which proceedings were invalidated.[24] To go to law was to refer the case to someone appropriate to hear it "with good ears," and the obligation to do so was matched by "'the right to speak' or to be heard." So strong was this right that the Akan proverb *Woamma manka masem, busufo a, Fanti* equated being silenced with being killed: "He [*sic*] who prevents me from speaking or telling my story is a murderer [and] carries the curse of evil death."[25] The person assigned to guarantee these rights and obligations was the *Okyeame*, a speech specialist who mediated all verbal communication with Asante rulers and gave judgment on all cases brought before them. The *Okyeame*—usually a man, but holding an office that oral tradition recounted was first held by a woman in the early seventeenth century—was bound by a solemn oath to speak the truth, bore a ceremonial staff of office, and needed to be "a walking storehouse of [the] proverbs" that encoded Asante custom and law.[26] Their regulation of speech guaranteed the rule of law and the continuation of the polity, and it established the boundary between humans and beasts.

Although it is difficult to identify their traces in the archival record of colonial societies, such legal practices, and the ways they were shaped by status and gender, would have informed enslaved Africans' interpretations of English law in the Caribbean. There were, of course, significant differences, at least by the eighteenth century. The *Okyeame*, unlike English judges or lawyers, were held directly accountable for cases overturned on appeal. Also, Asante courts held both plaintiff and defendant liable for judgment and depended on a wider range of witnesses and forms of speech than did their English counterparts, including suitably tested "supernatural testimony obtained through oracular consultations [and] spirit medium-ship." Yet there were also many ways in which these quite different conceptions of law might at least overlap. The fact that, in Asante law, dreams that disclosed bad "things in the mind" were, when told to anyone else, considered as reprehensible and open to prosecution as actions themselves, might have made comprehensible the Jamaican law punishing with execution slaves who imagined the death of any white person.[27] But the enshrining in law of the enslaved as commodities, the racialization of slavery through law, and the prohibition of slave evidence marked a clear difference from this and other West African forms of law.[28]

There was, therefore, a complex geography of law, which means that legal practices and the forms of speech on which they depended need to be understood at separate, albeit connected, scales from testimony given in court to the Atlantic's differentiated legal landscapes, both imperial and diasporic. The significance of law to the governance of colonies, to the identity of colonists, and to the lives of those they sought to govern meant that legal practice was a contested field, both locally and across the ocean. In what follows I want to demonstrate the processes whereby a differentiated set of legal subjects were made within this legal geography through the spoken word, examining how oath taking and evidence giving became an important ground in the formation of those who were subject to the law in different and unequal ways. What becomes apparent is that in this process of the constant and ongoing production and reproduction through the regulation and use of speech of these unequal, and often silenced, legal subjectivities, gaps and contradictions were also opened up. The claims of race, nation, liberty, religion, gender, and property intersected in ways that certainly drew strong distinctions among the sorts of subjects whose utterances might be tied to them. However, when these questions were part of a transatlantic discussion over colonial rule, and over slavery and freedom, these multiple axes also complicated the ways in which the distinctions among people could be drawn, and might even undermine them.

Taking Oaths

The official practices of government and law in seventeenth- and eighteenth-century Barbados and Jamaica were produced through and reproduced the social relations of race, gender, property, and liberty. This process was enacted through the spoken word, but it was also dependent on it as more than just words. For example, members of the islands' lawmaking assemblies were white, Protestant, male, free, and propertied. But although these were necessary qualifications, they were not sufficient ones. Assemblymen needed to swear a series of oaths before they could take office. After the restoration of King Charles II to the English throne, they had to swear the oaths of allegiance and supremacy required of every officeholder in England and the colonies, affirming that the monarch was the head of both church and state. The Barbadian legislation prescribed that this was to take place in person "in publick and open Court," and that those taking the oath must declare "that there is not any Transubstantiation in the Sacrament of the Lord's Supper, or in the Elements of Bread and Wine."[29] These oaths also had to be sworn by any "stranger" applying for a "Pattent for Naturalization" in Jamaica whereby they might "enjoy the immunities and Privileges of this Island in as ample manner as any natural born Subject."[30] In addition, Jamaican assemblymen had to swear an oath that they had sufficient property to qualify for office, and the governor and council of Barbados had to swear one before they could judge property cases in the courts of error and equity.[31] On the basis of contemporary ideas about who could effectively give their word—free, white gentlemen—such oaths bound together the social relations of a slave society and the public roles within its state apparatus to create a set of legal subjectivities which were at once materially embodied and linguistically constructed.[32] These oaths were of sufficient significance that one had to be instituted in 1692 to quell the "Doubts and Scruples . . . touching the authority of the [Barbados] Council to sit and judge" in matters of equity, and the oath guaranteeing that Jamaican assemblymen had sufficient property needed to be rewritten in 1760 to address the "many Doubts" that had arisen about its wording.[33]

These oaths were, therefore, part of the considerable armory of oath taking that underpinned government and law in the Caribbean and, along with colonial legislation, constituted specific sorts of legal subjects on the islands. This array of oaths provided a continuity of practice with the English legal and governmental apparatus from which the Caribbean versions derived some of their legitimacy and some guarantees against corruption and mismanagement, or at least a sense of accountability.[34] Oaths were of undoubted importance. On the occasion of the arrival of a new governor to the island in 1732, a contributor to the *Barbados Gazette* pointedly argued that oaths were used in "all civilis'd Countries and Nations" to "more firmly bind the Takers to declare the Truth, or to perform the Thing thereby promised to be done, for the Fear of offending invisible Powers, or the Dread of subjecting themselves to the Penalty which the Laws of the Society annexed to the Crime of swearing falsely." He used his readings of Sir Edward Coke, Samuel Puffendorf, Cicero, and Niccolò Machiavelli to remind his fellow islanders, and those who governed them, of "the binding Nature of an *Oath.*"[35]

A subsequent contributor to the same newspaper reinforced the point by informing readers that the British king's coronation oath obliged him to provide officers for the execution of justice who then spoke in the king's name the words of Magna Carta: "WE WILL NOT SELL JUSTICE, WE WILL NOT DENY JUSTICE, WE WILL NOT DELAY JUSTICE TO ANY ONE." He hoped that such oaths, sworn by magistrates along with the oaths of supremacy and allegiance, might be "set up in Capital Letters" in all courtrooms, keeping to their duty Judges "in a distant Colony" who could not be expected to "have the same great Talents and Endowments which are generally seen in those of our Mother-Country." Oath taking was also fundamental to the jury trial, "a very ancient Institution, [that] . . . has been continued down thro' a Succession of Ages as one of the greatest Privileges of *Englishmen.*"[36] Both judges and jurymen (women were excluded) had to have their oaths administered by the correct authorities in the proper wording in open court. Witnesses were also to give their evidence on oath.[37] Such oaths were to be sworn on the Bible to bring internal moral force, as well as external legal pressure, to bear. The author quoted Archbishop Tillotson that perjury is "Treason against human Society, subverting at once the Foundations of public Peace and Justice, and the private Security of every Man's Life and Fortune."[38]

Oaths were also central to the legal and governmental regulations that underpinned the doing of business on and with the islands. They might be used to bridge the Atlantic. In 1651, the Barbados assembly repealed legislation relating to royalist oaths and enacted that anyone "that expects benefits from the Courts of Justice" of Barbados, and anyone in public office, was to take the "Engagement," an oath accepting the sovereignty of Oliver Cromwell's Commonwealth.[39] Francis Hanson, in the preface to the first edition of *The Laws of Jamaica* in 1683, recommended that letters of attorney should be signed by those who traveled with them to the island so that they could "prove the same *viva voce*, which being recorded there, is good though the witnesses die or return."[40] Barbadian courts were to accept as sufficient in law documents "proved on Oath" under the seal of the Lord Mayor of London, or of any other city or town-corporate.[41] Persons wishing to leave either island had to "bring one or more of good Credit to make affadavit before a Justice of the Peace, that he hath known him or her to go by that Name, for one Year, or so long as he or she hath lived in the Island." Their name could then be put up in the secretary's office for twenty-one days before departure. Any slaves leaving the islands had to have a ticket signed by the governor after an oath of ownership had been made by a slaveholder before a justice of the peace (JP).[42] Elsewhere, those responsible for the activities and practices surrounding property ownership did so on oath. Each clerk in the Barbadian courts had to take an oath before a judge "for his true and honest dealing."[43] Jamaican land survey-ors were "duly sworn to do Justice in that behalf," and any encroachments were to be determined by a JP and two freeholders on oath.[44]

These were the words that the white, propertied, free, and predominantly male members of this slave society said, in public, among themselves. Some people found a partial inclusion. Quakers swore a different oath to secure their property, but this did not qualify them to give evidence in criminal trials, serve on juries, hold public office, or open schools.[45] Others were cat-egorically excluded. White women could own property but could not hold public office. In 1708, the Jamaican assembly debated, but did not pass, a motion that "no Jew, mulatto, negro, or Indian, shall have any vote at any election of members to serve in any assembly of this island."[46] However, a 1711 law excluded them from being "capable to officiate or be employed to write in or for any of the . . . Offices" of government in the island, and "mulattoes'" right to vote was removed in 1733.[47] This action was part of a differentiation of legal and political subjectivities on the basis of race, gen-der, liberty, religion, and property. Similarly, the Barbadian assembly ruled in 1709 and 1720 that "no Person whatsoever shall be admitted as a Free-holder [and therefore able to vote for or sit in the assembly], or an Evidence in any Case whatsoever, whose original Extraction shall be proved to have been from a Negro, except only on the Trial of Negroes, and other Slaves."[48]

Oaths were embodied, localized, and closely specified verbal perfor-mances which enacted power-laden identities and transformed social re-lationships. These ones could only be spoken by certain sorts of people—predominantly free, white, Anglican, propertied men—to enable those people to, for example, hold offices, guarantee statements, or transfer prop-erty. They were embodied and localized speech acts dependent on particu-lar felicity conditions. Their meaning and force drew on familiar, albeit historically and geographically contingent, notions that speech can be taken as a direct expression of thought and intention and bound subjects to their utterances.[49] In terms of race and gender—which were tied tightly to religion, liberty, and property—oaths could only issue forth from certain bodies. Some people were simply not permitted to say those words, or they had no force if they did. For those who could do so, these oaths were necessary because they had to be spoken for the social attributes of freedom, masculinity, whiteness, and property holding to be turned into active legal and political relations. Without swearing the right oath in the right place at the right time, the processes of institutionalization that were both intensely local and broadly transatlantic would not take place. This was certainly a process by which race, class, and gender privilege was consolidated and legitimated, even if only to the privileged themselves. However, as the discussion of evidence giving will show, this was never a completely clear-cut and uncontested set of exclusions based on simple and essentialized categories and identities, especially when viewed in a transatlantic frame. Neither were these the only oaths sworn on these islands.

Oaths were also sworn to make peace between the colonial government and the Maroons, within quasi-legal proceedings on plantations, and to bind conspiracies and revolts of the enslaved. White officials and planters might, at various times, depend on the efficacy of these oaths, deny their binding force, or punish them by death. While it does not fully explain the forms and purposes of oaths sworn and heard by Atlantic Africans in the slave societies of the Caribbean, it is significant that they or their forebears came from societies that gave great importance to them. To use the Asante example again, it was the *ntam kese*—the great oath, or, more precisely, the threat to say the forbidden words associated with the death in battle in 1717 of the first Asantehene, Osei Tutu—that both bound together their expanding polity and, through the need of the central authority to deal with the collective existential danger of those words being spoken, provided the means by which disputes were brought to law tribunals and judged as matters of life and death.[50] Within Akan political culture more generally, official and personal oaths served as "inviolable contracts for states, state and commercial actors, or armies." They were sacred acts involving the eating or drinking of spiritually powerful substances with the ability to kill if the oath was broken, a practice Europeans described as "eating fetish." As "loyalty oaths," they were, as Walter Rucker argues, "universally employed throughout Atlantic Africa—particularly in regions with expansionist states that needed the security of inviolable contracts to lengthen their political and commercial reach."[51] Their power and use therefore increased as the transatlantic slave trade transformed the political economy of West Africa.

Such oaths, undoubtedly transformed by Atlantic crossings, Caribbean brutalities, and the need to bind people of diverse languages and traditions, were crucial in cementing the bonds, primarily among fighting men, that underpinned conspiracies and revolts of the enslaved (as we will see in chapter 2).[52] They can be heard only in the words of those they threatened. Edward Long, writing of Gold Coast "Coromantins"—itself a forbidden word for the Asante, as the place (Kormantin) where Osei Tutu had been killed—whom he argued were the most rebellious of the enslaved, described how he imagined such oaths were administered by ritual specialists, their binding force, and the danger they posed to planters like him:

> When assembled for the purposes of conspiracy, the obeiah-man, after various ceremonies, draws a little blood from every one present; this is mixed in a bowl with gunpowder and grave dirt; the fetishe or oath is administered, by which they solemnly pledge themselves to inviolable secrecy, fidelity to their chiefs, and to wage perpetual war against their enemies; as a ratification of their sincerity, each person takes a sup of the mixture and this finishes the solemn rite. Few or none of them have ever been known to violate this oath, or to desist from the full execution of it, even although several years may intervene. (*HJ*, 2:473)[53]

In such contexts, swearing an oath was to commit to dealing death to others and to risk suffering death for either breaking the oath or for having sworn it. Those who administered oaths were particularly feared, aggressively hunted down, and spectacularly punished. The Jamaican laws outlawing obeah had their origins in the aftermath of the massive revolt on the island in 1760, which obeah men were understood to have bound with oaths and supported by distributing powders that offered protection against buckra's bullets (discussed further in chapter 2).[54]

Yet oaths and the spirit powers they invoked had everyday uses too. As in Africa, oathing ceremonies were used to detect thieves and other transgressors on Caribbean plantations, often with the tacit or explicit support of the slaveholders and overseers.[55] As Long described it, in such a "solemn oath," some "grave-dirt" was mixed with water in a calabash, and the person administering the oath "dips his finger in the mixture, and crosses various parts of the juror's [the accused's] naked body," asking at each point for a spoken affirmation that if they were guilty, "may the grave dirt make my bowels rot! may they burst and tumble out before my face! may my head never cease to ach! nor my joints to be tortured with pain! &c." Finally, the juror must drain the calabash to "*swallow the oath*." Such oaths were to be conducted with the proper ritual, ideally by an obeah man, but, as Long put it, "their superstition makes them hold it in great reverence and horror, even when administered by another Black, especially by an old man or woman." Significantly, even though white people might benefit from the effects of such oaths, if thieves were found and order restored, they could not effect them themselves. As Long said, when properly administered, the terrible power of the oath was a great force among the enslaved, "but they do not apprehend any ill consequence will arise from breaking it, when tendered by a white person" (*HJ*, 2:422–23).

There are, therefore, important points of similarity and divergence between oath-taking practices here. These comparisons produced a sense of the danger and potential of the oaths of the enslaved, which were certainly recognizable to the white planters as binding utterances, even when administered by women, but also a denigration of them as superstitious and barbaric. This combination of potential power and denigrated difference was particularly important when it came to questions of commensurability, as it did with the Maroons in Jamaica. As free people whom the British were unable to defeat in a war which lasted throughout the 1730s, and which threatened the stability and security of Jamaica's economy and polity, the Maroons commanded respect.[56] The key moment in their relationship with the island's British colonizers was the 1739 treaties that gave both the leeward and windward Maroons freedom from enslavement, their own lands, and significant legal and political autonomy in return for fidelity to the Crown and practical support for the plantation system, returning escaped slaves and fighting rebels.[57]

The leeward treaty was agreed between Captain Cudjoe (an Akan day-name, Kojo, for those born on a Monday)—a crucial figure in both Maroon and colonial history—and Colonel Guthrie, the commander of the British troops. Since the Maroons had fought the British to a standstill and enforced a cease-fire, this was a much more tense and symmetrical meeting than is suggested by the embodied differences represented in an image of their negotiations from 1803 (figure 7). Here Guthrie's polite and open martial masculinity contrasts with Cudjoe's squat defensiveness as they exchange hats. That Cudjoe was Guthrie's equal is evident both in the terms of the treaty, which established a form of restricted sovereignty for the Maroons, and in the practices by which it was agreed. Both sides signed a document, participated in a ritual mixing of blood, and swore the associated oaths. The Maroons gave no especial significance to the documents, although their leaders marked them, and the British did not rest their faith in the treaty on the mingling and drinking of blood and rum, although they are said to have participated in it.[58] However, both sides gave credence to the speech acts through which their representatives made peace, and the practices of each were recognizable to the other even if they understood them differently.[59] To secure the treaty, after initial negotiations had begun, Guthrie told Cudjoe "that he wished to come unarmed to him with a

Figure 7. *Old Cudjoe Making Peace*, from R. C. Dallas, *The History of the Maroons* (London, 1803). Courtesy of the John Carter Brown Library at Brown University.

few of the principal gentlemen of the island, who should witness the oath he would solemnly make to them of peace on his part, with liberty and security to the Maroons on their acceding to it."[60] Their word was to be the bond.

Significantly, however, both sides needed to attend to the felicity conditions of this speech act to satisfy themselves that what was being sworn to

would hold. British trust in the treaty rested on matters of political authority. Although Maroon politics was far from being that simple (as we will
see in chapter 2), Guthrie sought to reassure his own political masters that
Cudjoe was the Maroons' absolute leader, that he would hold to the treaty,
and what he swore to would be binding on others: "as to his Captains,
they pay him the greatest Defference imaginable, they are entirely under
his Subjection, And his Word is a Law to them."[61] In contrast, Maroon attempts to bind the British to the treaty were grounded in the supernatural.
They refused to agree to it on a Sunday—"'Tho Col. Guthrie and all the
Gentlemen with him made the most Solemn Asswerations to keep Religiously to the Treaty"—because of what they knew about the British relationship to their god, "alledging that no Articles of Friendship, could be
Sincere or valid, or any Oaths binding that were made on the Lords Day,
which we were enjoyned to keep holy and to do no manner of thing on."[62]
In this way both sides tried to make their different oath-taking practices
commensurate and binding when they came together.

Oaths, then, for all who swore them, made utterances attributable to
speaking subjects and bound speakers to their words. They were performative speech acts that, in their different ways, and with different forms of
legitimacy, drew on notions of gender, status, race, freedom, and a world
of gods, spirits, authorities, and various sorts of ancestors to constitute a
variety of subjects. In all cases, oath taking was a matter of specific, localized, and embodied speech acts with immediate or anticipated local effects
for those who swore them in terms of creating social relations, making
community, guiding violence, defining property, or effecting governance.
Yet they also had another, transatlantic, geography. This geography drew
on various imperial and diasporic forms of power and authority and on
understandings of the past, however much transformed in the Caribbean,
which gave these oaths force. The power of such oaths was understood
across these islands, as was the fact that all oath-taking practices were matters of inclusion and exclusion and were ultimately backed by violence.
This made the oaths sworn within the official legal practices of the islands
a powerful mechanism for defining the boundary between the free and the
enslaved and aligning it with notions of status, gender, and, particularly,
race. Yet, as has also been shown, other oaths were also sworn that, particularly in the case of the Maroons, made other oathing traditions—involving
blood, rum, grave dirt, and different forms of words—part of the process
of defining freedom and undermined a clear equation between race and
slavery. Indeed, an examination of the related practice of evidence giving
shows that this form of talk was structured to institutionalize radically

unequal social relations and was the subject of ongoing contention and transformation as the crosscutting social relations of plantation society were debated across the Atlantic.

Giving Evidence

As examination of oaths demonstrates, law and governance on Britain's Caribbean islands depended on a differentiation of legal subjectivities based, primarily, on the categories of race, gender, freedom, status, and religion. Within this system, the enslaved have been presented in contrasting ways, as either existing in a situation of "bare life" entirely beyond legal protection or, in contrast, as interested and active participants in the islands' legal processes.[63] As Elsa Goveia argued, they are perhaps best understood as having very heavily circumscribed and highly contingent legal subjectivities, such that "in so far as the slave was allowed personality before the law, he [sic] was regarded chiefly, almost solely, as a potential criminal."[64] On the one hand, the islands' judicial systems—including separate courts for the free and the unfree—acted to reinforce planter power and to terrorize the enslaved rather than to knit them into a unitary society.[65] On the other hand, the legitimation of slavery within the transatlantic constitution meant that the enslaved were part of the public realm of law rather than just the private realm of personal property. For example, one eighteenth-century historian of Jamaica repeated the planters' self-serving claim that the enslaved were better off in the Caribbean than in Africa since they were provided with food, clothing, and shelter, but also with legal rights since "their liberty in their Own Country" was not "any more than nominall or imaginary." As he put it, "for as many of them were Subject to the Arbitrary will and Pleasure of their Kings or chief Men who disposed of them as they thought proper, and had an Absolute Power of life & Death they may justly be said to be less Slaves in our Plantations than they were in their own Country. Because in our Plantations their Masters are allowed no Power of life and Death over their Slaves, and are even Restrained from maiming or Dismembering them upon any Pretence whatever without a legall Tryall."[66] In line with this reasoning, the Jamaican and Barbadian assemblies passed laws enacting that wantonly killing "a Negro or Slave" was judged to be a felony, and, in Jamaica after 1738, a second offense was to be treated as murder.[67] Yet punishing by fines and imprisonment was radically different from sentences of execution for any act of imagined or real violence directed by a slave to white person.[68] There was also a disjuncture between what was on the statute books and the actual practice of violence and prosecution on the plantations and in the courts, which meant that few whites ever came to trial for such crimes and even fewer were convicted.[69] It was enacted that there was no legal liability if those killed had escaped the plantations, were found stealing, or were "out of the owners Ground, Road, or Common Path, and refuseth to submit," and the inadmissibility of slave evidence against all white people fatally undermined clauses in the slave laws which claimed to offer protection from mistreatment, particularly from their own masters.[70] Despite this, however, the situation of the enslaved does not simply accord with the pure absolutism of Giorgio Agamben's *Homo Sacer*—a figure who can be killed without it being sacrilege, execution, or murder. The unlawful killing of a slave was a possibility, although that offered almost no protection in practice to enslaved men and women.[71]

These forms of severely circumscribed legal personality were evident elsewhere in the legal system. As Marisa Fuentes argues, the islands' laws "(un)gendered" the enslaved in that there was little differentiation between men and women in terms of crime and punishment until the late eighteenth century, thus obscuring, compounding, and making more deadly the subjection of black women to white slaveholders.[72] British colonial slave laws were, in order to punish slaves for their crimes, based on the notion that the enslaved were endowed with will. Yet their treatment under the law was to be different from those individuals who fell into other categories. The Barbados assembly argued in 1699 that "Brutish Slaves, deserve not, for the Baseness of their Condition, to be tried by the Legal Trial of Twelve Men of their Peers or Neighbourhood."[73] The enslaved in both Barbados and Jamaica were tried for small crimes by the magistrates and for capital crimes by two justices and three freeholders who had sworn an oath "to Judge uprightly, and according to Evidence."[74] In 1788 in Jamaica, under pressure from the imperial state, these arrangements were replaced by "slave courts" with three JPs and a jury of nine freeholders.[75]

Jamaican courts trying the enslaved could order any punishment, including transportation off the island and the death penalty. In such trials it was ordered that "the Evidence of one Slave against another . . . shall be deemed good and sufficient Proof," and the enslaved were to be punished, albeit more harshly than others, for perjury.[76] Indeed, for some crimes, particularly those relating to resistance against the plantation system, slave evidence was crucial to prosecutions.[77] It was also necessary in other circumstances. In 1750, the Jamaican assembly tried to deal with problems arising with an earlier act against piracy, which "cannot admit the Evidence of Slaves against Slaves, without Oath," leading to many crimes going unpunished. They enacted that these slaves' evidence could be taken without oath, but that was only to apply to crimes on the high seas and only to slaves who were acting as pirates, not to "Offences, committed, or to be committed by Slaves in their Passage to this Island, as Merchandise."[78]

In Barbados other, similar, contradictions arose as a result of needing some space for the evidence of the enslaved. There was a conflation of categories whereby it was enacted in 1688 that "any Negro, Slave or Slaves" would be tried by two JPs and three freeholders considering the "Evidences, Proofs, and Testimonies, or . . . violent circumstances" for and against them. Race and bondage were fettered together via the assumed brutishness of Africans.[79] This combination of racist assumptions and acknowledgment of the conditions under which the unfree gave their testimony provoked suspicion about the use of "slave" or "negro" evidence.[80] Richard Hall, a lawyer and member of the Barbadian assembly, glossing the 1688 act in 1764 and highlighting its inclusion of "violent circumstances," revealed these anxieties in his hope that other evidence would be found. As he said, "a violent presumption, amounts in Law to full proof, that is, where circumstances speak so strongly, that to suppose the contrary would be absurd. And where that presumption necessarily arises from such circumstances, they are more convincing and satisfactory, than any other kind of Evidence, because facts cannot lie."[81] Yet the spoken testimony of "Negroes" and "Slaves" could not be avoided; neither could the differences between those conflated categories.

In 1739, concerned about the "inconveniences" presented by "free Negroes, Indians or Mulattoes," the Barbados assembly enacted that the evidence of slaves, "where the same is supported with very good and sufficient corroborating circumstances" presented against any "free Negroe, Indian or Mulatto, whether baptized or not" was to be deemed "as good, valid and effectual in Law . . . as if the Slave giving evidence or testimony, was free, baptized, and not under servitude or bondage to any person whatsoever." Hall was quick to spot the contradiction. Since the Barbados assembly had ruled out evidence from anyone "whose original Extraction shall be proved to have been from a Negro, except only on the Trial of Negroes, and other Slaves," then it was unclear whether their evidence was in fact valid against free people of color or not.[82]

Such contradictions were heightened in the late eighteenth century when the institution of slavery came under attack from abolitionists for not affording legal protection to the enslaved (as we will see further in chapter 5). It was in this context, in Barbados in the 1780s, that Joshua Steele put the question of evidence giving right at the heart of his critique of plantation society.[83] Indeed, he organized his efforts toward what he saw as the reform of slavery, including the ideas with which this chapter began, around the problems that precluding "Negro evidence" against all but enslaved Africans created for slave societies. Steele's writings show what was at stake in the regulation of the spoken word when it came to oaths and evidence.

In his published work, Steele presented a series of conversations between the imagined clergyman and his planter acquaintances over cards, food, and drink at their homes and at Mr. T's store. Their debates on "Negro evidence" played on the transatlantic constitution, contrasting Barbados with England, "where all Ranks of Men, Slaves, Bondmen, and Apprentices, are admitted as legal Evidences." Steele also imagined a newcomer to the island "surprized to find, among English People, a mode of thinking, and a System of Laws, so new to me." Barbadian planters faced, Steele argued, a set of "Evils . . . principally derived from some impolitic System woven into the Web of your motley Laws." Since, sixty years before, "in the Frenzy of a contested Election, a Law, disgraceful to Humanity, was passed in the Colony," that "no Descendants of Negroes" shall give evidence in any case except the trial of slaves.[84] This was the law of 1720.

However, picking away at the question of evidence giving meant unraveling whole swathes of the system of racial slavery, calling fundamental powers and beliefs into question. Steele's clergyman queried the legal basis on which the island's legislature claimed the authority "to disqualify generally any Class of Subjects, of whatever Complexion, born within the British Dominions, and under the Allegiance of the Crown!" He suggested that these laws were repugnant to the laws of England and asked whether they shouldn't be repealed. He also attacked assumptions about Englishness and racial purity, questioning how one can draw a "black Line" between white and black on this issue, when, on the one hand, there were "worthless and sordid white People" who don't understand, or who ignore, "the Nature and Obligation of that Oath," and, on the other hand, "all the Descendants of Negroes, some of whom may now, perhaps, be seating among the Nobles and Commons in Parliament; (for without doubt, there has crept a little Negro Blood among them, thro' wealthy matrimonial Connexions, in the Course of near two Centuries, since our first Communication with them)." Indeed, he went even further by arguing that since all men are descended from Adam, then either Adam was a "Negro," and therefore all evidence must be disqualified, or "we pronounce against the Authority of the Holy Text, and say there were two Originals."[85]

Steele also presented opponents' objections and offered his heretical solutions to them. He had the slaveholders, perhaps echoing Edward Long, accuse the clergyman of advocating an "absurd System" which would "think of putting a Species of Creatures, very little above Baboons, in their intellectual Capacity, upon a Level with ourselves," and he presented their key objection that there was no oath that could be sworn in court by these "African-born Savages" since not one in three hundred of the enslaved was Christian. It was in their arguments that these imagined planters demonstrated the strong grasp of the power-laden geographies of talk on the sugar islands outlined in my introduction to this book. It was Steele that had them say that "such visionary Notions might be ornamental in the Oratory of the Pulpit; where, by Act of Parliament, no Man dares to contradict the Preacher; but in a Curing-House, a Counting-House, or a Court-House, he would find the Statutes of the island were flatly against such dangerous Doctrines." The "Freedom of Speech" implied in allowing "Negro evidence" against any but the enslaved would be, he ventriloquized, "opening a Door to a Train of unknown and inconceivable Evils!"[86]

The clergyman's replies drew support from the published and unpublished writings of Steele's own Barbados Society of Arts when he argued for "swear[ing] Negroes, as *legal Evidences*" by the sun, the moon, their gods, or on their ancestors' graves. They were also to be allowed to hold property under a reconstituted form of plantation slavery, a revived feudalism with the enslaved as copyhold tenants according to the "slave laws of our Saxon and Norman ancestors." This, he argued, would be "the most probable Way of civilizing Savages, and training them gradually to a rational Observance of, and Submission to, equitable and fixed Laws."[87] Yet, in arguing for this system, a direct outcome of his critique of the laws on "Negro evidence," Steele had questioned the fundamental tenets of plantation society. The chain binding race, freedom, Christianity, property, and evidence giving would be broken.

The links in this chain were also called into question by the legal status of the free black, "Mulatto," and Indian population of the islands. Their status was particularly significant in Jamaica, where free people of mixed race numbered about 3,000 in the 1760s, compared with about 145,000 enslaved Africans and about 15,000 whites. As shown already, although free and potentially propertied, these people suffered restrictions on their rights to representation in the assembly or to hold government office. Indeed, they were increasingly restricted across the eighteenth century.[88] A 1717 act required those without sufficient property to carry a certificate signed by a JP and to wear a blue cross on their shoulder; the 1733 act restricted their right to vote; and, after the revolt of the enslaved in 1760, the Jamaican assembly passed a law requiring that all "free Negroes, Mulattos, and Indians" without real estate register themselves with the vestry, get certificates, and wear their "Badges of Freedom." Later restrictions limited inheritances and landownership by free people of color to two thousand pounds, on the grounds that "such Bequests tend greatly to destroy the Distinction, requisite and absolutely necessary to be kept up in this Island, between white Persons and Negroes, their Issue and Offspring of Mulattoes, not being their own Issue born in lawful Wedlock."[89]

Their rights in the Jamaican courts were also circumscribed, albeit less closely than the enslaved. After 1730, they could give evidence against the enslaved but were not "admitted as an Evidence or Witness in any case where the life, liberty or property of a White Person is or shall be in question or in any wise Concerned." When they themselves came before the courts, punishments could include the removal of their freedom and transportation off the island, penalties never applied to whites.[90] Until 1748, those who were formerly enslaved but had been freed were to be tried according to the same procedures as for the enslaved, "and the Evidence of a Slave against them [was] to be good and valid to all Intents and Purposes."[91] Those who were born free had more rights, which created problems of who could give evidence against whom that were resolved by enacting that all "free Negroes, Indians, and Mulattoes" baptized as Christians could give evidence against one another regardless of previous status (although manumissions had to have been in effect for six months), and the courts were to "receive their Testimony on Oath . . . in the same Manner as if they were White Inhabitants of this Island."[92]

Such legislation required definitions of who was in these categories. There appears to have been little concern about who the free "Negroes" and "Indians" were. Those of mixed race were much more problematic for white planter society.[93] In Jamaica, as in Barbados, their exclusion from representation in the assembly was made law by an act regulating the "Freedom of Elections." However, the Barbadian exclusion of anyone "whose original Extraction shall be proved to have been from a Negro" would have been unworkable in Jamaica. Instead, "Mulattoes" were defined as "any Person who is not above Three Degrees removed in a lineal Descent from the Negro Ancestor exclusive." Thus it was enacted in 1733 that "no one shall be deemed a Mulatto after the third Generation . . . but that they shall have all the Privileges and Immunities of His Majesty's white Subjects of this Island, provided they are brought up in the Christian Religion."[94] This definition was used as the basis for other legislation, including the 1748 act regulating their evidence giving.

Yet these categories and the legal subjectivities assigned to them were more mutable than might at first appear. Tellingly, the 1748 act precluded free people of color from giving evidence not only against white people but also against any "Negroes, *Indians*, or *Mulattoes*, that have the Liberties of White Persons, by any Law of this Island."[95] As soon as there was public legislation enacting categorical racialized distinctions, then there were private acts shifting individuals between those categories and reallocating rights. There was the manumission of slaves, often in slaveholders' wills.[96] There was the declaration, in order to prevent the kidnapping of Amerindians from central America, "that all Indians brought to the island since 28th December 1741 be declared to be Free People."[97] Also, the printed volumes of the laws of Jamaica contain several instances each year from 1733 onward of the propertied and free children of white men and women of color—a wealthy mixed-race elite—petitioning for private legislation to accrue to themselves, as it was written in the titles of the acts, some of "the Rights and Privileges of Englishmen, born of white Ancestors."[98] After 1760, with the increased constraints on inheritance, the subphrase "under certain restrictions" was added. But there soon appeared other private acts exempting people from the provisions of this law as well.[99]

Most significantly, there was the case of Francis Williams—the free black Jamaican lawyer, mathematician, and poet who was denigrated as nothing more than a "parrot" by David Hume—who directly contested this construction of legal subjectivities according to racial categorizations of who might speak, albeit from a very particular position.[100] Francis was the son of a wealthy free(d) black slave owner, John Williams, who, in February 1708, and probably with the support of a political faction among the white planters, had brought in one of two bills (the other being for Manuel Bartholomew) "to enable him to be tried by a jury, as a white man." At exactly the same time as the motion was being debated to exclude free people of color from elections, John Williams was asking for exceptional status: to be a black person against whom the evidence of the enslaved, the slaves he held, would not be admitted. The assembly and council passed these bills with only one notable amendment, to "leave out the words 'as an Englishman.'" The exception was granted, therefore, but without a categorical change in the relationship between Williams's legal and racial identities. The day these bills were passed, there was a motion by another free black, John Callendar, and another a week later from Robert Bass. Sensing the opening of the floodgates, the assembly resolved that no more such petitions would be received during the session, and there is no sign that these two bills survived the prorogation of the assembly at the end of February

1708.[101] No other bills were brought in, but John Williams's bill passed through both houses of the British Parliament and was made law by Queen Anne in 1711.[102]

This legislation was certainly contentious, as was the extension of these rights by a private act to Williams's wife and three sons in 1716.[103] A nineteenth-century historian of Jamaica recorded that Williams came close to losing his newly acquired privileges "through having incurred the displeasure of a member of the house of assembly." The white man had called Williams a "black negro." Williams had returned the insult, calling the assembly member a "white negro." And, as the historian reported, this incident "was thought of sufficient importance to engage the attention of the learned legislators, by some of whom it was proposed to revoke the act of 1708, so far as it related to the culprit." While no contemporary corroboration of this slanging match has yet been found, it does foreshadow the well-documented and widely debated "bad language" of John's son Francis.[104]

Francis, it was said, was "a boy of unusual lively parts" who, according to Edward Long, went to England under the patronage of Lord Montagu as part of "an experiment" as to "whether, by proper cultivation and a regular course of tuition at school and the university, a Negroe might not be found as capable of literature as a white person." For Long, who wanted to prove that Williams's abilities did not demonstrate the equality of black and white, Francis's education shadowed that which the historian imagined could teach even an "oran-outang" to speak (*HJ*, 2:476).[105] However, the evidence there is of Williams in London shows a more illustrious career than Long allowed. In October 1716, he was proposed, at probably less than twenty years old, as a fellow of the Royal Society, presumably on the basis of his mathematical ability; and, in August 1721, he entered Lincoln's Inn to train as a lawyer.[106] His Royal Society nomination was made by the mathematician Martin Folkes as the last item of business at a regular weekly meeting of the society. This gathering included Isaac Newton, then the president; the physician (and Newton supporter) James Jurin; the astronomer Edmond Halley; Sir Hans Sloane, well known for his Jamaican connections and collections (described in chapter 3); and "Mr Smith, Professor of astronomy at Cambridge," who was not a fellow but had been given leave to attend the meeting.[107] However, it was recorded in the minutes of a council meeting a fortnight later—also chaired by Newton, and including Sloane and seven other fellows—that "Mr Williams a black Native of Jamaica was balotted for and rejected."[108] Folkes, a friend of Newton and Montagu, and later to lose out to Sloane in a fiercely fought contest for the society's presidency, was described after his death in 1754 as "in matters of religion an errant infidel & loud scoffer. Professes himself a godfa[the]ʳ to all monkeys, believes nothing of a future state, of the Scriptures, of revelation. . . . He thinks there is no difference between us & animals; but what is owing to the different structure of our brain, as between man & man."[109] As this description suggests, like Long's, Folkes's engagement with Francis Williams may have been an attempt to make a particular point about the nature and capacities of humans, regardless of whether the Jamaican mathematician's fellowship nomination was accepted.

Francis Williams returned to Jamaica following his father's death in July 1723. Once again, as with his father, it mattered who said what to whom. Williams was certainly insistent on his status as a man of learning, refinement, and property (figure 8).[110] Although he was excluded, as a black man, from either a legal or a political career, he set up as a teacher of reading, writing, Latin, and mathematics in Spanish Town, and, according to Long, "affected a singularity of dress" and wore "in common a huge wig" as well as a sword and pistols. Most significant, he would not be quiet. On "the measure of his wisdom" and other qualities, Long complained, "he had not the modesty to be silent, whenever he met with the occasion to expatiate upon them" (*HJ*, 2:478).[111] He was soon put to the test. In 1724, William Brodrick, the island's former attorney general, presented a petition to the assembly setting forth the insults he said he had suffered from Francis Williams. As a committee of the house reported, the men had argued over the wording of a document:

> And that the said William Brodrick offered to take his oath, that Francis Williams . . . coming towards him in a passion, he, to keep him off, put his hand to his breast, and that, without striking at him, he, the said Francis Williams, struck the said William Brodrick; that it fully appears the said Francis Williams called the said Mr. Brodrick "white dog" several times, and many other opprobrious words, and said he was as good a man as ever stood on Brodrick's legs; that he did not stand upon the act of this country, that exempts him from such trial as other negroes; that Mr. Brodrick's shirt and neckcloth had been tore by the said Williams. . . . The committee likewise find, the said Mr. Brodrick called Williams "black dog" several times, and that Williams's mouth was bloody.

In response to Brodrick's petition, Williams offered proof that he and his father had taken the oaths of naturalization, although he had nothing to prove he was a Christian. He then asked for time to make a proper reply to the charges.[112]

Figure 8. Francis Williams, *c.* 1740. Reprinted by permission
of the Victoria and Albert Museum, London.

Williams's particular legal status, and Brodrick's previous public position, meant that this was more than just a private dispute. Significantly, at
the same time the bill granting Williams these rights was going through
the House of Assembly, Brodrick had been at the sharp end of a reversal in
his political fortunes. He was accused of exercising illegitimate influence

on the new governor and of rigging assembly elections; he was described as "a forger of lies, a disperser of false news, an incendiary, and common disturber of the peace of this island." Indeed, a letter to King George I requesting that Brodrick be prevented from holding any public office was carried to England in the same ship as the Williams bill, which is likely to have been supported by Brodrick's political opponents.[113] Brodrick's accusations against Williams a decade later also aimed at larger targets. He argued of Williams's insults that "if such a precedent should pass by uncensored, it might be of ill consequence to the island in general." Indeed, the assembly's committee not only agreed but spelled it out in racialized terms: "that in regard the said Williams's behaviour is of great encouragement to the negroes of the island in general, and may be attended with ill consequences to the white people thereof, that leave be given to bring in a bill to reduce the said Francis Williams to the state of other free negroes in this island, &c." They set about crafting the legislation "for rendering free negroes more serviceable to the island," obliging them to wear badges of distinction, prohibiting them from wearing or using swords or other arms, except on militia duty, defining where they could or could not live, and bringing Francis Williams back into line.[114]

This bill was not made law by the assembly until 1730, when the flaring up of hostilities with the Maroons focused the attention of white planters and politicians on how various sorts of free black Jamaicans might combine against them.[115] Williams was "reduced" to the same state of trial and evidence as other free black inhabitants of the island. But he was not finished. He petitioned the colonial government in London against the act on the basis of the hardships that he and his family would suffer now that the enslaved men and women he saw as his property could testify in court against him, and as a spokesperson for other wealthy free(d) blacks. He subsequently won the judgment from the Crown's lawyers "that this Act is impolitick in its Tendency with respect to the Interest and Welfare of Jamaica as well as unequitable towards the Persons abovementioned [Williams's family] and to the whole order of free Negroes since it manifestly tends to discourage the Integrity of the Slaves in that Island, as well as the Industry of those who are become free."[116] Once this dispute over the power of speech and the rules of evidence had become part of an Atlantic discussion, the complexities and contradictions of legal geography became apparent. The binding together of property and freedom—Williams's slaveholding status and his legal privileges—as viewed from the imperial center was given more weight than the increasingly necessary alignment of race and rights as understood in Jamaica—reducing a troublesome black man to the same legal status as other "free Negroes." The 1730 law was repealed on order from London.

Francis Williams's "bad language" certainly troubled Jamaica's white political elite, along with others who were framing racist notions of difference. His moral and intellectual capacities—his command of mathematics, his Latin poetry, and his eloquence—needed to be denied or explained away by philosophers, historians, and proslavery pamphleteers (as will be discussed further in chapter 5).[117] Yet the ambiguities of his position are troubling in other ways too. Long, certainly not a disinterested commentator, reported that Williams "looked down with sovereign contempt on his fellow Blacks," treating his slaves and his children severely, and that "He defined himself 'a *white* man acting under a black skin'" who "endeavoured to prove logically, that a Negroe was superior in quality to a Mulatto, or other cast." Long glossed Williams's Latin poem to Governor George Haldane in 1759 with what he saw as his telling "apophthegms": that "'a simple white or simple black complexion was respectively perfect: but a Mulatto, being an heterogenous medley of both, was imperfect, *ergo* inferior,'" and "a proverbial saying, that was frequently in his mouth; 'Shew me *a Negroe*, and I will shew you *a thief*'" (*HJ*, 2:478). Yet these harsh maxims, understood through Francis's father's own pithy turn of phrase, can reveal a more consistent speech position. Other translations of Williams's poem than Long's note that the poet emphasized that his heartfelt sincerity and his "sweet eloquence in a learned mouth" should not be denied because of the color of his skin. What Long translated as Williams's imagined white body ("*Candida . . . corpore*"), John Gilmore suggests can also mean his sincere argument. For Williams, felicity in speech came from freedom. As he put it, "Every kind of eloquence is lacking to slaves."[118]

In his less poetic apothegms, Williams can also be understood as denying the devaluing of blackness via the association with slavery that white planters were so keen to construct. This interpretation makes sense of John Williams's retort as well. If he was judged a "black negro," then an assemblyman could be a "white negro," because, for him, to be "black" was not necessarily to be a "negro." Blackness and slavery—the status of the "negro" in law—needed to be kept separate for the slaveholding Williams family. While a black man saying that all "Negroes" are thieves is ridiculous in the mind of Edward Long, it makes perfect sense for the black slaveholder Francis Williams, particularly when he might also want to assert that they are liars too. Moreover, the Williams family grounded their status on the law, and on the rights to speak that it contained and constrained—on the oaths they had given and the words that the law decreed could not be said against them. Any claims based on race as a category of nature or blood—that white or "Mulatto" might be naturally superior rather than because of rights given in law—were to be vehemently denied. For Francis Williams, the lawyer who was not allowed to practice but had won his case, what mattered was whom the law allowed to speak. That was what defined freedom.

This contestation of the power of speech—of who could speak against whom—was also evident when Jamaica's white population, like Williams, sought to maintain the privileges and identities underpinned by the laws on evidence. The significance of these privileges for white Jamaicans is well illustrated by the fury unleashed in 1748 by the assembly's consideration of two bills: what would become the act on the evidence of free people of color, and a bill "for preventing the castration or other mutilation or dismemberment of slaves, without the authority of the magistrate." The latter would, it seems, have allowed slave evidence to be used in specially constituted courts to prosecute slaveholders, attorneys, or overseers who mistreated the enslaved in ways defined by the bill. It was promoted by Governor Edward Trelawny, who had, in 1741, proposed a scheme "for the abolishing of slavery, & putting the Negroes upon some such foot as the ancient villains of England were," which has clear parallels with Steele's later suggestions. In 1750, Trelawny was also to express concerns about the liberties slaveholders took with their slaves and the inability of the law to hold them to account.[119]

The debate over this second bill provoked a storm of protest beyond the legislative chamber. On May 18, 1748, the Jamaican assembly was treated to a reading of "a false, scandalous, malicious, and seditious libel" that had previously been declaimed in public outside the Kingston courthouse by Archibald Willock, a wharfinger. It purported to be a petition in support of the bill from the "negro slaves" of the island addressed to Edward Manning and the other representatives of Kingston. Displaying classical and religious erudition, parodying forms of legal documentation, and written in a high oratorical style, this satire's main thrust was to present the bill as one that would turn the slaveholders' and overseers' worlds upside down. The supposed petitioners—"unjustly and inhumanely detained in thraldom and bondage, in the island of Jamaica"—were presented as supporting "an act to make us evidences sufficient, on our bare word, (as we know not the meaning of an oath), to take from our unjust masters, the white men of this island, their liberty and property, and to take from them that shelter to which, for so many years they have fled, a fair trial by juries and lawful witnesses." Furthermore, the imagined enslaved petitioners argued that they themselves should have the privilege of a jury trial, that the island would only prosper when white masters were subjected to the black slaves, and that "all arms and ammunition" on every plantation should be "lodge[d] in the black drivers or favourite negroes house."[120] What had begun with extending the rules of evidence would end with armed uprisings of the enslaved. As Joshua Steele's imagined planters would later agree, allowing such evidence "would be opening a Door to a Train of unknown and inconceivable Evils!"

The satire's invented petition was sworn to be true and "signed" with "The mark + of CUDJOE," and carried the threat that there were "sixty thousand ready to carry the passing of the bill, or more, if required."[121] This, then, was another Maroon oath—like the treaty agreed with Governor Trelawny—that contradicted the correlation of whiteness with freedom and haunted the imagination of white Jamaicans and the wider Atlantic world.[122] Cudjoe's mark on the satire was a clear sign of the challenge that the legal rights of black people posed to the sovereignty and identity of free white Jamaicans. The satire used that challenge to object to the extension of the right to give evidence that the island's governor envisaged.

The agitation over the reading of the bill also raised divisive questions about Jamaican politics. The petition told Manning that the bill he had tried to get through with minimal debate would establish a "high court of inquisition . . . for reducing, by our means, artifices, and informations, the abominable pride of any white man, who shall ever hereafter complain at home of you or your patron [the governor]." There were also reports that Robert Dallas, a prominent physician, had gathered people together in Kingston and read to them "what he called a copy of a bill." Dallas had described the bill as the hidden work of Governor Trelawny, presented to the house by his creature Robert Penny, the attorney general. This action, Dallas suggested to his presumably enfranchised audience, made Penny a "person unworthy of being their member."[123] Identification of the satire's author—Dr. James Smith, another Kingston physician—also provoked complaints about a pamphlet he had published in London in 1747 entitled *A Letter from a Friend at J[amaica]*. The petition and the pamphlet certainly have similarities. Both attempted to use satire to skewer factions and factionalism in Jamaican politics, and both were concerned with the damage that could be done on the island through misusing oaths and evidence. However, they aimed at diametrically opposed political targets. In "Cudjoe's petition," Trelawny, Manning, and Penny were cast as the villains trying to pass a bill that would criminalize attorneys and overseers on the word of slaves. In *A Letter*, these politicians are the heroic victims of a "lurking" faction trying to undermine the Jamaican government and supported by "People whom neither Birth nor Education had intitled to better Principles than are found among the general Run of Overseers to Plantations."[124]

The members of the Jamaican assembly required to read both texts emphasized their parallel dangers. They judged the pamphlet to contain "several expressions which are false and villainous, and a high indignity to several members of this house, and tending to disturb the peace and quiet of this island." And they ruled that the "negro petition" was "injurious to the honour and authority of this house; tending to alienate the minds of the people from, and to incense them against, their representatives; to excite mutiny and disorder; and to destroy the well-being of this island, and the inhabitants thereof."[125] Willock was judged to be the "publisher" of the libel, by reading part of it out. However, since he had done so without evil intent, he was discharged. The author, Smith, was to be prosecuted for it. In the end, however, having apologized and paid his accumulated legal fees, he was released. His attacks on both sides made him difficult to prosecute by either.

As James Robertson has argued, this satirical petition reveals "deeply rooted assumptions about law, freedom and privilege" that were central to defining the identities of white Jamaicans.[126] Most notable here is that whereas the satire absolutely opposed the proposal to use slave evidence, it was far less critical of altering the rules for free people of color. Smith did seek to exaggerate the provisions of the 1748 bill by claiming that these free but nonwhite people would be accorded the same rights as the Jews but was more ambivalent over the "extirpation of the Jews" that he argued would follow. Of utmost significance, therefore, was the maintenance of the distinction between whites (who were by definition free) and all others, and the use of the laws on evidence to ensure that distinction. The 1748 act could pass because it permitted no extension of who could give evidence against white defendants. The other bill was not passed because it suggested the most significant extension of all: the use of slaves' evidence against their masters. Robertson is right that this was a matter of "local politics." Agitation beyond the walls of the assembly shaped what went on within it. However, this was also an issue of Atlantic political and legal geography. The publication and circulation of Smith's pamphlet shows that Jamaican political factionalism was a transatlantic issue, discussed in London (as we will see in chapter 2). Moreover, the argument in "Cudjoe's petition" that granting "a petition of the white men" against the bill "will prevent our getting possession of the promised lands . . . contrary to our natural rights as men, contrary also to your darling constitution of Britain, contrary to equity and good conscience" was, like the suggestion that the slaves be given guns, an invitation by Smith to his audience to recoil from something self-evidently unthinkable.[127] In this case, it was the suggestion that there was any basis for granting the enslaved either natural rights or the legal rights of freeborn Englishmen which underpinned liberty and identity in Britain's Atlantic empire. However, for it to be rejected, "Cudjoe" had to voice the possibility.

Conclusion

As this chapter has shown, oath taking and evidence giving were powerful sites of contested meaning and practice in the early modern Atlantic world. Through many particular forms of words they were embodied performances and localized speech practices that drew on longer histories and wider—imperial and diasporic—geographies to produce direct local effects by binding speakers and their utterances, and producing and reproducing relationships to other people, other places, and other times. In relation to formal systems of law on these islands, they operated both to underpin the particular social relations of Caribbean plantation societies and to knit them into Britain's Atlantic empire. They were part of a powerful oral culture of government and the law that was made across the Atlantic in the early modern period, and which placed a high value on the spoken word, or at least some people's spoken words. Oaths and evidence, and the institutions that they were part of—particularly lawmaking assemblies and jury trials—were seen to guarantee the freedoms expected by Englishmen, even those who found themselves in the Caribbean. They also bound together Atlantic networks of rights and property across a differentiated legal geography. Fundamentally, these speech acts depended on, and worked to establish, the identities and privileges of certain sorts of social subjects as legal subjects, ones which were primarily male, white, Protestant, propertied, and free. They were spoken words that sought to enact the closing off of the powers and meanings of law around these particular categories of people. Their freedom to act, and to use and be backed by the law, was based on speech.

This closing off is clear from the differentiation of rights to take official oaths and give evidence in court. One last example is useful here. In the mid-1790s, the Jamaican government went to war with the Maroons once more, led by their new governor, Alexander Lindsay, the Earl of Balcarres.[128] On his return to the island's capital from the war, Lord Balcarres received a petition from "a free Black man and Signed by the whole body of the Free people of Colour Resident about Spanish Town." This man had been assaulted by a white person, and although there were "Several people of Colour" who were witnesses, they could not give evidence in court. Governor Balcarres had, given the "recent and Eminent Services" of free people of color in the war, recommended the petition to the assembly and overseen a law allowing those baptized as Christians to give evidence in cases of violence against themselves. They could not, however, give evidence in cases involving other defendants. Indeed, in recommending this very limited extension of the right to speak, and in encouraging the assembly to do as little as possible, the governor recorded his general disapproval of granting such privileges. He noted to his superiors in London that "the Admitting of Petitions of this nature is extremely Dangerous in these Colonies & the line drawn between the Whites and Negroes and all the Intermediate Gradations of Colour cannot be too Strongly marked and preserved."[129] The laws of evidence, what could be said by and about whom in court, maintained the boundaries of race, gender, religion, and property. The identity of free, white, propertied Englishmen in Jamaica and Barbados was made through their power to speak and their power to deny that speech to others. Most significantly, ruling out the evidence of the enslaved against any but other enslaved people made the law an instrument of oppression and deadly terror that was wielded against them. Slavery was made of silence in court enforced by state-sanctioned violence.

However, it is also clear that these definitions of who could speak, about whom, and under what circumstances were continually contested, in ways that require acknowledging other modes of tying utterances to speaking subjects. There were other oaths on these islands, which drew on other traditions that had traveled and been transformed through the black Atlantic. Here "Koromantin" might name a powerful collective identity rather than being a word it was forbidden to say; and Akan loyalty oaths, the *Okyeame*'s swearing of witnesses, or the invocation of the *ntam kese* to convene a court of law might become the binding of a conspiracy or the finding of a thief via oaths administered by obeah men and women. How these new practices altered or maintained forms of status, gender, and political power as they traveled and changed is unclear, but planters and judges like Edward Long knew that they formed a realm of community and identity making within which powerful speech acts had no effect "when tendered by a white person."

This sense of the power of other oaths and of the oaths of others is there in the necessity of oath taking to make peace with the Maroons, and the admission of the fragility of colonial power that saying those words implied. It opened up a space that the equation of blackness and slavery, and the enforcement of silence, always sought to close down. It was a space where other forms of speech emerged, which troubled planter power. Francis Williams's successful overturning of the 1730 act removing his privilege not to have the testimony of the enslaved heard against him, the dreadful predictions of Cudjoe's petition, and Joshua Steele's unpicking of the social relations of plantation slavery via the contradictions of "Negro evidence" each came from distinct social and political positions. Yet they all showed that if these distinctions—the lines drawn—on the grounds of race, freedom, and property were only made via the regulation of "speaking *legally*," then Caribbean slave societies might be remade or turned upside down when others could speak. They revealed, sometimes despite themselves, that the world is continually being made and remade through practice, including speech practices, and that it could be made differently. Moreover, they show that this making and remaking had a geography. These contestations all became part of long-standing transatlantic debates over rights and powers—over the legal geography of the British Atlantic world—which could undermine the relationships between race, property, and freedom that the Caribbean planter class sought to fix. In the Maroons' oaths, Williams's victory, Steele's heretical alternatives, and the world-turned-upside-down of Cudjoe's petition was a recognition that the identities and powers of free, white, propertied men were not foundationally established but were made through legal practices and performances, through speech acts such as oath taking and evidence giving. They showed what happens to the binding of speakers and utterances when legal speech has an Atlantic geography, and they indicated the emergence of other speaking subjects. The next chapter shows that that was also the case in relation to the other key site of the making of the freeborn Englishman in the Caribbean: the legislative assembly.

TWO

The Deliberative Voice: Politics, Speech, and Liberty

In 1765, an Irish lawyer and member of the Jamaican assembly defended the freedom of speech in terms which put the issues of law and consent that were discussed in chapter 1 at the heart of questions of liberty thus: "There is certainly no other distinction between Freedom and Slavery, but that a Freeman has his Life, his Liberty, and his Property secured to him by known Laws to which he has given his consent . . . : Whereas a Slave holds every thing at the Pleasure of his Master, and has no Law, but the Will of his Tyrant."[1] Nicholas Bourke, a slaveholder himself, was using the political rhetoric of slavery to argue for the "privileges" of the island's legislative assembly—including the freedom of speech—against the actions of the governor William Henry Lyttleton and within the tense imperial politics of the 1760s, as the British government tightened its fiscal grip on the American colonies in the wake of the Seven Years' War.[2] Bourke's equation of slavery with the denial of the right to "consent" to the laws which governed Jamaica, and his defense of the assembly against what he saw as the illegitimate power of the British imperial sovereign's representative, raises the questions that this chapter addresses concerning the constitution of polities, political subjects, political spaces, and political relationships through speech in the slave societies of the Anglo-Caribbean. Considering "politics" on these islands in these broad terms, and through speech, makes it possible to bring together matters that are usually treated separately: the tensions between imperial authorities and colonial assemblies over the government of the islands; and the conflicts between slaveholders and the enslaved evident in suspected conspiracies and periodic revolts among the oppressed, and in their violent suppression. As a result, this chapter treats historically what Bourke conjoined rhetorically: the nature of political speech and resistance against enslavement.

Questions of communication have been central to debates over the nature of politics in the eighteenth century since the translation into English in 1989 of Jürgen Habermas's *Structural Transformation of the Public Sphere*. Habermas argued that new forms of discussion, undertaken by new participants in new ways, revolving round innovative uses of print, and happening in new spaces—especially the coffeehouse—produced a public sphere between the private individual and the state where, through critical-rational debate, "the better argument could assert itself."[3] Or, more commonly, Habermas's normative political theory read as history is taken to be insufficiently attentive to race, class, and gender, and therefore to the exclusive nature of the "public sphere" and the production and consumption of print, the rarity of critical-rational debate, and the existence of multiple publics or counterpublics.[4] As a result, despite notions of the public sphere shaping historical debates over European and North American politics in the eighteenth century, they have found little use in contexts of colonial slavery. Any public sphere here, particularly one based on print literacy, is so obviously not "public" in the ways just described, particularly when the discussion is about imperial politics or the resistance of the enslaved rather than movements for abolition and emancipation.[5]

However, while acknowledging the severe and violent delimitation and deformation of any kind of "public sphere" in Caribbean slave societies, it is possible to take from Habermas the idea that forms of "communicative action"—in this case, different forms of speech—constitute different sorts of politics.[6] This idea can then pluralize the forms of speech and politics and demonstrate the particular place occupied within them by the critical-rational debate of the coffeehouse.[7] In understanding the implications of these multiple forms of political speech, it is useful to turn again to Bruno Latour's ideas of regimes of enunciation to examine what he considers is happening when people "speak *politically*."[8]

Political speech, for Latour, is not defined by its content, but as a "manner of speech," whose prime characteristic is that it is always "disappointing" if judged by the demands of straightforwardness or directness, or of a Habermasian "reasonable reason," "enacted through the ideal conditions of communication." Evaluating political talk by the veridiction criteria of science or law means that "political utterances will be accused of concealment and lies, of corruption or fickleness, of falsity and artifice." That, Latour argues, is to misunderstand the particular felicity conditions of this regime of enunciation. Instead, political speech "aims to *allow to exist* that which would not exist without it: the public as a temporarily defined totality. Either some means has been provided to trace a group into existence, and the talk has been truthful, or no group has been traced, and it is in vain that people have talked." It is, he says, a matter of "the continuous creation of the public": "To study political talk we need to abandon the idea of a guaranteed existence of groups. These are continuously being formed and reformed, and one of the ways of making them exist, of making them 'take' as we say of a sauce, is by surrounding them, grasping them, regrasping them, reproducing them, over and over again, by 'lassoing' them, enveloping them, in the curve of political talk. Without this enunciation there would simply be no thinkable, visible, viable and unifiable social aggregate." In short, "there is no group without (re)grouping, no regrouping without mobilizing talk."[9] Moreover, this mobilization of political talk has a particular set of effects in relation to matters of sovereignty and representation. Latour, speaking of "democracies," identifies a circle of enunciation whereby the alchemy of political talk involved in these "regroupings" simultaneously effects the impossible transmutation of the many into the one (the moment of representation) and the one into the many (the moment of obedience, or sovereignty).

Although abstract and focused on understanding the workings of Western democracies, this identification of the specificity of political talk as a regime of enunciation that makes groups, or performatively constitutes temporary publics, can offer a way of hearing what is happening when people "speak *politically*" in the seventeenth- and eighteenth-century Caribbean. For example, in December 1719, the governor of Barbados, Robert Lowther, addressed the island's grand jury. He gave them a history lesson on the Glorious Revolution of 1689 and constitutional monarchy to establish the point that in Britain "Laws are the Test both of the Royal Authority & of the Subjects' obedience" but argued that there were many, particularly in the Church of England, who agitated against the monarchy by "preach[ing] the Pretender & themselves & not Jesus Christ" and created divisions by brabbling, quarreling, and berating; and cursing, swearing, and lying. Drawing on Montaigne's essay on lying, he told them, "We are tied to one another only by Speech, our Understanding being directed by no other way than that of words, he who speaks falsely betrays the public Society: It is the only Instrument whereby we impart our thoughts, it is the Interpreter of our Souls; if we are deprived of it, we are no longer tied to, nor know one another; if it deceives us, it breaks all our Correspondence & dissolves the bond of our Policy."[10] Here, speech—and speaking the truth—is the basis of the bonds of "public Society" and the social glue which holds the polity together. It is clear, given chapter 1's discussion of oath taking and evidence giving, that the idea of "public Society" here is severely constrained by race, gender, freedom, and property ownership. There is also the danger of suggesting that this eighteenth-century politician was motivated only by high-minded principle, however closely circumscribed. So it is important to note that this speech was a plea from Lowther, a Whig, to the jury to convict the Tory Reverend William Gordon, long a thorn in the governor's side, of treason as someone who had "advanced . . . doctrines from the Pulpit that tend to create jealousies and divisions among H[is] M[ajesty's] Subjects, or to incite them to Sedition & Rebellion."[11] Gordon, in turn, argued that he was simply holding the governor to account for the corruption with which Lowther was later charged.[12]

However, this complex factional political context only further demonstrates that political speech is, as Latour has it, a mode of continuous (re)grouping: the ongoing but temporary production of social aggregates through speech and other modes of communication which construct forms of sovereignty and representation. So it is significant that Gordon and Lowther's dispute worked through a range of means and across a range of spaces. Gordon had a pamphlet—A Representation of the Miserable State of Barbados (1719)—printed in London and circulated in Barbados, and hijacked the thanksgiving service the governor had ordered to mark the suppression of the 1715 Jacobite rebellion by preaching what Lowther, fuming in the congregation, heard as "a Virulent Satyr against the King's best Subjects & friends." In response, the governor collected sworn depositions of Gordon's bad character—one called him "a very busy, medling & active person in all party matters & publick differences in this Island"—and had a lengthy "Declaration" that he had written refuting Gordon's claims "published in the Bridge Town by Beat of Drum" by the provost marshal and ordered to be read out "with a Distinct and Audible Voice" simultaneously in all the island's parish churches. However, even those who stayed to hear it—Gordon's supporters reporting that when it started, "the Ladies taking up their Fans, and the Gentlemen their Hats, hasten'd out with so much precipitancy, that they trod on each others Heels"—would also have seen, fixed to every church's doors, a "paper writing" by Gordon, refuting the refutation. Perhaps they even read and discussed it before the assembly had it burned by the common hangman as "Impudent, false & Malicious."[13] At each turn, speech intersected with other modes of communication and political groupings were formed and reformed.

This example also indicates that as well as differentiating between "regimes of enunciation," such as law, politics, science, and religion, it is also necessary to differentiate within them to understand, in this case, how political processes of sovereignty and representation worked through particular forms of talk and in their relations with script and print. Consider again "Cudjoe's petition" from chapter 1, in terms of the modes of political speech it involved and the (re)groupings it effected in mid-eighteenth-century Jamaica. There was the imagined setting out of a political program by "Cudjoe," and the threat that there were "sixty thousand ready to carry the passing of the bill." There was the reading out loud of the petition outside the Kingston courthouse by the wharfinger Archibald Willock, a sort of "politics out-of-doors" designed to convene a public that could put pressure on the assembly. There was the assembly's deliberations on this "false, scandalous, malicious, and seditious libel" within its own chamber. And there were the transatlantic discussions of Jamaica's factional politics effected by James Smith's pamphleteering.[14] What is required, therefore, is a historical geography of political talk attentive to its variety of forms, the spaces it draws on and constitutes, and the particular historical contexts for the making of modes of speech and the conflicts between them.

This chapter shows that tracing these distinct historical geographies of political talk, from royal proclamations to the conspiratorial talk of the enslaved, means that the forms of legitimation—what sovereignty and representation might mean—and even the entities that are involved in the realm of politics, may not be the same as they are in the Western democratic contexts to which Latour's anthropology of "the Moderns" attends. In part, this is a matter of what even counts as political talk. As noted in the introduction, it is useful to consider Jacques Rancière's reflections on Aristotle's distinction between human "speech" and animal "voice," which Rancière calls "the beginnings of politics," since speech can communicate not just what is painful but what is unjust. Yet, to repeat, for Rancière this is not some kind of natural grounding of the human and the political but a key element of his idea that the political is differently made through different "distributions of the sensible," each one ruling on what is "speech" and what is "noise."[15] Once again, however, the politics of silence, speech, and sound in the Caribbean are ones that require more attention to questions of race, gender, and freedom than Rancière offers. Paying such close attention to these dimensions of social difference also questions whether the liberal and Enlightenment category of political speech is even appropriate to the conditions of violent oppression imposed by slavery.[16]

What follows examines the implications of starting from speech to understand politics in slave societies as the continuous creation of publics. This examination begins with the forms of political speech characteristic of white planter society and its place within the empire—sovereign proclamations, official oaths, and the debating practices of the assemblies. As will be

demonstrated, these were the shared basis of forms of governmental legitimacy and authority and, throughout the period, formed the tense ground
for conflict between empire and colony. As a result, these ways of speaking
formed the basis for a strong set of political identities, or subject positions,
which ruled in and out various forms of political speech. They also formed
the basis on which the speech of the enslaved was heard, or misheard, as
white colonists imagined and were confronted by conspiracies and revolts
that also drew on a variety of Atlantic forms of political talk.

Proclamations, Oaths, and Debates

Barbados and Jamaica were both governed through a political structure
said, at least by some, to replicate that of England. Each island had a governor, who was appointed by the monarch as their representative; a council—
with "freedome of Debate & Votes in all Affaires of publick concerne"—
recommended by the governor and appointed by the monarch to advise
and restrain him; and an assembly made up of representatives from the
islands' parishes, voted for by free, white, male property owners.[17] As chapter 1 showed, the key issues in the "transatlantic constitution" through
which these islands were ruled were ones of autonomy and control. There
were conflicts between the governors and the assemblies over raising taxes,
and where and how that revenue was spent; over the place of the islands in
imperial policy, and what sort of protection (both military and economic)
they might receive; and on who had the final say in how the islands were
governed. Official modes of political speech—proclamations, oaths, and
forms of debate—were shared between these powers and were also the
means and subject of disputes.

Proclamations are an obvious form of "sovereignty talk": the voice of
the sovereign in the colony speaking of matters of life, death, law, and
war.[18] For example, on August 15, 1660, the news arrived in Jamaica of the
Restoration of King Charles II, only five years after the island was taken
from the Spanish, "and the seventeenth [of August] was appointed for the
solemn proclamation of the king, and then done with all the expressions
of joy that the place could afford." This was mostly a delegated voice, with
the governor acting as the monarch's mouthpiece. So, on February 27,
1666, "Sir Thomas Modyford caused a war against the Spaniards to be
solemnly proclaimed by beat of drum and proclamation at Port-Royal."
This kind of proclamation was something of a regular occurrence, with
peace proclaimed on June 14, 1670, and war again on July 2, 1670, and on
throughout the long eighteenth century.[19] Indeed, proclamation was part

of the process for delegating that voice and power, particularly the reading of the governor's royal commission "with all due solemnity" as part of installing him in authority on the island, and the processes for instituting and dissolving assemblies.[20] Here, to "publish"—to make public—was achievable by speech and did not require writing or printing, although a combination might be used. Particular forms of words were to be spoken by specific people in combination with the use of objects, such as the large gilt mace given to Jamaica in the early 1680s by Charles II "to make the government appear more great and formal," and as a mark of the authority of the governor.[21]

Proclamations were part of the ongoing government of the colonies. The sound of them being read out filled the streets of the islands' towns as a signal of sovereignty that paralleled the dreadful sounds of judicial violence against the enslaved.[22] The mode of proclamation was specified in each case. In Barbados it was stated whether it should be by the secretary or deputy secretary and provost marshal, usually with soldiers sounding trumpets or drums, and at the marketplace, the stepping-stones, or the customhouse, or in the churches of the islands.[23] For example, in 1693, "to prevent insurrections" of the enslaved, "which are again threatening," the governor of Barbados issued a proclamation to be read in all the island's churches on the next two Sundays, reminding owners and managers of plantations of their duties under the *Act for the Better Governing of Negroes* (1661) and informing the clergy of their responsibility to "publish" the act in church twice a year.[24]

There were strict limits on who could speak like a sovereign. In 1656, in Barbados, it was ordered that no "writing or writings, printed book or books" be published in church except by command of the governor and council. Any "Minister or Clarke" who did would be "Deemed a Disturber of the Publicke Peace." There were also concerns in December 1697 whether a proclamation of peace was authentic, it having some errors, being "not printed by any manner of Licence, nor in the Kingdom of England," and no order to proclaim it having been received.[25] And, in November 1667, Nathaniel Kingsland and Robert Hackett, two military men recently evacuated from the English colony of Surinam when it was taken by the Dutch, had hired George Allen, Bridgetown's bellman, to inform the public that they should not buy any of the enslaved Africans from the mainland over whom Kingsland and Hackett claimed ownership. Allen told the Barbados council that they had given him two sixpences to "publish & cry with his bell," and "that he did go from the said market place, towards the prison & through the usual streets, & publish the words" they had paid him to say. In turn, the court told Kingsland and Hackett that as well as causing discontent among the other soldiers, they had, by making the proclamation, "manifestly usurped upon the authority belonging only to His Majesty, and those that govern under him."[26] Yet, it is telling that in 1673 Barbados's governor and council allowed a printed declaration from the Royal African Company—which had a monopoly over the slave trade to the islands—to "be published in the several parish churches in this Island; or at such convenient place, & at such time, as to the said Royal Company's Agents shall seem meet."[27] The company, headed by the Duke of York, the future King James II, was at the heart of an imperial political and economic strategy based on sovereign power over the colonies and was, as a result, allowed a voice.[28]

These proclamations were important as illocutionary speech acts: words that made something happen. For laws, they were "deemed . . . essential to the existence of an act." So, as Nathaniel Lucas, the early nineteenth-century chronicler of Barbadian government, reported, when the Barbados assembly wished to print the laws of the island in 1697, it found that one act of 1667 was not recorded in the book at the island secretary's office. However, the assembly had the original manuscript that had been signed by the governor, and "it having been Manifested at [the council] . . . by persons of undoubted Reputation, that the same was publisht in the Churches," it was declared law. Lucas's point in recording this story, and offering examples of the reading in English churches of English acts of Parliament, was to establish, against the ideas of the new bishop of Barbados, that "there is nothing Heathenish in our Publication of Laws in Churches; it is undoubtedly of English Extraction & of every day use."[29] Speaking the act was a speech act, and an English one at that.

Proclamations can, therefore, be set alongside oaths (which were discussed in chapter 1) as speech practices that activated sovereign power in the colonies. All members of the political process were bound by and dependent on oaths of various sorts.[30] So, when the governor of Barbados, Sir Francis Russell, died in the night in August 1696, the council met the next afternoon. The council had his royal commission read, and, following its orders, appointed the eldest councilor, Francis Bond, as president. They then issued a proclamation that all officers should remain in their posts until further notice. There was to be no discontinuity of governance as a result of Russell's death. Bond and the council members then all took the appropriate oaths and subscribed to the Test Act (renouncing Roman Catholicism), enabling them to govern. They then took the oath for the security of trade—swearing to uphold the Navigation Acts—and another, voluntary, oath to prevent frauds and abuses in the plantation trade. The assembly members all had to take the oaths of allegiance and supremacy. Then the deputy secretary, provost marshal, and clerk all swore their oaths of office, as would the judges, surveyors, and customs officials.[31] This oath taking, along with the reading of commissions and much speech making, marked the installation of all new governors on both islands. Such oath taking naturally imposed limitations on who could be part of the political and administrative process. In the early years of the Jamaican colony, the requirement that all officeholders and assembly members swear the oaths of supremacy and allegiance was waived to ensure that the posts could be filled, although it was still insisted on for the council. Instead, governors were instructed by the crown to find "some other way of securing yo[ur] selfe of their Allegiance to Us and Our Governm[ent] there."[32]

However, sugar island politics was not simply rule by proclamation. Indeed, those proclamations were lampooned as empty ceremonies conducted for disinterested audiences, particularly in the case of contentious issues such as repeated proclamations of martial law that put a stop on trade and increased the obligations of the islands' planters and merchants (figure 9). Such concerns raised significant questions of representation and, in answering them, what the appropriate forms of talk for the islands' assemblies were as they participated in the process of making laws. As Nicholas Bourke argued, and as chapter 1 has shown, rule by known and appropriate laws was crucial to white islanders' sense of their Englishness and of being part of an extended imperial polity which accorded them the gendered rights of freeborn Englishmen. They accepted that sovereign power established the basis of colonial government and agreed that the sovereign should have a veto over laws passed. They were the monarch's laws, after all, and the islands depended on English law for the security of their property and the ability to transfer it. But, as to the making of the laws by which they were governed, that was a matter, the islands' planters and merchants argued, for the representatives of property owners on the islands themselves.[33] It was, therefore, a matter for the white, propertied men who met as the islands' assemblies to make laws through debating and voting: giving their "consent" to the laws by which they governed themselves. Indeed, the mode of speech used in the assemblies was vital for legitimating not only the decisions made, and the laws passed, but also the political authority of those assemblies.

In large part, the assemblies modeled themselves on the English House of Commons, even calling themselves, in Jamaica, "the Representatives of the Commons of the Island."[34] Indeed, Edward Long argued that since

Martial Law declared by beat of Drum, and the
Standard hoisted in St Jago de la Vega.

Figure 9. Proclaiming martial law in Jamaica. Detail from [A.J.],
Martial Law in Jamaica (London: William Holland, 1801).
Courtesy of the National Library of Jamaica.

the governor was the monarch's representative, and the council was not
the equivalent of the House of Lords, it was only the assembly that was a
perfect copy of an English political institution (*HJ*, 1:10–11).[35] More prac-
tically, legitimating the assemblies' actions involved setting out formal
regulations governing how members spoke that would guarantee that a
bunch of rich white men arguing about politics was a "representative assembly." Thus, in 1674, the Barbados assembly listed its rules.[36] These
rules stipulated that the assembly be governed by a "Speaker"—described
later in Jamaica as "the mouth of the house"—chosen from among themselves by themselves. The presence of fifteen, out of twenty-two, representatives would mean that the assembly was quorate, and decisions would
be by majority vote, with the speaker having the final say. For any act to
be passed, it had to be "Read & Voted" three times in at least two separate
sessions, with at least one Saturday between them. The speaker was to or-
chestrate the assembly's deliberations. The rules ordered "that the Speaker
have a wooden hammer or maule lyeing by him & whensoever he Knocks
or strikes on the table therewith, all the members are to keep silence" or be
fined a shilling. It was also ordered "that in all Debates in the House, every
member who is Intended to Speake signifie itt by his Rising up & onely

address himself in all he saith to the Speaker" or be fined two shillings. The decision on who should speak first on a motion was by majority vote, and anyone interrupting a member (unless to alert the assembly's speaker to a breach of the rules) was to be fined two shillings and sixpence. Discursive order was also to be extended over the content of what was said, and how it was said. It was ordered "that whatsoever member shall speak to any other thing than is in Debate, or Begin any new Question, untill that in Debate be resolved[,] ended or Determined," was to be fined two shillings and sixpence, and "that whatsoever member shall utter any ill or unhandsome Language in any Debate, or use any Reflecting & Reviling speeches to each other shall pay tenne shillings." There were also penalties for absence and the threat of expulsion for misbehavior.

As with the laws governing slavery, the rules of the Jamaican assembly were very similar, although they also stipulated that no one could speak more than twice in the same debate, unless invited to do so by the speaker (a practice later adopted in Barbados), and "that every man place himself as he comes, that there may seem no disparity."[37] That is, assembly members were ordered to take the next available seat rather than sitting in more favored places or together in parties. These rules endured over time, being debated, added to, and amended, but with the basic principles maintained. In 1688, in Jamaica, it was ordered that a "table" of them be "hung up in the house."[38]

These rules can't be treated as descriptions of what happened in the assemblies but rather as presenting a normative political theory of speech. Establishing how political speech should be conducted aimed to prevent some of the contentions seen in early assemblies riven by conflicts between royalists and commonwealthmen. That of Jamaica in 1664, for example, was described as one where "business went on but like bells rung by boys, all jarring, and every day caused more ill blood." Indeed, it had ended with one member killing another at a final dinner where "the plenty of wine made the old wounds appear" and "all went together by the ears."[39] Yet the application of these rules and norms remained contentious.[40] In part, this contentiousness concerned how far the assemblies lived up to the model of the English House of Commons when there was no simple mirroring of procedure. For example, in 1694, the Barbados assembly was censured by the governor for voting in a manner "not practiced in England." While they agreed to desist, they did resolve "that it is their Privilege to make their own Rules, as the Parliament of England proceeds to make their Rules."[41] There were also some affinities to the idealized Habermasian coffeehouse, such as the injunction to take the next available seat, albeit with a presiding authority to see order and punish transgressions that the coffeehouse did not have.⁴² Overall, the rules aimed to ensure decision making through debate among "equals"—an equality achieved by excluding anyone from the assembly who was not free, white, male, Anglican, and with sufficient property—and via the rule of the majority after proper deliberation. The geography that this theory of speech implied is also striking. These rules defined the space of legitimate and determinate political speech. In their early years, the places where assemblies met were less important than how they met, and how members spoke when they met, in defining that place, with those people speaking according to those rules, as an assembly. For example, the Jamaican assembly had met in 1678 in the courthouse and in 1679 in a church, while the Barbados assembly in 1672 had first met at William Wilson's house and then adjourned to the Roebuck just outside Bridgetown, and in 1673 met at Paul Gwynn's tavern.⁴³ It was not until much later that particular buildings were more regularly used or constructed specifically for the assemblies. In mid-eighteenth-century Jamaica, a grand building housing the assembly and the law courts was constructed in Spanish Town across a neoclassical square from the governor's residence (figure 10). Yet the assembly chamber was, in keeping with these rules, a simple and unostentatious wood-paneled room.⁴⁴

This political theory of speech also underpinned the assembly mem-

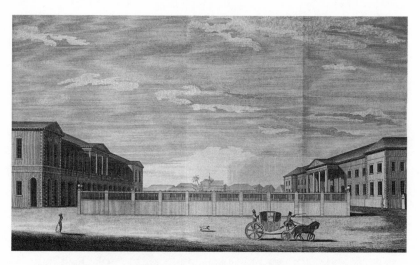

Figure 10. The King's House (*right*) and Public Offices (*left*), housing the Jamaican Assembly, Spanish Town, 1770s. [Edward Long], *The History of Jamaica* (London, 1774). © The British Library Board. 146.c.8.

bers' understanding of their relationship to sovereign power. In Jamaica, at the opening of each assembly session, and after taking their oaths, hearing a speech from the governor, and giving one in return, the speaker would ask the governor to confirm their "privileges" in relation to arrest (guaranteeing their ability to meet without interference), their "freedom of speech," and access to the governor's person.[45] In Barbados, these privileges were taken for granted until the late eighteenth century, when the ritual of asking for them began.[46] This ritual, for the assembly's representatives, confirmed their centrality to the legislative process and their ability to make, and therefore consent to, the laws that governed them. These were "privileges" that were keenly defended to ensure "the liberty of speech of this house," not least within the mid-1760s conflict with the Jamaican governor that prompted Nicholas Bourke's arguments.[47] However, although the monarch, the governors, and the assemblies generally admitted a place for both proclamation and deliberation, they did not always agree on what that was, and where and how political power should be located and exercised. This lack of agreement is what led to conflict. What was at stake in relation to the various forms of political speech can be seen in a foundational crisis in late seventeenth-century Jamaica.

Deliberative Power

The 1670s saw a concerted attempt to reorganize England's empire, particularly in the Caribbean, so that it was increasingly based on centralized, sovereign control and worked for the benefit of the Royal African Company's slave traders and in the monarchy's economic interests.[48] The islands' planters—who provided the market for the enslaved Africans supplied by the company and the taxation revenue to run the colonies, and who asserted their right to govern their affairs through the legislative assembly— saw things differently.[49] In Jamaica this quickly became a political conflict between the right of royal prerogative (rule by proclamation) and the rights of the representative assembly (government by deliberation) within which the very existence of the assembly was a stake. As the conflict developed, the Lords of Trade and Plantations, who oversaw imperial policy for the king in London, argued that Jamaica had "never had any other right to Assemblies than by permission of the Governour's & that only Temporary," while the assembly itself argued that "they never desir'd any power but what your Majesty's Governours assur'd them was their Birth Rights."[50]

The immediate substance of the conflict was that, in 1678, the Earl of Carlisle had arrived as the new governor of Jamaica with a set of laws drawn up by the Lords of Trade based on those previously passed by the assembly under the governorship of John Vaughn but not ratified by the king at that time.[51] What was new was that the assembly was asked to pass the laws as "originally coming from" the sovereign, on the model of Poynings' Law in Ireland.[52] It was a denial of the active role of the assembly in making law prompted by "His Ma[ty] taking offence at these irregularities which did seem to arise from the deliberative Power of the Assembly in framing & passing of Laws." This change, if accompanied by a perpetual revenue bill to fund the island's government, meant assemblies would only be called when the king desired it, or by the governor in exceptional circumstances such as rebellion or invasion. They would certainly not be regular deliberative chambers making the island's laws.[53]

Conflict over political speech was how the crisis worked through as well as its cause. Carlisle was also to end the earlier practice on oath taking and to insist that "for the prevention of future inconveniences, and greater assurance of loyalty towards yo[ur] Ma[jesty]," no assemblymen should be allowed to sit until they took the oaths of allegiance and supremacy. The governor's power over debate was also reinforced by a rule that if he dismissed a member of the council, they could not join the assembly.[54] Carlisle also tried to make the council members and assemblymen swear an oath "that they would submit & acquiesce to the said form of Government until his Majestys further pleasure should be known concerning the same." Most refused, saying they had taken the pledges prescribed by law, and that because this new oath was an "oath of his own invention," it was one "for which there was no authority."[55] Finally, Carlisle was instructed by the Lords of Trade to continue the laws in force by proclamation if there was any doubt about their validity, acting in the name of the king and over the heads of the assembly to keep the colony running.[56] It was, however, clear that these measures risked losing the legitimacy to govern in the name of any form of collective good at a time when the island was threatened by slave revolts and war with the French, and the treasury was almost empty. Moreover, it seemed evident to the planters that Carlisle's personal political and economic interests were in opposition to their own.[57]

In turn, the assembly tightened control over collective deliberation, the source of their legitimacy and power. Despite Carlisle's encouragement that "upon their discoursing it openly and freely, they might bee the better convinced of the necessity of their being dutiful therein," they refused to debate the constitutional changes with him, or even in his presence, declaring it was a breach of their privileges for him to enter the assembly. They also strengthened their collective voice by ruling that members of the house were prohibited from "divulg[ing] any of the secrets of it, or declar[ing] what is there debated, before it comes to a vote; or mention[ing] the name of any person that moved or spoke in the house." Once a matter was decided collectively, it "should not therefore be discoursed of or debated . . . by any private members, lest the whole house receive disadvantage by it."[58] What was debated in the chamber was to stay in the chamber, or risk undermining the assembly's power. This requirement for secrecy then produced conflicts over who was entitled to know what had been said, with Carlisle enforcing what he claimed was the king's right to appoint the clerk to the assembly, and the Lords of Trade reiterating their requirement of a quarterly journal of the assembly's proceedings, which they told the governor would "hinder the concealment [he] complained of." Yet the assembly could not be made to say what the governor, the Lords of Trade, or the king wanted them to say. As they put it to Carlisle, "they will submit to wear but never consent to make chains" for their descendants.[59] However, by not passing a revenue bill or the laws the governor had brought, they were putting their own property, and that of those they represented, at risk as there was no legal framework to guarantee its possession and transfer.

This conflict did not simply set the assembly against the sovereign power of proclamation. It would have been very dangerous to deny the royal prerogative. Instead the assemblymen argued that the king's word was "the best means of preserving our own privileges and estates" because it had proclaimed their liberties. In particular, they cited a 1661 proclamation by King Charles II, who, to encourage settlement, had "publish[ed] & declar[ed], that all children of our natural born subjects of England to be born in Jamaica . . . shall be free denizens of England, and shall have the same privileges . . . as our free-born subjects of England." And they pointed to the governors' royal commissions that had established the lawmaking assemblies they claimed as "their birthrights."[60] Citing these royal words supported an argument that they distrusted powerful governors, while remaining loyal to the king, and that "as his [Majesty's] English subjects they ought not to be governed as Irish Men" but by laws they made themselves according to "their ancient form of government." The crucial issue was consent and the forms of deliberation on which it depended. As the assembly put it, "It is no small satisfaction that the people, by their representatives, have a deliberative power in the making of laws, the negative and barely resolving power being not according to the rights of Englishmen, and practiced no where but in those commonwealths where aristocracy prevails."[61] The sovereign—or the governor, acting *in loco regis*—could say only yes or no. The assembly had to deliberate.

Carlisle deployed all the strategies and resources of political speech, but he had to report that "having used all methods possible with the several Members apart, and jointly with the body of the Assembly for ye passing the Laws, I was, after many conferences, debates, and several adjournm^ts, frustrated, and they threw them all out."[62] The assembly took its lead from Samuel Long, Edward Long's great-grandfather, a council member and former speaker of the assembly, who, as Carlisle put it, the assemblymen "esteem[ed] . . . the Patron of their Rights and Privileges as Englishmen," to the extent that "they spake his words and were solely guided thereby." Carlisle could not keep Long quiet, even after removing him from public office. In the end he brought Long to London to answer the charges of leaving the king's name out of a revenue bill and obstructing royal policy. As Carlisle put it, this would mean that Long and "some of the stubbornest of ye General Assembly [were] to before your Lo[rdshi]ps receive those reasons as sufficient inducements for their obedience in England which they will not bee persuaded to admit of in Jamaica."[63] The matter would be decided before the Lords of Trade.

Having sought information from former governors on the history and status of the legislative assemblies, from "substantial merchants trading to Jamaica" on how to regulate the island's trade, and advice from the solicitor general, attorney general, and the judges, on the constitutional position, the Lords of Trade aimed to determine the question of "Whether by His Matie's Letter, Proclamation, or Commissions [to governors] . . . , [he] hath excluded himself from the power of establishing Laws in Jamaica, it being a conquered Country and all Laws settled by authority there being now expired."[64] This was no foregone conclusion. Former governors Thomas Lynch and John Vaughn affirmed the long-standing legislative role of the assemblies, and others confirmed that the 1661 proclamation "was published in Jamaica." The Jamaica merchants reported that the crisis had substantially reduced investment, and the legal authorities could not produce a definitive ruling.[65]

Over a series of meetings during the winter of 1680 in the Whitehall council chamber, the Lords of Trade sought to establish which laws were in force, and the constitutional mechanisms for making Jamaican legislation. They heard evidence read, including the 1661 proclamation, and they questioned witnesses. Samuel Long and the other Jamaican politicians appeared before them and stood their ground. They argued that laws were not legitimate if made "by the Governor and Council without an Assembly," and they got overturned, on a technicality, a ruling that the revenue act of 1663 was still in force on the island.[66] With the king beset by political difficulties elsewhere, the inability to get a definitive legal judgment supporting the prerogative, and the Jamaican planters and merchants—including Long and the speaker of the assembly, William Beeston—petitioning for "remitting them to the ancient form of Government and regulating the Negro-Trade," a deal was struck.[67] A seven-year revenue bill would be passed in exchange for the king agreeing to the laws the assembly made for the same period (later extended to twenty-one years), and the Royal African Company would agree to new terms of supply, price, and credit for enslaved African labor that were more favorable to the planters. Most important, the imperial authorities agreed, in the language of the prerogative, to "his majesty's favor to the assembly of Jamaica in restoring to them the Deliberative Voice" in making laws. They had won back their right to speak, which was to be established on the same basis as for the Barbados assembly.[68]

In both the laws passed and the king's acknowledgment of the assembly's legislative role, this was what Thomas Lynch, who replaced Carlisle as governor, described as a new foundation for Jamaican politics. It was to be recorded in writing in the colony's records so "that you may discourse it to your children, they to their posterity, so that the generations to come may know it, bless God for it, and recur to it as to another kind of *magna charta*," a declaration of rights from the monarch. Yet it was also intended to stop other forms of political talk. Lynch, speaking to the assembly in 1682, warned them to avoid "refining on things that are not pertinent, or reviving those detestable names of caballers, prerogative and property men," divisions that had been widened and deepened by the constitutional crisis.[69]

The Bad Language of Politics

As a structural tension within imperial politics, conflicts between prerogative power's proclamations and the "Deliberative Voice" extended across the long eighteenth century and all the Anglo-Caribbean islands. For example, in 1747, an exasperated Governor Trelawny of Jamaica wrote that "managing a Colony Assembly" was "of all farces . . . surely the greatest, & most stupid." As he went on, "they do not see, except 2 or 3 of them, beyond their nose; & yet they will prate an hour together upon liberty, of which they have no notion, mistaking it for licentiousness, (a liberty for every one to do as he lists, & his humour inclines him) or they will harangue as long against a Governor's power, who really has not enough."[70] Presenting the countercase in Barbados in 1767, the assembly argued that its privileges, including freedom of speech, "must be the natural inheritance of any representative body of free People, or every such body will be found upon the trial, to be a form without a substance, whose powers are inadequate to the end for which they were created."[71]

As each major conflict emerged, those involved drew on the history of previous confrontations. In Jamaica, the establishment of the voice of the assembly at the heart of planter identity in 1679–80 was revisited in the conflict with Governor Lyttleton over the assembly's privileges in the mid-1760s. When Bourke's speech about liberty and slavery was printed, it carried an appendix containing the most important late seventeenth-century documents.[72] Both episodes were then retold in detail in Long's *History of Jamaica* (1774), celebrating both his own and his ancestor's place in the island's history as defenders of the assembly's rights. Indeed, Edward Long had also briefly been speaker of the assembly during his time on the island.[73] Additional seventeenth-century archival materials were later uncovered in London by the Jamaican assembly member, planter, and historian Bryan Edwards, and reprinted in full in the first volume of *The Laws of Jamaica* in 1792 and in Edwards's history of the West Indies a year later as a proslavery defense of the planters' right to make law in the face of the growing abolitionist challenge. The use of the past for political ends was repeated, even more pointedly, in the early nineteenth century as the imperial center again threatened to make law for the islands in the name of the "amelioration" of slavery (as we will see in chapter 5).[74]

In these conflicts between assemblies and governors, colonies and empire, forms of speech were worked together with print in ways that made these disputes transatlantic ones, with active and vocal political audiences in both the Caribbean and London. While all sides involved themselves in official and private correspondence, there was also a broader audience to be reached through books, pamphlets, and newspapers. Much of this material was printed in London, even after the arrival of presses in the islands in 1717 (Jamaica) and 1730 (Barbados), and it circulated back and forth, prompting further debate and more paper.[75] For example, Reverend Gordon and Governor Lowther's clash in early eighteenth-century Barbados was accompanied by a number of lengthy pamphlets on both sides and concerted effort to prove who had printed and circulated them.[76] One, *A Letter from an Apothecary in the West-Indies* (1720), reported that a satire on Lowther as Don Quixote the second—and "How he fancy'd himself the greatest Monarch under the Sun and commanded Obeisance from every one accordingly"—had been posted "on several of the most publick Places" in the island, and that a Barbadian poet was turning Lowther's self-justificatory *Declaration* "into Doggrel or Hudibrastick Verse" to make it "more useful and entertaining" when "read amongst the Congregations of the Upright."[77]

What mattered was the effect of such writing within dynamic forms of political speech in both public and private. In 1714, Hugh Totterdell and Peter Beckford, in Jamaica, were informed from London that the pamphlet *The Groans of Jamaica*, which attacked the faction including attorney general William Brodrick (see chapter 1) as "a Triumverate of vile, profane, mercenary, and ignominious upstarts" with undue power over governor Archibald Hamilton, "has made an abundance of Noise here among the Ministry & all Persons anyway concerned in that Island." Since it had not been "publickly sold" but was sent to each member of the British Parliament by the penny post, they were told that "you cannot imagine how greedily that Pamphlet is hunted for." It had raised doubts about the executive's arguments that "all your Assembly's Addresses & general Complaints were but things to be expected of Course from a Factious Brood that can never be contented with any Governm[en]t," and it threatened the governor's future. Consequently, their correspondent sought more material to press the case and suggested that the man currently charged with doing it would not be effective since "he seems as little fit for Action as an Old Woman of ffour Score."[78] For his part, Totterdell asked his correspondents for "the most valuable pamphlets and witty Ball[ards]" which appear daily in London.[79] He may have been sent *A View of the Proceedings of the Assemblies of Jamaica* (1716), which called *The Groans of Jamaica* "a most egregious, false and scandalous paper" that, once distributed on the island, had affected the assembly elections and produced a body which engaged in "absurd ridiculous Lying" and "gross abject Flatteries."[80]

These wars of words developed a basic grammar of illegitimate political speech that—contrasted with official proclamations and the rules of debate—could be used in accusations against opponents and to effect more or less temporary forms of political grouping and regrouping. Governors were presented as overbearing. Gordon accused Lowther of using the "sort of Language and Behaviour which *Westminster-Hall* is a Stranger to," including "Scurrilous and Ungentleman-like Language at the Publick Council Board" that was reminiscent of Oliver Cromwell's pronouncement that he would not be controlled by the "Magna Farta."[81] Supporters of Governor Byng, also of Barbados, complained that he was accused of being a "*haughty imperious Governor*" who treated everyone with "Disdain, Anger, Distance, or indecent Language."[82] The danger of political control by "arbitrary Power" was that it produced silence; a situation where "these supposed Directors, have the very Tongues of the People, and the Voices of the Assembly in their pockets."[83]

In contrast, assemblies were charged with being unnecessarily argumentative, squabbling, and prone to prate for hours on their supposed liberties. All sides accused each other of forming factions to shape the political process in their interests and using manipulative language to make illegitimate groupings. As *The Groans of Jamaica* put it, identifying Richard Rigby as the head of the "Triumverate," the problem was that the governor had been "implicitly captivated, ensnar'd, and deluded, by the fawning and sycophantick Addresses, the smooth artificial Smiles, the fulsom designing Flatteries, the close and parasitical Attendance, the servile and indefatigable seeming Obsequiousness, the frothy, obscene and prophane Witticisms, the many false insinuations, the malicious Suggestions, the artful, crafty, and most injurious Contrivances, and with the fallacious, equivocal, gawdy and affected Oratory of so over fashionable a Demi-Foreigner, and so grossly dis-ingenuous a Sophister, as our *Machiavel Junior*."[84] In the same vein, James Smith's 1747 *Letter* (see chapter 1) also denounced political opponents for "servile flattery," scurrilous abuse, the love of scandal, taking false oaths, haranguing, and attempts to manipulate assembly debates and decisions. One political fixer was described by Smith as being "insinuating and artful," so that "his very Approach to a Gentleman warns what is to be expected of him; the Knowledge of himself, scarce ever permits him to look into the face of those he speaks to, which is seldom in public, and hardly exceeds a Whisper in private."[85]

The boundaries between forms of political speech were also fought over in contests about what could be said where and when: the practicalities of freedom of speech, and how far those freedoms extended. These disputes involved whether speeches in the assembly could be prosecuted as libels, how far the intention of the speaker mattered, and whether words spoken when the house "had no business before them" were the same as when they had.[86] When Hugh Totterdell was prosecuted by the Jamaican governor in 1704, for speaking "several words, very much derogatory to her majesty's authority and government in the island," the prosecution was clear that "the liberty of speech" did not apply since the case was brought not "for any of his transactions in a parliamentary way, or any thing spoken publicly in the house, or reasoning concerning any matters of parliament, but for words privately spoken to some of the members, endeavouring thereby to create ill will against the governor, and contempt of the government . . . and tending to stir up sedition and rebellion." In response, in asking for their privileges, the speaker subsequently requested the governor's "favourable construction of their words and actions."[87]

These contestations over the legitimacy of political speech were also profoundly shaped by the ways its forms were gendered and understood through racialized notions of freedom and slavery. The exclusion of women, people of color, and the unfree from formal political positions, and therefore from making proclamations and from the debates of the council and the assembly, was crucial to constituting free, white propertied men as political subjects whose voices would be heard in the ways described so far in this chapter. Forms of political speech were also denigrated by feminizing them as seduction, flattery, or mere gossip. In part, such feminization asserted differences between public and private spheres. However, these differences were far from fixed. The legitimacy of public figures could be condemned on the basis of their private lives, and it was clear to all that domestic spaces were sites of political discussion among white men and women that could, under certain circumstances, become public matters.[88] There were also spaces for political discussion, such as the coffeehouse, which had an uncertain position, and where talk might be taken seriously or dismissed as "a Coffee House or Gossiping Story."[89] Indeed, these instabilities of public and private spaces could also be used to construct positions from which to speak truth to power. *The Barbadoes Packet* (1720), part of the attack on Governor Lowther, included a letter to him from a woman, naming her as "Marcia." Readers were told that while they "may probably be supriz'd to hear of a Woman in *America*, capable of exerting herself in so uncommon a Manner, against *abused Authority*," she had done so because "her Courage in the Behalf of virtuous Liberty would not suffer her tamely to sit still." She was, the letter said, taking "the Advantage which my Sex gives me"—although whether the letter really was from a woman was uncertain—and using the means of a private letter, albeit one now in print, to "allow me to use such Freedoms as I know you are not wont to receive" and "which would be Provocation enough to whip or set in Pillory the Man that dar'd say half as much to your Face." With that, "Marcia" delivered some home truths, telling Lowther that his lies would soon be revealed.[90] Women's speech might find a place as part of politics out-of-doors.

Finally, the freedom of speech was intimately linked to freedom itself. It constituted freedom and slavery as political conditions that were grounded on who spoke and who did not. This distinction was evident in Samuel Long and the Jamaican assembly's refusal to consent to making chains for their children and grandchildren. It was also evident in Gordon's supporters' claim that their inability to have their complaints heard "means acknowledging ourselves, who are born to such Liberties and Privileges as no People under the Sun have besides, the most despicable and miserable of all Slaves!"[91] And it was at the heart of the political controversies of the 1760s with which this chapter began. For the Jamaican assembly, the defense of its privileges against Governor Lyttleton was a matter of "giving the People of this Colony that Protection against arbit[r]ary Power, which nothing but a free and independent Assembly can give." The specific moment of asking—or refusing to ask—for their privileges became a highly charged one, in both Jamaica and Barbados, as the colonists sought to demonstrate that "the Assembly consider their privileges, as derived to them from their Constituents; and that they are not concessions from the crown, but the right and inheritance of the people."[92] "The privilege of speech" was, as Samuel Frere put it, "a constitutional privilege, inherent in that body." Asking for it was "a mere act of manners" that the sovereign could not refuse. To not stand up for those rights was, therefore, to become a slave. As Bourke insisted, "IT IS THE POWER, WHICH ANY MAN HAS OF TAKING MY LIFE, LIBERTY, OR PROPERTY WITHOUT MY CONSENT, THAT CONSTITUTES AND DEFINES SLAVERY."[93] And it was only slaves who would "submit to Oppression." It was better to be forced into submission than to consent to such a condition. Slavery was, therefore, a political state of war.[94] Edward Long, quoting Montesquieu, called slaves "the natural enemies of society" (*HJ*, 2:431), and, for Bourke, it was the duty of free men to resist slavery's imposition. Indeed, as he argued, "His Majesty, and every Honest Man in Britain will think the better of us, for shewing a Manly resolution and constancy, in defence of our Privileges."[95] As John Dickinson of Philadelphia put it, accusing Barbadians of being "a people *chusing* to be *slaves*" when they failed to resist the Stamp Acts through which the British empire sought fiscal control over the colonies, "The *British* nation aims not at empire over vassals. And must, I am convinced, be better pleased to hear their children speaking the language of freemen, than muttering the timid murmurs of slaves."[96]

Planters might argue, even in the face of clear evidence to the contrary, that enslaved Africans did not have "that idea of liberty which European nations have" and were happy in their "ignorance."[97] Yet other commentators—considering slavery as a political condition and its existence as a choice for politicians looking to the "publick Benefit"—used Puffendorf and Locke to argue that people could not be legitimately enslaved by force, since there was no "Right to the perpetual Service of every one he is able to beat, with whom he quarrels, or picks a Quarrel."[98] This meant that by the mid-eighteenth century, the assemblies' talk of liberty and slavery was seen as hypocrisy, with, as Long noted, "the enemies of the West-India islands" asserting "the impropriety of their clamouring with so much vehemence for what they deny to so many thousand Negroes, whom they hold in bondage. 'Give freedom' (say they) 'to others, before you claim it for yourselves'" (HJ, 1:5). While distinctions between political subjects might increasingly be drawn on racialized grounds (see the introduction), the continued construction of potent political identities—of freedom and slavery—from something as impermanent, mutable, contested, and open to appropriation as speech practices meant that attempts to tie those practices to particular sorts of people were both constantly transgressed and met by the violent policing of the boundaries that this organization of speech sought to establish in all that hot air. This was particularly the case since—as with the law—political identities, or subject positions, were always constituted in relation to race and freedom while, as with Francis Williams and the Maroons (discussed in chapter 1), these categories were never simply coterminous. Neither were these the only forms of political speech on the islands. The next section explores the ways in which the assembling of polities through political speech was spoken and heard among the enslaved and free black populations of Barbados and Jamaica as they sought their own liberty.

Speaking against Slavery

Could the men and women of color on these islands have a political voice? As noted at the beginning of this chapter, one important argument is that the politics of the enslaved under such violent oppression is simply beyond speech and should be understood as undercutting the liberal Enlightenment categories of political identity that the speaking subject is heard to enunciate. In this way, in her examination of the violence visited on enslaved women, Marisa Fuentes argues that "shrieks and cries are a rhetorical genre of the enslaved"—"the sound of someone wanting to be heard." She describes it as a genre that both "communicates an historical and human condition in response to routinized violence" and, quoting Fred Moten, "resists certain formations of identity and interpretation by challenging the reducibility of phonic matter to verbal meaning."[99] Similarly, Elizabeth Maddock Dillon, examining politics and performance in the Atlantic world, draws on Glissant's discussion of the "counterpoetics" of creole to argue that "speech thus takes on a materially sonic form, abrogating the sense-making qualities of language in the name of shaping a new sensorium—'a verbal delirium [at] the outer edge of speech'" made up of "'improvisations, drumbeats, acceleration, dense repetitions, slurred syllables, meaning the opposite of what is said, allegory and hidden meanings.'" She also argues that the meanings of such sounds are performed in conjunction with nonlinguistic forms of expression and representation, particularly textiles and dance. There is, she notes, using Rancière's term, a "division of the sensible" here between "*phonê* and *logos*—sound (mere noise) as distinct from language," which makes some noises, and the people who make them, politically meaningful and others not.[100] This argument is certainly evident in ideas of singing, drumming, horn blowing, and dancing as elements of an autonomous culture of the enslaved that could become part of more direct forms of resistance. It is also apparent in the prohibition of such sounds and activities by colonial authorities for those very reasons, particularly in the wake of revolts of the enslaved.[101]

There are, then, quite different positions on speech and politics taken by scholars who want to break the links between voice, identity, and resistance for people of color on these islands and those who want to affirm those connections, albeit through other modes of communication. The usefulness of Latour's broad sense of "speaking *politically*" as the temporary convening of publics via speech—or, indeed, any form of collective, through any sort of communication—is that it leaves those questions open. It is then possible to examine what spoken words and other sounds do politically rather than assuming in advance that there are particular relationships between speech, identity, and politics. It is, therefore, possible to look for forms of political speaking among both the enslaved and free(d) people of color such as the Maroons that simultaneously hold a mirror up to and disrupt the categories of political speech that were so important to the islands' white politicians: oaths, sovereign proclamations, and modes of deliberation.

The legal and political functions of oaths are entwined. As chapter 1 has shown, these speech acts tied individuals into groups through forms of law, and therefore into polities, as they bound speakers to their utterances. They have a long and rich African and Caribbean history. Swearing an oath for or against the political order was much the same act in terms of its invocation of fidelity, life, and death. Planters such as Edward Long were certainly convinced that the oaths administered by practitioners of obeah were crucial to the making of slave revolts—particularly the extensive Jamaican revolt of 1760, which he had witnessed—and that these oaths signaled an alternative political order on the islands. This, he argued, meant that revolts of the enslaved could be traced back to words said several years earlier:

> It was . . . well known, that several Coromantins, who had actually been in arms during the late commotion, whilst their cause wore a promising aspect, slunk away afterwards, and returned again to their duty, affecting great abhorrence at the behaviour of their countrymen, and even pretending that they had been exerting themselves in opposition to the rebels. With good reason therefore it was suspected, by many persons in St. Mary's, that these deserters, who had taken the *fetishe*, or oath, which they regard as inviolable, would dissemble their genuine sentiments for the present, and wait a favourable opportunity to execute their bloody purposes. Some time in July, 1765, there was a private meeting in that parish, of several Coromantin headmen, who entered into a conspiracy for a fresh insurrection, to take place immediately after the Christmas holidays; they bound the compact with their fetishe, according to custom, and received assurances from all or most of the Coromantins in the parish, that they would join. (*HJ*, 2:465)

The oath, properly administered, was a lasting bond—guaranteeing secrecy, ensuring fidelity to their leaders, and promising death to their enemies—which, for Long, assembled an invisible underground polity or "cabal" and meant that even "if defeated in their first endeavours, they still retain the solicitude of fulfilling all that they have sworn; dissembling their malice under a seeming submissive carriage, and all the exterior signs of innocence and chearfulness, until the convenient time arrives, when they think it practicable to retrieve their former miscarriage" (*HJ*, 2:473).

There were also proclamations on life and death, law and war, issued in the name of powers other than imperial sovereigns and their representatives. While, in making the treaties in 1739, the Maroons may have accepted a form of provisional sovereignty, their existence continued to unsettle where sovereign power lay.[102] Several years after swearing the treaty oath for the leeward Maroons, Cudjoe, it is said, discovered that some of his "head men" were dissatisfied with it and, perhaps, that "They had not the same notice taken of Them" by the British as he did. In response to this dissent from his power and his political strategy, Cudjoe sought to demonstrate his "inviolable Regard to the Treaty"—and to produce in practice the sovereign power over his followers with which the British had credited him—by sending four of these head men to the governor, accused of having "Entered into private Caballs with the Negroes in the neighbouring Plantations, and incited them to Revolt." The courts, finding them guilty, proclaimed death sentences on two of them and ordered transportation for the others. The governor, taking into account that these were first offenses, granted them pardons and sent them back to Cudjoe. He, in turn, "insisted upon his Own Authority," executed the two condemned to death, and sent the others back to the governor for transportation. His actions certainly made it unclear who had the sovereign power of deciding on the exception: of pardoning, or not, in relation to the law.[103]

Indeed, Maroon powers of sovereign proclamation might be distributed in ways that were troubling to white male observers. Philip Thicknesse, one of the officers sent to negotiate the treaty with the windward Maroons, later reported that he had the "mortification" of seeing the "under jaw" of the "poor *laird of Laharrets*"—a previous treaty negotiator, with particularly distinctive teeth—"fixed as an ornament, to one of their hornsmen's horn." Seeking to find out what had happened, Thicknesse was told by the Maroon leader, "Captain Quoha," that he had been willing to agree to the same terms as Cudjoe but said that "when I consulted our *Obea woman*, she opposed the measure, and said, *him bring becara for take the town, so cut him head off.*"[104] The sentence was duly carried out.

Thicknesse was recalling events fifty years later, and many of the details of his self-serving account cannot be corroborated. He certainly wished to stress his courage in the face of danger, having been accused of cowardice in a previous encounter with the Maroons. Yet he still allows a glimpse of what seems to be the sovereign power of life and death wielded on the word of a woman whose political authority seems to have rested on her longevity and her skills as a ritual specialist. As a woman, she may have acted as a complementary power to Quoha, a male military and political leader, in ways suggestive of diasporic transformations of African forms of shared sovereignty.[105] Indeed, gendered forms of African sovereign authority, and their material trappings—transformed in diaspora, and remade under conditions of slavery—were drawn upon as part of revolts of the enslaved, particularly those involving Akan ("Mina" or "Coromantee") "nations."[106] It is, however, now impossible to know what words were said, and how they were spoken, when "one *Coffee* an Ancient Gold-Coast *Negro*" was chosen as "King" and crowned in June 1675 in Barbados "in a Chair of State exquisitely wrought and Carved after their Mode."[107] Or when "a wooden sword . . . with a red feather stuck into the handle" was carried in a procession in Kingston in 1760 as "the Signal of war" (*HJ*, 2:455). Or by what verbal formula in the same year "the Coromantins of the town had raised one Cubah, a female slave belonging to a Jewess, to the rank of royalty, and dubbed her *queen of Kingston*," where she attended meetings with robe and crown, seated under a canopy.[108] We do, however, know that she was not just a mute symbol. Arrested for "some rebellious conspiracies" during the 1760 revolt and ordered for transportation, she "prevail[ed] upon the captain of the transport to put her ashore again in the leeward part of the island" where there was still fighting. She was only silenced when she was finally captured and executed.[109]

Other powerful statements that convened rebellious collectivities were made at the threshold between life and death. Thomas Thistlewood noted in his diary in September 1760 that "Cardiff," sentenced to be burned to death at Montego Bay, "before expiring informed the spectators 'that Multitudes of Negroes had took Swear that if they failed of Success in this Rebellion, to rise again the Same day two years'" hence.[110] The power of obeah practitioners to act on this boundary between worlds was part of the making of political groupings. As Long wrote of the 1760 revolt, "these priests administered a powder [to those who swore to fight], which, being rubbed on their bodies, was to make them invulnerable." They also told the combatants "that Tacky, their generalissimo in the woods, could not possibly be hurt by the white men, for that he caught all the bullets fired at him in his hand, and hurled them back with destruction to his foes." One "old Coromantin" captured in St. Mary's parish during the revolt was tried and hanged, surrounded "with all his feathers, teeth, and other implements of magic," a punishment intended to terrorize the enslaved (HJ, 2:451–52).[111] Yet this obeahman would not pass quietly. As it was later reported, "At the Place of Execution, he bid defiance to the Executioner, telling him, that 'It was not in the Power of the White People to kill him.'" This proclamation on sovereignty, life, and death was taken by those "White People" as a mistaken belief, born of superstition and trickery, that the powers of obeah could prevail against the death-dealing violence of the colonial state, so that the "Negro (Spectators) were greatly perplexed when they saw him expire."[112] Long also argued that "many of his disciples . . . soon altered their opinion of him," "notwithstanding all the boasted feats of his powder and incantations" (HJ, 2:452). Yet this proclamation might equally be taken as a statement of the illegitimacy of a murderous act: that "the White people" did not have the right to execute him, as they had no sovereignty ("power") over him. Or it might also be an ontological statement: that extinguishing his physical life would not destroy how he would still inhabit a world powerfully moved by ancestral spirits. The spectators may have demonstrated their "perplexity"—in another version, they were "astonished"—because of the revelation of this collective of humans and spirits at the point of death rather than because of its denial.[113]

But what of the "deliberative voice"? Long knew that slaves talked, and not just to swear oaths and make proclamations. He argued that the underground polity that underpinned the 1760 revolt was geographically extensive as well as enduring over time, so that there was "scarcely a single parish, to which this conspiracy of the Coromantins did not extend." There had been, he wrote, a "strict correspondence which had been carried on by the Coromantins in every quarter of the island" and an "almost incredible secrecy in the forming their plan of insurrection." This, he argued, was a particularly dangerous instance of a more general set of island-wide oral communications. The relationships of family and fictive kin that the enslaved enacted through relayed messages and personal visits were, he said, "in almost every parish throughout the island. . . . Thus a general correspondence is carried on, all over the island, amongst the creole blacks; and most of them become intimately acquainted with all affairs of the white inhabitants, public as well as private." (*HJ*, 1:114, 2:451, 455–56). Although Long mistakenly argued that revolts were only the work of "Coromantins" and not creoles, these were networked "publics" convened through speech that inhabited planters' fears.[114] It is important to note that they were not separate from planter politics, either. As Long suggested, when "party feuds" took hold in Jamaica, "even our very Negroes turn politicians" (*HJ*, 1:25). And, in 1776, John Lindsay, the rector of Spanish Town, complained in a letter to the historian William Robertson in Edinburgh that a decade of "our late constant disputes at our tables (where by the by every person has his own waiting man behind him)" and involving toasts to colonial heroes, such as the American rebels, who had received "Immortal Honours for Encountering Death in every form, rather than submit to Slavery let its chains be ever so gilded," has meant that "Dear Liberty has rung in the heart of every *House-bred Slave*, in one form or other, for these Ten years past." For Lindsay, loose talk of liberty, such as that engaged in by Nicholas Bourke, was dangerous because it cost white lives.[115]

While Lindsay's concerns connected the North American colonies and the Caribbean through the speech of white planters, merchants, and politicians, these were only a small part of the flows of intelligence, news, rumor, and gossip which, carried by sailors, slaves, peddlers, and traders, circulated among the sugar islands and around the Atlantic world. What Julius Scott has called "the common wind"—a regional network of oral communication binding the societies of "Afro-America" and developed through inter-island and transimperial trade, especially after the 1760s—blew particularly strongly between Jamaica, Cuba, and St. Domingue in the 1780s and 1790s. These islands, although within different empires, were separated only by easily navigable stretches of water and were bound to each other, and other parts of the Atlantic world, by regional and oceanic trades in enslaved people and other commodities. As well as providing slavery's infrastructure, these connections formed the basis for multiple fleeting conversations within which news and rumor—of changing imperial policy, opportunities for trade, and prospects for freedom—were impossible to separate.[116] Indeed, the process of passing and parsing rumors—a form of speech "poised between an explanation and an assertion"—was an active political engagement with the power-laden worlds the enslaved inhabited and a way of collectively imagining and assessing the possibilities for freedom.[117] So, when a massive, and ultimately successful, revolt of the enslaved erupted on St. Domingue in 1791, their counterparts on Jamaica's north coast knew about it before the slaveholders did. Black Jamaicans had found out through what one plantation owner called "some unknown mode of conveying intelligence among Negroes," prompting discussion on all sides of what events across the water might mean for Jamaican slavery.[118]

All these discussions of concerted oral communication among the enslaved were underpinned by the specter of such revolts and fed by the inability of anyone to differentiate between intelligence and rumor. This uncertainty produced fear on all sides.[119] A planter's report from the thick of the 1760 revolt noted that "the Acc^{ts} are so Various there [is] scarce any dependence upon them but I think these are the most Authentick I Can Learn."[120] The enslaved would have felt the same about what they were told. Yet such uncertainty could also be exploited. In 1796, following the Second Maroon War, Captain Gillespie of the Twentieth Dragoons reported to Jamaica's Governor Balcarres that three years earlier, on St. Domingue, Deputy Sonthonax—the official who had declared the end of slavery in the French colony—had told him that he hoped that the "'Children of the Blue Mountains woud shortly throw off the shameful yoke, under which they then groaned: and establish the Tree of Liberty on the same Basis of their Brethren of St. Domingo.['] Hinting at the same time, that there were then Emissaries laboring for that desired event."[121] Sonthonax clearly aimed to frighten the British with stories of revolutionaries working in Jamaica, and perhaps to embolden the enslaved there too; but Balcarres, seeking justification for his questionable actions against the Maroons, could also make good use of rumors of revolution.[122] Indeed, such ideas frequently justified extensive and often summary executions of the most horrific kind, as well as other forms of extreme violence against rebels. This was also the case when all that had happened was talk. Thus, the Reverend Lindsay was writing in response to the prosecution and execution of enslaved men and women for involvement in a "conspiracy" to revolt that had been "discovered" in the northern parish of Hanover in summer 1776.[123] This "conspiracy" had shaken the planters since it involved the island-born creoles who ideologues like Long had said would be a source of stability. The evidence for such slave conspiracies was, quite literally, all talk—about who said what to whom about who would rise up, and when and how they would do it. As such, these "conspiracies" raise significant questions about speech, politics, slavery, and freedom, both for those who take them to be significant articulations of resistance and for those who see them as expressions of planters' fears woven out of extorted testimony, the whisperings of the condemned, and the nightmares of slaveholders and the enslaved, brought together in life-or-death courtroom examinations.[124]

The judicial process surrounding such "conspiracies" certainly demonstrated a desire to reveal the deliberative voice of the enslaved: a site and moment, or series of sites and moments, of collective decision making that would make a conspiracy woven of words real as a political event, as it convened through speech a temporary insurrectionary polity—one of Edward Long's "cabals." Thus, for the 1692 conspiracy on Barbados, which Jason Sharples interprets as an originary moment in the formation and articulation of planters' fears, the investigating officers reported that the confessions they had extracted showed that the conspirators "have been most active and industrious in gaining men to their Party within these three months" and that "the matter had been long ere now putt in Execution" had not news of victory over the French and the plan to send two regiments to Martinique "putt a stop to it." However, "finding that some of their confederates were in chains upon Gibbetts & believing their time drew nigh for being made Examples of, [they] were resolved on yᵉ mischief immediately (which they had sometime before defer'd for yᵉ reasons aforementioned)."[125] They had, it was concluded, collectively discussed it and resolved to act.

Another example, from Jamaica in 1776, involves the testimony under threat of death of an enslaved man, Sam, before the Hanover magistrates led by George Spence:

> The said Sam being brought before Us and Interrogated Saith that he . . .
> with the following Negroes, Malcolm Charles and George of the Baulk Estate, Adam and Zachery of Richmond Estate, Mingo at Batchelors Hall Estate,
> Creole Cuffey and Young Cesar at Prosper Estate, Christmas at Unity Estate,
> Bottom Quaw and Billy at Venture Estate [etc.] . . . had Conspired together
> to Rise in Rebellion and that at their several Meetings they had appointed

the above people to be their Captains and Officers, each to Act in their separate districts—That the first general Meeting was on Sunday last five Weeks
at Lucea Bay and that they met again last Sunday three Weeks at the same
place, Both times Blue Hole Harry was with them. That at the Second meeting they agreed to meet again the Ensuing Sunday as before at Lucea Bay
And agreed that in Case they were all ready they were to Rise the next day
but if they found they were not then prepared that they were to postpone it
and that at that Intended Meeting they were to Discover to Each Other what
Guns Powder &c They had Collected that they might from thence know or
Judge how far they were in Readiness to Rise, And that they Carried on all
their Deliberations in the Open Street on Lucea Bay but still very cautious
who they trusted.[126]

This account of meetings, participants, "Deliberations," and decisions reads
like the journals of an alternative house of assembly for Jamaica, planning
a very different political future. Such accounts are repeated across conspiracies and revolts, and in the official archives, published reports, and private
letters, along with the conspiratorial words overheard, retold to masters, or
confessed to by the enslaved. Evidence of the enslaved meeting in "Councel" to "Consult & Contrive" was produced through the power-laden questions of magistrates full of vengeance and fear, and the answers of terrified
witnesses speaking in a context where such meetings, and such words and
intentions, were punishable by death. They were, indeed, exactly what the
magistrates had to hear in order to prosecute men and women who "did
conspire, combine, confederate and agree together . . . to be concerned in
a rebellious conspiracy."[127] This is particularly clear where effort was made
by political authorities to find an outcome other than the confirmation of
a widespread plot. For example, the desire to prove that the Maroons were
not involved in the 1776 conspiracy, because of the political difficulties
that would entail; or that meetings in St. James, Jamaica, in 1823 were not
evidence of "rebellious conspiracy" but only of "an active spirit of inquiry"
into the prospect of freedom (as we will see in chapter 5).[128]

Yet is it sufficient to conclude that the "deliberative voice" of the enslaved was a fabrication of official judicial and political processes working
to legitimate the deployment of deadly violence? For example, while it is
important to hear evidence from these conspiracy trials as ways of talking
that exerted extreme caution over attributing definitive statements or actions to those speaking or being spoken about, the extensive material they
provide can also be taken to indicate that what was going on in the clandestine places of talk on the plantation was a more fluid, open-ended, and

indeterminate form of talk than the active planning of an uprising.[129] As Walter Johnson has asked, "How . . . did enslaved people set about forming social solidarities and political movements at the scale of everyday life? How did they talk to one another about slavery, resistance, and revolution? How did they sort through which of their fellows they could trust and which they could not?"[130] Here, for example, is the evidence of Adam from Jamaica in 1776:

> Adam further Saith that the said Leander on the Sunday aforesaid Acquainted him that Coromantee Sam belonging to the Baulk Estate promised to get hands from Fatt Hog Quarter, Cacoon, Top River, Richmond, and the Baulk Estates and at the same time the said Leander declared their Intentions of Rising some time this Week, And the said Adam further saith that Previous to the Conversation had with Leander he had met With the said Sam who Informed him that he had been Ill used by the Overseer and that the Hardships on Negroes were too great and that if all the rest of the Negroes Hearts were like his the Master would have been finished long ago, And the said Adam further Saith that the said Leander endeavoured to prevail on him to give him his Overseers Gun and to procure Powder and Bulletts.[131]

This testimony might weave a picture of the enslaved traveling between plantations to build a rising, or, perhaps, of them sharing experiences of hardship in a time of drought, and, through news, rumor, and gossip, imagining with others what might be done—an imagining that could include tales of cruel overseers, where guns might be had, how the Maroons would act, whether "buckra" was strong, and whether the regiments had left the island. These deliberations figure an uncertain collective assembled through particular forms of speech, and one that could easily remain just words if not fabricated into a conspiracy by the magistrates or into a revolt by (some of) the enslaved who felt they trusted what they had heard enough to act. If political meaning can be given to such talk—part of everyday life in the "slave houses"—as the temporary making of a "public" through speech, there remains the question of what specific forms of speech could have bound such a social aggregate.[132]

There were, for example, the gendered forms of hierarchical leadership and militarism among the enslaved that generated sovereign proclamations. These forms of politics might, more or less directly, deploy the idioms of West African sovereignty within Caribbean slave revolts. Yet, determining which forms were adopted and adapted is by no means straightforward; neither is understanding how they were changed by their new circumstances.[133] In addition to notions of absolute sovereignty, there is evidence of ideas of geographically dispersed leadership hierarchies, non-statist forms of organization, leaders guided by councilors, and the forms of shared and gendered sovereignty evident in the practices of the windward Maroons and the coronation of the queen of Kingston.[134] Moreover, as John Thornton has argued for the kingdom of Kongo and the influence of its political theory on the Haitian Revolution, there was a long-standing debate in Kongolese political philosophy, just as in other African traditions, between absolutist notions of kingship and more collective versions of power.[135] Thus, it was reported in 1743, albeit in a disparaging way, that the leeward Maroons had elected "Cudgoe" to be their leader. That choice having been made, he was granted sovereign authority: "His Words being absolutely all the Laws they have." Yet, as with the windward Maroons, this political authority was, it seems, shared with spiritual specialists. The same report noted that the Maroons' "Obia Man" was one whose "Words Carried the Force of an Oracle with them, being Consulted on every Occasion." However, like the challenges Cudjoe faced to his authority, the obeah man was now "disregarded," having assured the Maroons of the safety of a town that was subsequently destroyed by British soldiers.[136] This suggests that where sovereignty and authority were parceled out—just as with the monarch and the colonial assembly—what bound the polity, but might also split it, were discussions over where power was located and how best to organize it. The speech practices used in such discussions, and in the oral exchanges that might become "conspiracies" or "revolts," would certainly have mattered and were necessarily both diverse and dialogic.

For example, in 1678, following "a rebllyon of Negroes" on Captain Duck's plantation near Spanish Town, the rebels, led by "a notable bold subtle Negroe" called Captain Agiddy, had escaped into the woods and were trying to get those enslaved on Colonel Modyford's plantation, "they living formerly togeather in Barbadoes," to join them in "thire rising." It was reported that a "Negroe Woman," who was passing through the woods in the early hours of the morning, had come across "some Negroes," all of whom she knew by name, "sitting in ye path wth a bole of watter betweene thire Leggs." She saw them toss the water into the air "to see by thire witchcraft whither they should be fortynate in thire proceedings or not." We cannot know how the patterns of falling water were interpreted, and what part Agiddy, or even the woman who reported it, took in this "counsell," but such signs needed to be talked over to decide what action to take.[137] These deliberations would have been given direction by particular oratorical forms. A century later, in Hanover in 1776, the Maroon

Asherry—a "tall yellow fellow" who wore "beads on his hand" and an "Officer's Badge" from the island's government—was said to have told those gathered at Philander's house on George Gilchrist's plantation that "Negro man fool for thinking white man better more than Cudjoe," and argued that the Maroons and "negroes" should make common cause against the "white people." In saying this, he was reported to have "struck the ground with great violence, in the manner of his Country (for he said he had Coromantee blood in his body)."[138] Here, then, was a trace of Akan oratory used, or maybe just heard and imagined, in Jamaica as part of an attempt to constitute a temporary polity seeking liberty.[139]

Walter Rucker has argued that Gold Coast diasporas in the Americas exhibited a "commoner consciousness," which was produced by the class and gender dynamics of who was taken into Atlantic slavery and involved a reworking of spiritual, cultural, and political technologies (including oaths and symbols of sovereignty) from hierarchical West African societies in ways that were egalitarian and revolutionary in the Americas.[140] This set of social relations included the "normative" practice of obeah as a more democratic involvement of spirits in the world and a strong political role for women. However, focusing on formal politics in West Africa, Rucker does not examine whether "commoning" itself—the practicalities of managing common resources—provided forms of deliberation and determination that might find other uses in the new world of enslavement.[141] There were as many forms of West African "commoning" and collective deliberation as there were forms of sovereign authority, with institutions and spaces such as the "palaver tree" suggesting their production within long histories of encounters with Europeans and others.[142] Yet, alongside the modes of royal authority that can be traced to the Americas from the Gold Coast, it is worth considering the political practices of the "secret societies" or "power associations" of the Upper Guinea Coast and their ways of organizing political life.[143] As the slave trader–turned–abolitionist John Newton put it in 1788:

> They have no men of great power or property among them; as I am told there are upon the Gold Coast, at Whidah and Benin. The Sherbro people live much in the patriarchal way. . . . But the different districts, which seem to be, in many respects, independent of each other, are incorporated, and united, by means of an institution which pervades them all, and is called The *Purrow*. . . . [It] has both the legislative and executive authority, and, under their sanction, there is a police exercised, which is by no means contemptible. Every thing belonging to the *Purrow* is mysterious and severe, but,

upon the whole, it has very good effects; and as any man, whether bond or free, who will submit to be initiated into their mysteries, may be admitted of the Order, it is a kind of Common-wealth. And, perhaps, few people enjoy more, simple, political freedom, than the inhabitants of Sherbro, belonging to the *Purrow*, (who are not slaves,) further than they are bound by their own institutions. Private property is tolerably well secured, and violence is much suppressed.[144]

Newton was keen to demonstrate the law-bound, peaceful nature of West African societies against proslavery accounts of the violence of their kings and everyday life (discussed in chapter 1). However, later accounts of the Poro society also emphasize its role in organizing the use of common resources—when and where to plant and harvest, and how to deploy collective labor—as well as mediating with ancestors and spirits, arbitrating chiefdom disputes, declaring war, and deciding on capital punishment. As what Newton saw as a "Common-wealth," the Poro sought, through various modes of deliberation, both in seclusion and via palavers in the *bari* or meetinghouse, to create what the Mande people termed *ngo yela*, or "one word."[145] Significantly, while Newton only described an all-male Poro society, it operated in conjunction with the women's associations of the region, particularly the Sande, to provide the basis for forms of shared and gendered sovereignty. This gendered sovereignty included ways in which older women with spiritual powers and a place in the Sande hierarchy might become members of the "male" Poro society and assume positions of political leadership.[146]

Marcus Rediker has argued that the Poro's practices—reworked on the slave ship and in North American courtrooms—provided a means for the *Amistad* rebels to contest their enslavement and fight for freedom in the early nineteenth-century United States.[147] The political investment of Jamaican Maroons in the organization of the commons as a space of life, death, and the ancestors—and crucial to their existence as an enduring polity—is also telling.[148] However, it is important not simply to replace one set of diasporic political practices with another—emphasizing collective solidarity and not political leadership—since both were necessarily involved in convening groupings of the enslaved.[149] Indeed, while reports of slave revolts and conspiracies afford little direct evidence of such declamatory and deliberative forms of talk in action, the diversity of trajectories and experiences that rebels brought together—as Vincent Brown shows for Jamaica in 1760—can only have led to both confluences and clashes of ways of speaking politically as those involved discussed what to do next and how to judge the military, political, and geopolitical forces within which they were operating as they rode the waves of global warfare, tapped into interisland communication networks, and assessed local geometries of power in the 1690s, 1730s, 1760s, 1770s, and 1790s.[150] These discussions might convene temporary polities in which the spirits and ancestors were invoked, or which used forms of royal ceremony, proclamation, oratory, or storytelling that had resonance within a black Atlantic diaspora created by slavery. These polities might be deliberative assemblies more or less open to different voices: of men and women, of old and young, or of those who spoke with the spirits or had military experience. Or those collectivities might involve other forms of experimentation with Atlantic political practices. Thus, on the island of St. Christopher in 1770, the authorities, suspecting a plot, found "nothing more than a Meeting every Saturday night of the Principle Negroes belonging to Several Estates in One quarter of the Island called Palmetto Point, at which they affected to imitate their Masters and had appointed a General, Lieutenant General, a Council and Assembly and other Officers of Government, and after holding a Council and Assembly they Concluded the night with Dance."[151] The governor decided that no conspiracy had yet occurred as a result of these meetings, but that it certainly might.

Attempts to delineate modes of political speech among the enslaved offer an alternative Atlantic history of the "deliberative voice" to place in dialogue with Enlightenment discussions of liberty and the arguments made by colonial assemblies over their freedom of speech.[152] Indeed, they cannot be kept separate. These interconnections are evident both in the slaves' "Assembly" on St. Christopher and in the uncertainty among the islands' political masters about what they were hearing when the enslaved spoke politically. Here is the governor of Jamaica in 1776:

> It appeared from the Examinations of the Negro Conspirators, the chief of whom were Creoles, who never were before engaged in Rebellions, and in whose Fidelity we had always most firmly relied, that they were invited to this attempt by the particular Circumstances of the times; well knowing, there were fewer Troops in the Island, than had been at any time within their Memory, that the number of Slaves were near doubled in that time, that Great Britain was exerting all her Strength in reducing the Rebellious Colonies of North America, that the Shipping were all leaving the Island, for the Protection of which the Man of War were likewise going; they could not withstand such a tempting opportunity, to throw off the Yoke of Slavery; now, or never, they thought was the time, to make themselves Masters

of this Country; and altho' this train of Reasoning may seem above Negroe comprehension; Yet it certainly does appear from all the Examinations, and informations, written and Verbal, that they were encouraged, and did Act upon these foundations.[153]

The governor set out the conspirators' considerations—the local and global economic and political circumstances that made them decide that sovereignty was "now, or never"—and, while constituting them as deliberative political subjects, attempted to simultaneously both affirm and deny them that status. Interrogating modes of speech and reason, British, Jamaican, and Barbadian politicians heard, and needed to hear, evidence of political speakers not unlike themselves to understand what was going on. While this may have been a mishearing (or silencing in the case of women) of other Atlantic diasporic speech practices with different roots and routes than those of the colonial governors and assemblies, and while it was certainly used to visit violent retribution on those speakers, it also worked, through the politics of speech, to constitute some subaltern voices as ones with a powerful and recognizable political agency.

Conclusion

This chapter has investigated how slavery and freedom (or "liberty") were constituted as political conditions—matters of power, consent, and resistance—through forms of speech. It has demonstrated how ideas and practices of political speech—of proclamations, oaths, and collective deliberation—were the means and cause of contestation and political action in the relationships between the empire and the colonies and in the revolts and conspiracies of the enslaved against those who would keep them in bondage.

Forms of rule by distant sovereigns were enacted through oaths and proclamations that were combined, and came into conflict, with the privileges of colonial legislative assemblies that built their political identities—and ideas of liberty contrasted to the condition of slavery—around notions of the "freedom of speech" and the practices of the "deliberative voice" that ensured them. Freeborn Englishmen, they proclaimed, would never be slaves if they spoke out to defend their liberties. This view, in turn, shaped the ways in which the voices of the enslaved were heard, and misheard, as they discussed and protested slavery's violent oppressions. There were proclamations, oaths, and forms of deliberation that can be heard as opposing the slaveholders' view of the world and which were gendered in how they were made and heard. Since the archival traces of what was spoken and heard are predominantly from courtroom and execution sites laden with threats and violence, they echo with the sounds of death. These ways of speaking and listening were both shaped by the violence of slavery and drew on sources that ranged from diasporic versions of West African political philosophies of sovereignty—including ways in which it might be shared by powerful men and women—and modes of collective discussion over the "commons," to performative reworkings of the colonial relationships between governors, councils, and assemblies for the purposes of the enslaved. Heard by magistrates, politicians, and spectators as echoes of the speech practices of European sovereigns and colonial legislative assemblies, they were understood, predominantly when made by men, as statements of political agency uttered by political subjects who thus appeared recognizable in European terms.[154] Such signs of the political agency of the enslaved, and the potential recognition of humanity that they demanded, were met with the violence of military action and execution.

In all cases, what has been traced here—through multiple forms of both sovereignty and collective representation among slaveholders and the enslaved—are the ways that "speaking *politically*" is a matter of grouping and regrouping, of making temporarily constituted "publics" that would not otherwise exist. In using the term *publics*, it is important to recognize Elizabeth Maddock Dillon's caution that, because of the dependence of slave societies on structural illiteracy and radically asymmetrical access to all technologies of communication, "accounts of the Habermasian public sphere, insofar as they focus primarily on print, risk reinscribing the technologies of 'social death' associated with race slavery."[155] Yet, an insistence on how speech makes publics, instead of a focus on print, can more effectively understand both the differentiation of forms of political speech in the violently asymmetrical context of Caribbean slavery and the shared political practices these differentiations belie. Here speech is a dialogic dance, with deadly consequences, between different speakers seeking to make, unmake, and remake polities and their publics.

One long-standing idea about the politics of speech and language in Atlantic slavery is that, to prevent resistance and revolt, slave ship captains and plantation owners deliberately mixed people from different places—usually defined by port of embarkation—who spoke different languages. These diverse people could not, then, it was supposed, combine against those who would exploit and oppress them. As the governor of Barbados put it in 1668, of the forty thousand enslaved Africans in Barbados, it was their "different tongues and Animosities in their owne Countrey [that] have hitherto kept them from Insurrection."[156] However, the effectiveness of this strategy has been questioned on the basis of the history of West African state and empire formation, which generated large areas of mutually intelligible languages and the strong and enduring demographic connections between particular regions of West Africa and the Americas produced by the slave trade.[157] As this chapter suggests, that strategy should also be questioned on the basis of the multiple, complex, and contradictory uses of forms of speech (and silence) to make collectives on these islands. It certainly took little time for new ways of speaking to convene new groupings. For example, a pamphlet from Barbados reported that, in the wake of the slave conspiracy of 1675, two men were condemned to be burned alive, with another twenty-five to be executed. Just before the sentence was carried out, the deputy provost marshal called upon them *"to Confess the depth of their design"* to break one collective and shore up another. In response, one man seemed prepared to speak and called for a drink of water, "which is a Custome they use before they tell or discover any thing." He would, it seemed to the spectators, testify to the truth of whatever he had to say about the plot by drinking the oath, as he might have done on the African coast or in swearing to rebel: calling the spirits to witness his words.[158] Yet before he could speak, the man chained to him, "one *Tony*, a sturdy Rogue . . . jogged him, and was heard to Chide him in these words, *Thou Fool, are there not enough of our Country-men killed already? Art thou minded to kill them all?"* The first man then kept silent and, after Tony had addressed the crowd in English to tell them, in rhetorical and proverbial style, *"If you Roast me to day, you cannot Roast me tomorrow:* (all those *Negro's* having an opinion that after their death they go into their own Countrey)," both were put to death.[159] In this tragic moment there is evidence of the uses of shared speech forms, and shared languages, to undermine resistant collectives (the "confession"). There are also indications of the ways of speaking of what governor Willoughby feared as "the Creolian genera[ti] on now growing upp" that "may hereafter Mancipate their Masters," who might deploy speech practices that combined West African forms of oratory, proclamation, and oath taking—which were familiar to but different from European versions—with a creolized English to build new groupings in both speech and silence.[160]

Considering the politics of slavery on the asymmetrical common ground of speech makes it possible to hear planter politics as part of a transatlantic oral culture of proclamation and deliberation. Within this culture, identities were formed, and often violently policed, in vain attempts at the demarcation of something as mutable as talking. At the same time, it also provides a way of attending to the speech of the enslaved as part of the intellectual history of the long eighteenth century wherein powerful ideas of sovereignty, slavery, and freedom were formed through the transatlantic political geographies of ways of speaking. The inclusions and exclusions of these Atlantic histories of ideas and of knowledge are also evident in the next chapter, which turns to the ways people spoke about plants and what uses they might have.

THREE

Master, I Can Cure You: Talking Plants in the Sugar Islands

William Wright, an eighteenth-century Scottish doctor and Jamaican bota-nist, was comfortable communing with plants. As he put it in the pref-ace of his *Hortus Jamaicensis*, "The Man who inclines to this happy turn is never at a loss for society, whether in the Garden, In the Field, on the bleak summit of the mountain, In the plenteous Vale, in the sweet range of the Hedge Row or in the cool umbrage of the Wood, He never fails to meet with numerous acquaintances, whether adapted to the purposes of Health, Food, agriculture or to gratify the Sight, Smell or Taste." Materialized into Wright's *Hortus*—via dried and pressed leaves, stalks, and seeds, and the accompanying handwritten text—were both the fruits of and the basis for other forms of communication with a variety of interlocutors. Its elaborate title page, which imitated that of a printed book, and its presentation to David Steuart Erskine, the 11th Earl of Buchan and a fellow of the Royal Society, opened up realms of gentlemanly conversation and scholarly de-bate about natural history and natural philosophy. Its plant descriptions also signaled other oral encounters. Wright's specimen of sugarcane (*Saccharum officinarum*), that most significant plant for Caribbean plantation slavery, took up a full page. The text opposite noted that "the Sugar Cane is probably a Native of Arabia as well as of Guinea and the Continent of South America. The new Negroes brought here well know its use and give an account of their boiling it into Syrup in Africa." Finally, in its discussion of the medical uses of plants, Wright's Jamaican herbarium let the doc-tor speak to his patients. For example, for lignum vitae as a treatment for "the Lues Venerea" (syphilis), he noted that "ten drams of Gum Guiacum Six Drams of Speices of Edinb[urgh] Treacle and thirty drams of Corrosive Sublimate infused in a Bottle of Rum is our mercurial Tincture. Two tea Spoonfulls morning and Evening in a pint of Decoction made of Sarsparilla & Lignum Vitae is a dose for a grown person." He would expect a cure in six weeks.[1]

Wright's *Hortus* begins to demonstrate the range of ways in which speech was involved in the making of natural historical knowledge in the long eighteenth century, and the ways in which spoken words flowed around and into and out of texts (in both script and print), images, and objects such as dried specimens and mounted collections. Much has been made of the norms of civil conversation that underpinned truth telling in the early Royal Society and of the contemporaneous chatter and clatter of coffeehouse lecturers and their audiences.[2] In addition, close examinations of both the production and the reading of printed texts in the construction of scientific knowledge have revealed worlds where what was said between printers and authors around the press and between diverse sorts of readers in a range of venues were of utmost importance.[3] Following the previous dicussions of law and politics, which have demonstrated how speech both differentiated and connected the enslaved and the free, this chapter argues that multiplying the forms of scientific talk considered, and attempting to assess the relationships between different forms of talk on the same subject, can address important issues of knowledge and communication in the sugar islands at the troubled intersection of appropriation and exchange between those enforcing and enduring slavery.

Such talk was part of new global relationships forged through scientific practice. As James Delbourgo points out, although natural philosophers in the Atlantic world increasingly argued that knowledge of nature came from direct observation, the "identities and properties" of natural objects—as they appeared to those separated from them by oceanic distances—"were not self-evident but dependent on the webs of spoken and written testimony in which they were embedded."[4] Such approaches to the historical geography of science draw on the best-known part of Latour's theorization of forms of veridiction, that which focuses on the production of scientific knowledge. Here Latour emphasizes how the "universality" and "objectivity" of science has to be produced through complex chains of translations and mediations between people, instruments, images, and texts across networks of many kinds of actors, human and otherwise. When these multiple and networked mediations work as they should, they produce—in something like a boringly unsuccessful game of what modern-day Americans call "telephone" and the British call "Chinese whispers"—access to what is to be known.[5] As Latour puts it, "when the network is in place, access is indeed obtained; you put your finger on a map, a document, a screen, and you have in your hand for real, incontestably, a crater of the Moon, a cancerous cell deep within a liver, a model of the origin of the universe. You really do have the world at your fingertips. There is no limit to knowledge." Once again, these networked mediations are clearly not just a matter of speech practices, but they do provide a basic building block in a way of thinking about scientific knowledge that wishes to challenge the idea that it is all about "straight talk" in a world where "the facts speak for themselves."[6] Interpreting these modes of mediation and articulation becomes all the more important where they are entwined in imperial contexts with questions of exchange and appropriation.

What this chapter investigates, therefore, is talk about plant life in seventeenth- and eighteenth-century Jamaica and Barbados as part of a wide-ranging botanical endeavor that lay at the very heart of early modern European scientific practice in the Caribbean and elsewhere.[7] Understanding orality here can address issues central to debates over science and empire, and the changing relationships between particular forms of botanical knowledge and their producers. Richard Drayton, Londa Schiebinger, and Claudia Swan, in their surveys of imperial and colonial botany, emphasize long-term changes in the ways in which botanical knowledge was produced as "Europeans' respect . . . for traditional knowledges gradually waned over the course of the eighteenth and nineteenth centuries."[8] From the mid-eighteenth century onward, Africans, Asians, and Amerindians were seen less and less as the holders of valuable botanical knowledge. Yet, as Schiebinger also argues, this "respect" had always been qualified and exchanges limited by the conditions within which botanical knowledge was made in what she calls the "biocontact zones" of empire: the enormity of the task of describing new flora, the cacophony of languages, the inflexibility of European theory, the parceling up of transferable factual knowledge, the threat of violence, and the refusal of the indigenous and enslaved to cooperate.[9]

The complexities of the changing relationship between exchange and appropriation in colonial botany requires finding ways to work across and between interpretations of botanical knowledge making in the colonial Atlantic world that have used the same sources to come to different conclusions about knowledge and power. In her wide-ranging analysis of the culture of natural history in the colonial British Atlantic world, Susan Scott Parrish emphasizes what she calls "the diasporic African sources of Enlightenment knowledge" and the consequent reliance of Europeans on African botanical expertise. She notes how certain forms of botanical knowledge, particularly of poisons, required an African source to be credible, and she reconstructs the chains of exchange that led from the American woods to the metropolis.[10] In contrast, Pratik Chakrabarti's discussion of economic botany in eighteenth-century Jamaica emphasizes the role of violence, including "epistemological violence," in the process of extracting plant knowledge.[11] What this chapter argues is that considering talk as it is understood in this book—as a practice which effected both power and communication; that was used by all, albeit unequally; and that was organized into quite different forms of practice—allows, in the case of Caribbean natural history, the recognition of both exchange and appropriation at the same time and over time.

This chapter will explore these questions of knowledge, power, and communication through three forms of talk about sugar island plants—*botanical prescription*, *botanical conversation*, and *botanical oration*—which demonstrate contrasting social and spatial configurations of botanical speech. Each involved distinct forms of knowledge and distinct relationships of power and exchange, and each had implications for what differently positioned people could know and say about the islands' flora. Once again, investigation of orality is not restricted to the botanical cultures of the indigenous, enslaved, and unlettered, or their encounters with Europeans. European botanists were part of highly developed oral cultures shaped by gender and status, located in particular spaces, and employing specific modes of veridiction. For each form of talk, the aim is to show how it was productive of certain sorts of botanical knowledge and not simply reflective of it. The chapter also addresses the question of how the place of indigenous and enslaved knowledge in the production of colonial botany changed over time by first considering plant knowledge in the context of oral encounters between healers and patients (botanical prescription), where, it is argued, the form of talk itself opened the possibility for exchange, and then demonstrating how, over time, other modes of speech—those between botanical gentlemen in their gardens from the mid-eighteenth century (botanical conversation) and the types of public talk associated with botanical gardens in the late eighteenth century (botanical oration)—constructed more exclusive plant talk that restricted the space for the voices of the enslaved.

Botanical Prescription

In the entry on "Hog Plum" in his *Hortus Americanus*, Henry Barham, a Jamaican doctor and practitioner of natural history, wrote:

In the year 1716, after a severe fever had left me, a violent inflammation, pain, and swelling, seized both my legs, with pitting like the dropsy: I used several things, to no effect. A negro going through the house when I was bathing them, said, "Master, I can cure you," which I desired he would; and immediately he brought me bark of this tree, with some of the leaves, and bid me bathe with that. I then made a bath of them, which made the water as red as claret, and very rough in taste: I kept my legs immersed in the bath as long as I could, covering them with a blanket, and then laid myself upon a couch, and had them rubbed very well with warm napkins; I then covered them warm, and sweated very much: I soon found ease, and fell asleep. In five or six times repeating this method, I was perfectly recovered, and had the full strength and use of my legs as well as ever.[12]

Having benefited from the cure, Barham, in a characteristic move, then denied the source of the knowledge and skill that had helped him by "giving God thanks for his providential care, in bestowing such virtues to mean and common plants, and that the knowledge of them should be made known to so vile and mean objects as negro slaves and Indians."[13] Yet the moment of speech remains—"Master, I can cure you"—however much it is mediated in Barham's text.

Such statements can be located within historically variable modes of European medical talk (and silence) between doctors and patients, and their changing locations, from the early modern to the nineteenth century. Michel Foucault's *The Birth of the Clinic* identified "the minute but decisive change, whereby the question: 'What is the matter with you?,' with which the eighteenth-century dialogue between doctor and patient began (a dialogue possessing its own grammar and style)"—and defined, he says, by "a botany of symptoms"—"was replaced by that other question 'Where does it hurt?,' in which we recognize the operation of the clinic and the principle of its entire discourse."[14] In addition, Nicholas Jewson's theorization of the disappearance of the "sick man" from medical cosmology after the end of the eighteenth century argued that a "polycentric and polymorphous" "Bedside Medicine" within which "diagnosis was founded upon extrapolation from the patient's self-report of the course of his [*sic*] illness" was replaced by "Hospital Medicine" and then "Laboratory Medicine" with the patient (and the bedside) disappearing further at each stage.[15] Although historians have since blurred the clean lines of these epistemological and cosmological breaks, they still suggest a shifting historical geography of modes and sites of talk between healers and the sick as the practice of patients telling their histories was supplanted by physical examinations.[16]

The statement "Master, I can cure you" also needs to be located within Caribbean slave societies, where disease and death were a central fact of life.[17] Doctors' limited and dual roles in relation to masters and the enslaved meant that European distinctions, and historical shifts, between personal and institutional medicine were less clear on the plantation.[18] There were also both continuities and discontinuities with Britain in the way doctors were expected to talk to their patients. William Wright was said to be at ease with botanical conversation among his learned friends (as we will see in this chapter), but not fit to be a London doctor, "whether from a certain diffidence of manner in the presence of strangers, which, in his case, may be said to have been constitutional, and the total want of that *brusquerie* and self-assurance, so necessary to success in this bustling profession."[19] If this was not a problem for him in Jamaica, there were other issues to consider. For example, in *The Medical Assistant* (1801), Thomas Dancer, director of Jamaica's Bath botanical garden, provided extra reading for white women on sexual health because of "the prohibition which the sex lies under from delicacy, in seeking advice; particularly in the West-Indies, where the major part of the faculty are young bachelors."[20] Finally, although it is difficult to assess, it seems to have been as much Hans Sloane's confident bedside manner as a physician in late seventeenth-century Jamaica, rather than his medical success, that brought initially wary patients round. One wrote that Sloane's presence was "Very much wanted," while commenting of his replacement that "I never much beleave much of his Doctorin."[21] For his own part, Sloane claimed a plain medical language, at least in writing. He was dismissive of those who fetishized words, saying that "Neither have I seen any Effect of Gibberish or other Words used as Charms to cure or rather fright Diseases, tho' in ancient Times, and even now some have a great Opinion of them from a Belief they have in an axiom *herbis, verbis, & lapidibus, inest magna vis.*"[22] In an extensive set of case histories printed in the first volume of his *Natural History of Jamaica* (1707), a book best known as an account of Jamaican botany, Sloane replaced the magical power of words (alongside herbs and stones) in this formula with the observation of physical symptoms and, most important, talking with his patients.

Indeed, patients of all sorts would speak the variety of their symptoms. Thus, Sloane reported that "A black Man, of about forty Years of Age, told me he had great pains about his Navel, so that he could not sleep, he was in a cold Sweat, in great pain, and had not been at Stool in four days." There were, however, social differences. Sloane would sometimes talk to masters about the enslaved, and to mothers about their children. In the latter case, talking was so important that Sloane advised "tampering not with Physick too much with Children, where the Disease is not plain, being not safe, they not being able to inform the Physician of their Malady, but by frowardness and crying." More powerful patients might, in a medical system based on personal patronage, consult several physicians at once, moving between those "who promis'd his Cure." They might also enforce their own interpretations of condition and treatment. One Jamaican gentlewoman with a very bad bellyache "had been with all the Physicians of the Island" and could not be persuaded by Sloane to desist from drinking "Wine, Punch, and Vinous Liquors." She said she "would not abstain, alledging that her Stomach was cold, and needed something to warm it."[23]

Given its importance, doctors' talk was readily satirized. William King imagined reading Sloane's *Natural History* as a conversation with the doctor and contrasted his convoluted language with the actual content of a work he judged "more like a House-Wife's Receipt Book, or as Physick was said to be in its first Age."[24] Edward Long gave, for "amusement," "the history of a Jamaica *quack*" who substituted talk for knowledge and gave too little or too much, delivering lengthy jargon-filled monologues that put his patients to sleep (*HJ*, 2:583, 587). At issue in these satires of medical incomprehensibility and incomprehension were the ways in which male doctors' specialized talk of symptom, condition, and treatment related to everyday discussions among men and women of health and medicine, including medicinal plants.[25] Barham informed Sloane that the first volume of his *Natural History* had been criticized by those who had not read it but had just taken the word of "some of these men who are thought by the Rest to be more Capeable of Reading & Judging." In this harsh court of oral public opinion, Barham noted, "a book is oftimes Condemned in Generall before ever it is Particularly Read or Considerd . . . [and] you shall not meet with one in tenn that Spakes Slightly of it that ever Read itt."[26] The main complaint, which Barham himself made to others, was that plant names and descriptions were given in Latin rather than "Plain English" and too little was said about their "Common Medecinall qualityes."[27] For these readers, unlike the metropolitan satirists, Sloane's book should have been even more like a domestic cookbook.[28]

Understanding these ways of talking about health and natural history involves situating the statement "Master, I can cure you" and its utterer's work with plants within the world of household medicine. Accounting for the majority of medical treatment, but in an uneven dialogue with more or less professional medical practitioners and printed texts, household medicine was often under the care and control of women.[29] In Europe, depending on wealth and household size, they might store, often for many years, large quantities of medicines—particularly distilled waters and in-fusions made from botanical materials both grown and purchased, some exotic and some everyday, along with commercial preparations. Also on hand would be the ingredients, many of them foodstuffs—fats, oils, and vegetable matter—to make medicines that would not keep, and carefully compiled manuscript books of recipes necessary to prepare them. Here women's knowledge of plants, health, and the body intersected with the medical marketplace; such knowledge was deployed through talk about health and its restoration that was continuous with the relations of power and care in the household.[30] In the Caribbean, that store of knowledge, care, and power was redistributed among planters, doctors, and enslaved men and women. On the one hand, areas of medicine such as childbirth were often under the control of enslaved women, transferring oral and practical knowledge of perinatal bath rituals and the need for mothers' si-lence when giving birth across oceans, communities, and generations. On the other hand, slaveholders and doctors increasingly sought control of plantation medicine, which offered a role for script and print.[31]

It was here that Barham saw an opportunity to rework Sloane's book: "If there were an Abstract writt of the Nature and as Plaine as Culpepers English Physician and Referr them to the figure of the Plant: I Believe every Planter would have one of them in his House."[32] And he set about produc-ing what he later called "A Small Treatise of American Plants Setting forth their Experienced Medecinal Vertues or Specifick Qualityes . . . as I gained them from Spaniards Indians and Negroes." It was organized by "the Plain and Vulgar Name of every Plant in an Alphabetical Manner," referencing them to Sloane's descriptions and pictures. It was written "in the Plainest English I can put it in" and to be produced as a "Small Pockett Book."[33] Al-though it was not published in his lifetime, and the manuscript he sent to Sloane is now lost, it is evident that Barham's *Hortus Americanus* combined material on useful plants from his book collection, his own experience as a doctor and planter, and other "local" knowledges, including testimony from patients.

This inclusion of "local" knowledges certainly involved obtaining bo-tanical and medical knowledge from the enslaved. Barham determined that a bark used by a Jamaican "Practitioner in Physick" to cure intermitting fe-vers was not, as many thought, "bully-tree bark" but that of a "Locus-tree" because "I at last got out of a negro, that used to gather it for him, what tree it was."[34] Barham also questioned Sloane's account of Africans' medi-cal uses of clay and water plastered on the body by finding, through obser-vation and questioning, that they also included in the mixture "a yellowish

Root wch they Call Attoo."[35] In the process, Barham also gleaned something about their ideas of disease:

> Sometimes they only Illute the Head and Face, if thats affected. Sometimes their Stomach if their Heart is affected: for they attribute all inward ails or illness to the Heart Saying their Heart is Noe Boon not Knowing the Situation of the Stomach from that of the Heart: if their Limbs and Solid parts are affected they Illute themselves all Over Saying their Skin Hurts them: The Africans hath Such Confidence and Opinion of this Root, that if they Neither finde Relief by this Root outwardly Applied or Inwardly given Decocted they wholly Dispair of any Relief believing it to be the most Soveraigne Remedy that they Know Amongst all the Plants that comes Within their Knowledge.[36]

There was much to learn from talking to "the most Knowing or Skilfull Negroes Such as they Call (Obamen) or their Country Doctors." Indeed, one of Barham's own patients had been cured of a gonorrhoeal gleet by such a doctor and, as Barham wrote, "affirmed to me with Reflections that the Negroes Could doe more with their Herbs than Wee Practitioners with all our Art or Skill."[37]

Barham collected and identified plants and their uses through enslaved intermediaries who brought him specimens they thought might interest him. He also went out looking and asking around. Barham wrote of peanuts that "the first I ever saw of these was in a negro's plantation, who affirmed, that they grew in great plenty in their country; and they now grow very well in Jamaica." Likewise, for what Barham called "Oily pulse"— sesame—he recorded that "the first time I saw this plant, it was growing in a negro's plantation, who told me, they ground the seed between two stones, and eat it as they do corn." Finally, he sent enslaved workers out collecting, noting for the medicinal Nhandiroba nut that "the negroes I employed to get them for me called them *sabo*," and for lignum aloes (also known to us as lingnaloe, linaloa, or agarwood) that "a negro that I employed to get some of it, when he brought it to me, said the same sort grew with them in Africa, where they called it *columba*."[38] Such interlocutors could be enlisted in the tricky process of botanical disambiguation. One of the Jamaican naturalist Anthony Robinson's correspondents, keen to identify whether seed pods Robinson had found on a beach were the "Fruit of Sloane's Bichy Tree" that "the Negroes might have flung . . . over Board from some of the Guinea Ships," suggested that "it might be worth while enquiring of the New Negroes & also the Whites on Board these Ships shewing them some of the Fruit & asking if it is . . . what they call the Bichy or Buzz Tree in Guinea, according to Sloane Vol. 1 pa: 22 in his Introduction or rather his Bizy or Billy or abicey or abicy as they seem now to pronounce it." Africans emerging from the horrors of the middle passage might be recruited to shape natural history through what they knew and what they said, just as the captains and surgeons on those ships were.[39]

As this account shows, for Barham and others who addressed a Caribbean audience as well as a metropolitan one, these plants were part of local oral cultures of botanical knowledge made through the intersection of many transatlantic movements.[40] Here the naming of plants was not a matter of the cultural politics of Sloane's Latin polynomials or Linnaean taxonomic classification.[41] It was, instead, an amalgam of vernaculars. Names came from use—"broom-weed," "belly-ache weed," "pilewort," "ague-tree," or "self-heal"—basic physical description, such as "prickly white wood," or characteristics, like the "four o'clock flower" and "hell-weed, or devil's guts" for the invasive weed dodder. The names might be very local. Self-heal was also known in Jamaica as Pickering's herb, since a poultice of it applied by "an old negro man, well skilled in plants" had restored Captain Pickering's dislodged eyeball. Or they might be in use across various Atlantic worlds. Barham gave some Spanish, Portuguese, and Indian names as well as ones common to England and its colonial cultures. He, and others, also recorded, albeit unsystematically, some of the African names they had been told—for example, *sabo* and *columba*, and also *soonga* or *wolongo* for sesame. For "mad apples" (a nightshade), Barham recorded that "Angola negroes call them *tongu*, and the Congo negroes *macumba*."[42] Wright, discussing "The Mackaw Bush," reported that "Mackaw is an African word and indiscriminantly used for any thing prickly." And for the medicinal root "attoo," Barham noted that "I never could find any other name for this Plant and that I had from a negro."[43]

Plant names and stories could signal African origins and uses, and European botanists in the Caribbean noted a range of ways in which they, and others, had benefited from the skills and knowledge of the enslaved.[44] Yet, this facility to manipulate nature was a matter of threat as well as utility. Nowhere was this clearer than in the use of poisons. Even doctors were at risk. Discussing the "savanna flower," Barham related a case where "a practitioner of physic was poisoned with this plant by his negro woman, who had so ordered it that it did not dispatch him quickly, but he was seized with violent gripings, inclining to vomit, and loss of appetite."[45] Just as Barham had noted the medical skills of "Obamen" or "Country Doctors," these people were also at the heart of planters' botanical fears. For example, the concern that skilled poisoners among the enslaved were ready to covertly convey plant poisons into food and drink was often repeated. Barham claimed to have it from the enslaved themselves: "They say such wicked Retches that Practice it are use to in their own Country & when they see a Negro keep his Thumb Nails very long it gives a Jealousie that is an Oba Man or Sorcerer, Poisner or a Wizard." African medical practice might be denounced as highly dangerous or as useless "Hocus Pocus Tricks" in the same texts that discussed the medicinal plants in use "among the *Indian* and *Negro* Doctors."[46]

What European doctors and botanists were getting some limited access to was the dynamic complex of conjoined spiritual and medical knowledges and practices through which the enslaved sought to sustain their bodies and spirits under the death-dealing conditions of New World plantation slavery.[47] The oppressive conditions under which these practices developed, the secret nature of much of the knowledge, and the increasing fear and denigration of ritual practitioners in the Anglo-Caribbean, particularly after the 1760 revolt in Jamaica, mean that it is now hard to recover more than the broad outlines of what might have been involved. As far as talk about plants went, this was an oral culture of herbal remedies that connected body, mind, and spirit, and offered a range of practitioners and practices providing varying degrees of engagement with the supernatural world. There were herbalists, often women, who were continuing the long, and transatlantic, process of developing syncretic pharmacopoeias from weedy and herbaceous botanical materials that had already been exchanged between the Old and New Worlds for several centuries by the 1700s.[48] These herbalists overlapped with diviners and healer-sorcerers, practitioners of obeah, whose herbal healing was combined with the promise to intervene among the spirits for a fee, and whose formulae— both botanical and oral—might be used to protect enslaved rebels or make oaths binding (see chapters 1 and 2).

Dealing with health and sickness in particular cases required specific forms of speech. There were diagnostic rites, often conducted communally, between sufferers and healers, examining bodily conditions, past events, and disrupted relationships with the living and the dead. Divining, perhaps through dream interpretation, power objects, or using the Bichy tree's kola nuts, as was done in West Africa, sought the causes of conditions through what was revealed to the healer.[49] Such natural and magical objects might also, along with plant remedies taken internally or externally applied as smoke, baths, or poultices, form part of a sensory cure accompanied by ritual speech consisting of powerful, and often incomprehensible, words that demonstrated the healer's access to and control over numinous natural forces and engaged the spirits in the body.[50] Some of the enslaved might find some benefit in disclosing elements of this knowledge to survive or gain some small advantage within the plantation system.[51] Others might be threatened. What, for example, lay behind Barham's revelation of knowledge that he "at last got out of a negro"? However, either because what they disclosed was within the least sensitive realms of herbal remedies or because Europeans were uninterested in or threatened by plants' connections to the supernatural, even when they sought out such cures, what was exchanged carried with it little magic except as an overhanging and increasing sense of the danger that obeah posed.

The radically asymmetrical intersection of these forms of talk within a slave society also lay at the heart of what passed between white doctors and black patients, as evidenced in Sloane's recounting of the case of Emanuel, "a lusty *Negro* Footman" ordered to guide a dangerous early morning mission through the woods to seize a band of pirates. At midnight, Emanuel "lay straight along, would not speak, and dissembled himself in a great Agony, by groaning &c." The Europeans thought he was dying, the "Blacks" thought him bewitched or poisoned. Sloane thought Emanuel was faking it and "that he could speak very well if he pleas'd." If talk was the problem, it was also the remedy: as Sloane recounted, "To frighten him out of it, I told the Standers by, that in such a desparate condition as this 'twas usual to apply a Frying-Pan with burning Coals to the crown of the Head, and . . . to put Candles lighted to their Hands and Feet," and Sloane sent for the necessary items. In a quarter of an hour, the doctor reported, "he [Emanuel] came to speak. I question'd him about his pain, he told me that 'twas very great in his Back. I told him in short that he was a Dissembler, bid him go and do his business without any more ado, or else he should have due Correction, which was the best Remedy I knew for him, he went about his Errand immediately, and perform'd it well, though he came too late for the Pirats."[52]

There is much to examine in this account. Sloane's prescription is both a set of tortures devised to elicit a confession and a plausible cure within a medical regime designed to make slaves work. It is of a piece with the brutal punishments of the plantation system, where prisons and hospitals for the enslaved were interchangeable and where botanical works might note plants useful as "Switches" "to scourge the refractory slaves."[53] Sloane's deceit was also a parodic extension of a European medical practice built on blisterings, vomits, and purges that could terrify enslaved Africans. Indeed, medical knowledge, including botanical knowledge, could come from the pain, or threat of pain, experienced by enslaved African or Indian bodies.

For example, Barham's account of poisoning from the apple-like fruits of the manchioneel tree (figure 11, in which the tree name is spelled "Mança-neel") informed his readers that "I had a negro man that wilfully poisoned himself with them," enabling the doctor to see and hear the consequences: "I observed, he complained of a great heat and burning in his stomach, but could not vomit; his tongue swelled, and was burning hot, as he called it; he was continually calling for water; his eyes red and staring; and he soon expired."[54]

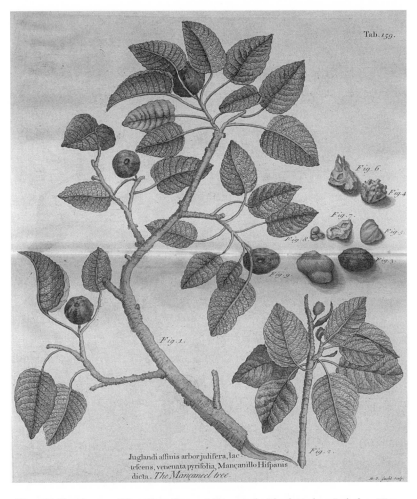

Figure 11. The Mançaneel Tree. Hans Sloane, *A Voyage to the Islands Madera, Barbadoes, Nieves, St Christophers, and Jamaica*, vol. 2 (London, 1725). © The British Library Board. 38.f.4.

However, the lesson of Sloane's case history is not simply the necessity of punishment for those who would not work. Emanuel could have been cruelly punished just for his refusal to follow orders. Instead, Sloane put all his efforts into getting him to speak. He required answers to both of Foucault's questions: What is the matter with you? Where does it hurt? He needed Emanuel to recount his symptoms, to enter into a dialogue, so that the prescription could be made. This account can be set alongside William Wright's later report of his treatment of an enslaved child for lockjaw. As Wright told it, "the Negress was my own; and, with her consent, I plunged her infant in cold water. It grew stiff as a board. A Mulatto rubbed it till it became warm and flexible. The child had no more tetanus." There is evidence that Wright's cold water treatments, which he also used on himself, were learned during a 1768 Jamaican smallpox epidemic from "a custom among the Maroons of Jamaica, as well as in some of the nations on the coast of Guinea, to cover the body with wet clay during the eruptive stage of the disease."[55] Yet, as significant as what he did was what he said. In law, Wright had no need to ask consent of the enslaved mother to plunge her child into a freezing bath, yet he did so. Indeed, in another, later, case the enslaved mother's opposition to the practice stopped it happening.[56] There is in both Sloane's and Wright's accounts recognition of the voice of an interlocutor—a question that needs to be asked and that needs to receive an answer, either describing symptoms or giving consent for treatment—that is rooted in the eighteenth-century healer-patient dialogue.

Returning to Henry Barham's swollen legs and the hog plum bark, the early eighteenth-century enslaved African who said "Master, I can cure you" inhabited, at least for a moment, the healer side of the dialogue as he set out and acted on his prescription. There was here a recognition of the "cunning" knowledge of nature's pharmacosm attributed to Africans, and the shadow of its threat. The exchange was made possible by the power of the patient as patron in the practice of eighteenth-century bedside medicine and the redistribution of the responsibilities for household medicine on the plantation. It was also enabled by a partial commensurability of speech practices between eighteenth-century European doctors, practitioners of household medicine, and the healers and diviners of the black Atlantic. They all shared the need to talk. It might have been vocalized and materialized in contrasting languages and communicative forms—the dream interpretations, ritual speech, animal parts, and grave dirt of the obeah practitioners (Barham's "negro doctors"); the medical preparations and recipe books of the "House wife" or her surrogate; and the bedside narrations, pills, clysters, printed books, and prescriptions of the Edinburgh and

London trained medical men—but there was in all these practices an attention to medicinal plants such as the hog plum and a need to talk to and hear from those they sought to heal. Yet, as the remainder of this chapter shows, there were other, increasingly prominent, forms of talk about Caribbean plants which involved a more limited range of voices.

Botanical Conversation

Viewed from the history of science rather than of medicine, perhaps the most familiar form of talk about Caribbean island plants occurred within conversations that were part of what April Shelford describes as modes of Enlightenment sociability. Here natural knowledge was produced through exchanges between "polite" gentlemen whose homosocial conversations were part of European botanical practices of collecting, identifying, and naming plants, and which resulted in the publication of works of natural history and their subsequent discussion.[57] Thus, as Shelford shows, Anthony Robinson taught Thomas Thistlewood the Linnaean system of botanical classification while staying as Thistlewood's guest, and showed him appropriate ways of drawing and describing plant and animal specimens. Robinson also shared with Thistlewood tales of the problems of producing natural knowledge for a broader audience.[58] Likewise, William Wright's correspondence with Joseph Banks, the linchpin of British imperial botany, demonstrates the range of uses of botanical talk between gentlemen. Such talk might, for example, involve instruction and close personal communication. Wright reported from Edinburgh that his nephew James had been well educated at the university and in the field in "the Medical and Philosophical Classes under my immediate Inspection." He also commented favorably on the young Mr. Thibaud's "competent knowledge of the English Tongue, and of the various Doctrines taught in this university," and noted that "he has been often with me and I find him a most agreeable young man, and hope he will make a conspicuous figure in Natural History and medicine." Wright later told Robert Brown, who was going to Australia on Matthew Flinders's expedition, that he would have opportunities to instruct others: "In a voyage of discovery, you will probably go in a Kings Ship, who will have a surgeon: If he is docile, you will instruct him in *re medica & in Historia Naturali.*"[59]

Talk was vital to the identification of new plants in the field, alongside work with specimens and books. In 1783, Wright reported to Banks from Jamaica about a plant he intended to name "Wallenia Juglandifolia." He had discussed it with others, and "Mr Wallen says he knows 4 other Species of this and that Browne [in his *Civil and Natural History of Jamaica* (1756)] has described it by the name of *Phælorium.*" There was also the talk that evaluated and reconciled collections of specimens, and botanists conversed to share their experience of collecting, to express their hopes for expeditions, and to exchange specimens. As long as people—either botanists or their emissaries—traveled, botanical knowledge could travel orally too. For example, Mr. McGrigor, a long-term resident "in the mountains of St Anns Jamaica" turned up on Banks's doorstep in London carrying a letter from Wright informing Sir Joseph that he could be trusted and that "he knows a number of the plants in that Island by the country names, and will furnish you with seeds or Plants for Kew Gardens such as you may direct."[60]

These forms of talk were embedded in particular practices of sociability. Conversations and correspondence between botanical gentlemen worked through a culture of "candid friendship" that related intimacy and truth-seeking. It is important to understand the significance of face-to-face meetings within these practices of intimate connection, which managed both botanical collaboration and the often tense personal politics of scientific priority.[61] Conversations conducted according to the norms of friendship helped botanists like Wright assess the work of others in Jamaica. They also provided the means to help visitors. Thus, Wright informed Banks that "I have seen Dr Oloff Swarts and shewen what civilities I could in person and by Letters to Kingston[.] He is gone to Hispaniola where I hope he will meet with many new things."[62]

Talk and text, conversation and correspondence, are hard to separate. They worked together to forge the links in chains of scientific sociability that stretched across the Atlantic via exchanges of letters and specimens. Botanical conversations involved the shared reading of letters from others, which, in turn, might report conversations.[63] Margaret Meredith notes that written correspondence was "understood as a conversation conducted through the medium of the letter between persons separated by distance, duty or protocol." She argues that prior meetings formed the basis for correspondence networks.[64] But writing could also come first. Barham corresponded with Sloane from Jamaica in 1712 without having met him.[65] Yet talk was still vitally important. When Barham came to London in 1717, he was keen to engage with Sloane in person and continued to write him long letters on natural history, sending them across the city rather than the ocean. On a later visit in the 1720s, Barham's efforts to take his leave from Sloane face-to-face speak to the cultural conjunction of intimacy and copresence. As Barham put it, "I am not a Little Troubled when I think of my Disappointement of having some Discourse With you & takeing my

Leave wch the Hurrying me a Way was the Occasion: A Thursday I Went to Pontacks [a well-known French eating house in the City] to See for you: but you were Just Gone & the Fryday Morning Early went to Graves End & a Saturday Embark^d."[66]

Eighteenth-century male friendship and its communicative forms were an important way of negotiating the demands of patronage through which natural history and so much else was organized. The mobilization of obligations and the forging of connections were crucial to plant collectors in the Caribbean and back in Britain, and they worked through talk as well as practices such as botanical nomenclature, publication, and illustration (see figure 12).[67] William Wright was keen to remind Joseph Banks that he had helped international collectors in Jamaica such as Swartz, because "I was happy to show him kindness on your account." In return, he sought plant specimens for others and solicited Banks's help organizing the necessary patronage to get his nephew a position as an East India Company surgeon. William sent James to converse with Sir Joseph about botanizing in Iceland, which both men had done. As Wright wrote to Banks, "I leave you to talk to him about his Northern Voyage with Mr Stanly. He was the only person that parted with him on good terms. The others having unfortunately quarrelled the whole Voyage."[68] Perhaps James talked his way into the position he subsequently obtained.

Carrying the weight of friendship and patronage, face-to-face meetings and polite conversations were as productive of enmity as intimacy, particularly where claims on the support of the powerful, hierarchies of competence, or precedence in significant discoveries were involved. Thus, Wright reported to Banks from Jamaica in 1783 on a couple of botanical colleagues (one of whom, Reverend Lindsay, was heard holding forth on political speech in chapter 2), "I told you in my last I had seen Dr Clarke he is in considerable Practice near Bath, has his Salary from the Assembly [as director of the botanical garden] and I believe as usual does nothing for it. Dr Lindsay Rector of Spanish-town went off in May last on the Verge of Dropsy. I have seen him, and he is no less a Blockhead in Botany & Drawing than in Theology, and common sense." Sometimes it was precisely the conventions of botanical friendship, as conveyed through conversation, that were at stake. When Wright saw what Swartz had produced from his Caribbean collecting, he regretted having helped him since it seemed to have undermined his own discoveries. Wright pointed out to Banks that "I hope he will do me justice, and that I shall have no reason to complain. I shewed him much attention personally and by Letters, and pointed out numbers of Plants he did not know or had not seen. But if he is disposed to

Figure 12. Palm tree dedicated to the Prince of Wales. Griffith
Hughes, *The Natural History of Barbados* (London, 1750), Plate IV.
© The Trustees of the Natural History Museum London.

take all those as discoveries There are many Evidences against him in your
possession and a Herbarium of mine in this City of Eleven years standing."
Wright later noted of Swartz, punning again on botanical acquaintance,
that "Candor and Gratitude do not seem to make a great part of his composition. He mentions me but once at the article Geoffræa. I once thought
there was some merit in being a Collector of Exotics." Wright himself was
said to have been courted in metropolitan botanical circles "from the

ample store of information he possessed, and from the talent for conversation which enabled him to make his knowledge at all times available," but others struggled to live up to the demands of botanical friendship.[69]

Beyond the botanists of Banks's international circle there was perhaps less at stake, and talk among friends might partake more readily of the pastoral culture of sensibility that Wright signaled in his idealized vegetable communion in field and vale. Thus, while there were also elements of the master–pupil relationship between them, Thistlewood referred to Robinson as his "beloved Friend." When Robinson died in 1768, Thistlewood recorded that he "walked out alone and wept bitterly," noting later that "when he lived in this parish he spent a good deal of time at Egypt with me, and I was never happier than in his company." This relationship involved the two men sharing their low moods, what Robinson called "the gloomy disposition of my Soul," as well as their enthusiasm for natural history and, it seems, a disdain for men whose verbal self-presentation exceeded their learning.[70] Yet this intimate friendship was also part of a wider circle of "gentlemen" interested in natural history. In the mid-1760s, Robinson drew together Thistlewood and both Edward Long and Long's brother Robert to describe and draw the birds they hunted or had brought to them by others, including the enslaved.[71] While it is unclear whether or not Thistlewood and the Longs met, Robinson did introduce the plantation overseer to the wealthy planter George Spence, who, as a magistrate, was later centrally involved in the interrogation of those accused of "conspiracy" in Clarendon in 1776 (see chapter 2). On the basis of a shared interest in Jamaican meterology but noticeably different results from their thermometers, Spence told Robinson that he "must get acquainted with" the "honest and Philosophical Man" whose letters Robinson had showed him. Thistlewood later carried botanical specimens from Spence to Robinson as all three worked to puzzle out, through books, letters, and conversation, which were new species.[72]

A key site for these conversations was the garden, which gave plant talk both subject matter and meaning.[73] It is significant that on May 12, 1768, when Thistlewood heard of Robinson's death from the merchant Samuel Hayward, he had that day "sowed above 30 sorts of seeds in the garden." Thistlewood's own sociability was often organized through his horticultural pursuits and those of others. On October 17, 1774, for example, he visited the Copes at Paradise and carried on to Mr. Haughton's place, where he found John Cope trimming vines in the garden. Cope then joined Thistlewood to visit Mr. Blake and walk together in his garden. In August 1786, Thistlewood and his neighbor Hugh Wilson rode to Paul Island Estate to visit Julines Herring and walk through his garden. They spent the night there and then went on to visit Mr. and Mrs. Anderson's garden and pen at Belle Isle. Visitors were also entertained in Thistlewood's garden. On February 16, 1779, Mr. and Mrs. Blake and Mr. and Mrs. Swinney paid him a visit, and "Phibbah made them tea, gave them porter, &c. under the guinep tree in the garden."[74] Unfortunately, Thistlewood himself was not at home, having gone visiting.

These visits were part of a wider set of exchanges of horticultural (and natural historical) books, seeds, plants, flowers, and fruit and vegetables—as gifts and in lavish dinners—through which Thistlewood established and sustained his relationships with other members of Jamaica's white society: men and women. This activity made him part of a circle of botanical gentlemen, which included, along with George Spence and Samuel Hayward, the doctors Richard Panton, Robert Pinkney, and Thomas King, and the wealthy planters John and William Henry Ricketts, Richard Vassall, and even William Beckford, one of the weathiest men on the island.[75] Yet it also involved supplying produce to neighboring families who were entertaining Governor Trelawny and his wife, although Thistlewood was forced to note that "Mr Weech says the Governor and his Lady several times expressed a great desire to come and see my garden, but were prevented by Mr Haughton's representing the road to be so very rocky and bad."[76] This was the exception that proves the rule that from the mid-eighteenth century onward, the extension of the plantation complex with its network of great houses, the development of the island's roads, and a notion of "conscious equality" between white men of different ranks within a racially ordered society had provided the basic social infrastructure which enabled the polite exchange of natural historical knowledge, and of which Thistlewood wanted his garden to be a part.[77]

Thistlewood's plot combined polite horticulture, natural philosophical experimentation, and agricultural production for profit. There is no direct record of his horticultural conversations, but a sense of their content comes from his diary entries and letters to Edward Long. His written account of "Plants growing in Thos. Thistlewoods garden at Bread Nutt Island Pen, June 1776" shows what he thought would interest enlightened visitors such as Robinson, Cope, Blake, or even the governor and his wife had they braved the road. His view could be compendious, identifying both cultivated and uncultivated plants and fungi, and what the slaves grew to feed themselves as well as what was in his own garden. His primary topic was what would grow in this place and what would not. Such plant talk would have had to be attuned for other ears since, as he noted, "People of a Philosophic turn" on the island were "So thinly sown." Like the nurserymen from whom he bought many seeds, he dealt in plants' common names and did not need to identify everything. Yet there were highlights. He claimed to have grown the first white narcissus "that ever flowered in this island" as well as the first English pink and honeysuckle. He would certainly have shown visitors what was newly planted, which might have been English trees, or flowers and vegetables for the table: crocuses, tulips, and narcissi; turnips, cabbages, and carrots; or granadillas, grapes, and christophenes. Perhaps even his greatest triumph: asparagus. He could have discussed where the seeds or plants had come from, outlining a series of local, regional, and transatlantic transplantations and exchanges—such as with Henry Hewitt of Brompton in London, or with other island gardeners—which brought seeds to his plot and sent his seeds to others. He might have discussed the care he took in arranging his garden, including the arbor constructed for granadillas and grapevines; his transplantation of species from the island or farther afield: nutmeg, coconuts, and coffee, or bamboo, mangoes, and the coral tree; his experiments as a livestock pen owner with "Fattening" grasses and fodder such as "clover, lucern, burnett, timothy grass, sainfoin, french furze, angelica & borage seed"; or, as a slaveholder, planting yams and rice for food and carefully recording how well they grew. No doubt his visitors would have had suggestions—about planting, grafting, and tending—just as his correspondents did.[78]

In his garden, Thistlewood was engaged in a small way with the grand work of imperial transplantation, naturalization, and improvement.[79] While he complained of the "hot Air, poor Soil, and bad Seasons" at Bread Nut Island pen, he argued that Jamaica had "such a Variety of Soils & Situations of diff^t Temperature, that am apt to imagine, most of ye Fruits, both of the Torrid, & Temperate zones might be cultivated with success, and that the latter might be Naturalized by degrees." He thought that spices would grow, and, presumably from his reading, "the Bread Fruit & other Productions of Otaheite . . . &^{c} in the Pacific Ocean."[80] However, as Douglas Hall calculated, of the two hundred kinds of seeds and trees he imported from England between 1768 and 1770, only about twenty were still growing in September 1770.[81]

Just like the British Empire for which the work of transplantation was undertaken, Thistlewood's garden was deeply rooted in the social relations of plantation slavery.[82] Enslaved workers, both men and women, dug, planted, weeded, and pruned under his orders and under the whip. They also tended their own gardens, or provision grounds, growing, as he listed them, "plantains (horse & maiden), bananas, maize or Indian corn, guinea corn, scratch toyer, coco roots, cassava bitter & sweet, pandas [pea-nuts], and sugar cane &c. &c."—a combination of African domesticates and American plants, for subsistence and medicine, that enslaved Africans had become adept at growing for physical and cultural survival.[83] It was en-slaved men and women who taught Thistlewood how to space corn plants when he was the new arrival in Jamaica in 1751, and who prepared his food, kept him healthy, and made his garden grow.[84] Their knowledge and skill underpinned his use of the garden as a space of sociability, not least as they carried gifts of seeds, fruit, vegetables, and flowers back and forth between Thistlewood and his horticultural friends.

The garden was Thistlewood's Anglo-Caribbean world writ small. The enslaved were always hungry, particularly when hurricanes destroyed their subsistence crops. They fed themselves from the plants they had tended on his orders—avocados, peas, plantains, and cassava—and he flogged them for doing so.[85] Yet there were also exchanges of gifts from these gar-dens that passed both ways between master and slaves. They all used their gardens to make a little extra and to deploy it in intimate social interac-tions to strengthen the connections that might help them live.[86] Thistle-wood supplied food for planters and the governor. Phibbah, he recorded in 1773, "made me a present of £10 18s 1 1/2d, all in silver; money she has earned by sewing, baking cassava, musk melons & water melons out of her ground, &c &c."[87]

Yet the conversations that surrounded these horticultural exchanges have left little trace. How was horticultural expertise and plant knowledge made and exchanged in the gardens between masters and the enslaved, and how was that exchange shaped by gender as well as race? What was said as gifts were given and received? Did Thistlewood and Little Jenny talk as they spent two hours in February 1779 gathering mimosa seeds at Poole's Rock to send to Henry Hewitt in London? We know that Thistle-wood was introduced to new fruits and vegetables grown and cooked by enslaved workers who told him the names they called them. These men and women also talked of plants' properties and botanical remedies, and they let him glimpse the riches of the storehouse of stories of plants, ani-mals, and spirits that inhabited the Afro-Jamaican landscape: *anansi* the spider, the fruit-gorging monster *cokroyamkou*, and the *duppies*, spirits of the dead that lived in the silk cotton trees.[88] Yet not only were slaves brutally punished for stealing food they had grown in Thistlewood's garden, but it was also a site of sexual coercion within a system built on terror. On March 20 and 23, 1786, Phoebe was at work weeding when Thistlewood raped her by the granadilla arbor over the cotton tree roots.[89] Indeed, the garden was a place where he committed multiple acts of sexual violence against numerous women, many more than in any other space beyond the plantation house.[90] If it was a place of polite conversation and sociability, the garden was also a place of violence and silencing enacted by a white man on many enslaved black women. And if, as Trevor Burnard emphasizes, Thistlewood did not imagine the "Negroes" with whom he lived and gardened as having anything like the same connection to him as his botanical friends, then this was a matter of gender as well as race.[91] Both Thistlewood's "gentlemanly politeness" and his botanical conversation were tightly circumscribed.

Botanical Oration

In contrast to these sociable, but essentially private, networks of gentlemen in conversation were more avowedly public pronouncements that sought to make Caribbean botany itself a matter of wider utility through establishing and maintaining botanical gardens funded by the islands' governments. This effort was part of a broader process of initiating botanical gardens, expeditions, and wide-ranging surveys from the 1760s onward as imperial policy makers sought to use economic botany as a practical and ideological solution to a series of imperial crises. This process led to the establishment of a botanical garden on St. Vincent in 1765 and to new gardens on St. Helena and in India in the 1780s and 1790s.[92] Under the banner of "improvement," these were sites for what Richard Drayton calls "a complicated theatre of virtue" played out by metropolitan and colonial interests.[93] In the late eighteenth-century Caribbean, with debate raging over abolition and fears of slave revolts, what was at stake was the nature of the relationships between the imperial government, West Indian sugar planters, and the enslaved (as we will see further in chapter 5).

Jamaica's first public botanical garden was established at Bath (in St. Thomas) in 1779 under Thomas Clarke.[94] However, a broader public interest in botany does not seem to have flourished as a result. Some observers, including William Wright, blamed Clarke's inactivity and were delighted when he was replaced in 1788 by Thomas Dancer.[95] But by July 1789, Dancer was writing to Edward Long in London that "after what I have already observed concerning the Taste of People in this Country for Science I need not add that your Proposal respecting a Botanical Lecture though a very proper one is not likely to meet with much Encouragement."[96] It is not clear what sort of talk Long had in mind. His *History of Jamaica* had suggested establishing a botanical garden stocked with plants "most distinguished for their virtues in medicine, or value for commercial purposes" (*HJ*, 2:259) near a proposed boarding school for the island's middling sort, and there were certainly botanical lectures at other gardens.[97] But Dancer persisted, and less than a year later, on March 1, 1790, he gave a botanical lecture of sorts by offering an "Oration" or "Address" on public botany in Jamaica, subsequently reprinted in the island's newspapers, which argued for "the necessity or importance of a public garden."[98] The contents and delivery of this speech, and the ways in which its audience was called into being, can define the contours of this form of botanical talk.

The occasion was Dancer's first address to the new directors who oversaw the garden for the Jamaican assembly, and he welcomed the fact that botany had become an arm of the state and a matter of collective concern. He then addressed a much broader public to justify government funding for a botanical garden: "Utility is the grand object of every public establishment, and if a Botanical Garden is not subservient to this end it should meet with no support or encouragement:—But can it be necessary, that in such an enlightened age and country, when amidst the many improvements that have taken place, Natural Science, and Botany in particular, has made such rapid advancement, that I should undertake to demonstrate the advantages arising therefrom to mankind?" It seemed that it was, so he reminded his audience of "the several establishments of this kind in every part of Europe and elsewhere" and noted that the French had made exchanges of plants between their botanical gardens a matter of imperial policy. He encouraged the audience to think to the island's past—when the now failed crops of cacao, cochineal, and indigo had been profitable—and to the future, when coffee, cotton, and even sugar might fail to make money. Surely this suggested the wisdom of investing in the "possibility that many of the plants introduced by means of a Botanical establishment, may, on a future day, become of great importance." And if an appeal to the island's economic base was not enough, there were other public purposes: "Is not the institution an ornament to a civilised country? Does it not tend to introduce a spirit of enquiry and improvement, and to enlarge the sphere of our rational gratification by adding to the elegance and luxuries of the table?" A "Public Garden" would, he assured the public, serve the public.[99]

However, these questions also needed to be asked of "the Public of Jamaica," Dancer suggested, because of the specter of private interests. As chapter 2 has shown, the existence of a planter "public" could certainly not be taken for granted given the planters' long-standing unwillingness to be governed. Moreover, by 1790, abolitionist activists were fully engaged in attempts to redefine public benefit by arguing that those who organized the slave trade and plantation slavery served no imperial interest wider than their own illegitimate private profit (an argument to which we will return in chapter 5).[100] Dancer sought to counter both charges by making the botanic garden "a symbol of the 'improving' plantocracy."[101] While he assured his audience, despite what he had written to Long, that "there is no country upon earth that abounds with a more intelligent people," he noted that "in every community there will be found men of narrow views and prejudiced sentiments; men destitute of Amor Patriæ, who estimate every thing by the standard of immediate self-interest, and become dupes to their own œconomy, by opposing indiscriminantly every plan of public improvement."[102] He offered his arguments about utility and public benefit to convince both them ("an arduous undertak[ing]") and those who were not entirely lost to self-interest, which meant showing the limits of private interest where public botanical good was concerned.

To do so, Dancer celebrated the actions of "certain Gentlemen"—a version sent to Long mentioned Matthew Wallen and Hinton East, who had both established well-known private botanical gardens—"for the zeal they have shown in enriching this Island with useful and ornamental plants."[103] These men had certainly aimed to naturalize Asian and Pacific plants for the benefit of the plantation economy, with Tahitian breadfruit foremost among them as a food for the enslaved.[104] Yet Dancer informed his audience that "I must take the liberty of suggesting, that the necessity of a Public Garden can never be superseded by their efforts as individuals—Private property is fluctuating, and the sentiments, taste, and persuits of different possessors, will seldom correspond." The introduction of foreign plants, if it was of any importance, "should not be trusted to the changeable humours of people in private life." He put it to his imagined audience, "the Public of Jamaica," that "You will therefore concur with me in opinion, that a private Garden can never be deemed a fit deposit of plants that are intended for the public use." A "Public Seminary"—and he also called it a "Public Depository"—"under the direction of properly qualified and responsible persons, is necessarily required." He assured the new directors that he would spare no effort in making the botanical garden answer these ends. He offered as evidence of his success "some Correspondence I have been honoured with by the *respectable President* of the *Royal Society, some Gentlemen of the Society of Arts, and Others*, respecting the Cinnamon of our growth," a potential new cash crop cultivated in the garden at Bath.[105]

Dancer's speech was, in both form and content, designed to construct a public botany for Jamaica. Doing so depended on a series of oral and material strategies to secure a space—spoken and horticultural—for public botanical practice and to structure that space as both imperial and masculine as a way to "render it popular and useful." Doing so was not just a matter of ensuring that useful plants grew and were propagated; it also depended on ensuring that "the Public of Jamaica," the taxpayers, knew about it. Dancer had to spread the word as well as the seeds. He complained to Long that, without an expert propagandist like the great man himself, "we have few Gentlemen here sufficiently embued with Natural Science to enable or incline them to estimate the Importance of natural Studies to Society and Commerce," but he sought to work with useful plants to "so far awaken that Attention of the People as to convince them that the pitiful Allowance of £300 per Annum for the Bot. Garden, *including Salaries* is a very inadequate one."[106] Long had devoted much of his *History of Jamaica* to a discussion of the island's botanical wealth, at least in part to argue that a more botanically diverse economy would provide the basis for a larger, more dispersed white settler population "able to quell the insurrections of their Negroes, or oppose any hostile invasion" (*HJ*, 3:691n).[107] He had no doubt that Dancer was well placed to engage the public with such a vision. He informed him that "it is not sufficient that the Gentleman entrusted with this office possesses Science, he must also possess your Ardor, Activity, and that Enthusiasm of Spirit in the cause, without which, no Seminary of this sort can be brought to perfection, or rendered popular in the island."[108] This was to be a seminary in both senses—a seed plot and a place of education—education of the public in the utility of public botany.

This version of botanical knowledge involved plants' public virtues as well as their uses. The Bath garden was only three acres, but Dancer made room for "an Obelisk in the Centre of our Garden over a Well in which is to be included an Hydraulic Machine for raising Water." He had it inscribed in honor of Admiral Rodney, who had saved Jamaica from invasion in 1782 and presented the garden with cinnamon trees and other valuable plants captured from French ships. The obelisk, and its Latin inscription, secured these plants to the garden through an act of imperial heroism—a virtuous botanical gift to the Jamaican public from the British Empire—that might also save the island. It stood against Dancer's privately expressed frustrations at being unable to persuade "Gentlemen at Home or the *Nursery Men*," thinking only of their short-term private interests, to send him plants, even though he promised "it would be in my Power to make them ample Returns from this Country."[109] Even more troubling were the problems Dancer had in ensuring that the garden got the plants it deserved through official channels. He asked Long to ensure that any plants sent by Banks for the garden, including the breadfruit being brought by Captain William Bligh, were directly addressed to him or landed near Bath, "otherwise it is a Chance if I ever get them." It took political clout to plant a public garden. Without it, "the Gentlemen of the Country will be scrambling for all they can get (as they did for the Seeds brought out by Lord Effingham [in 1789]) & the Bot. Garden, the *Public Depository*, will be forgotten or come in only for a Remnant."[110] The public nature of this botanical space lay in the desirable completeness and security of its provision, keeping a collection that served imperial purposes and was beyond the short-term private interests of greedy men.

The particular nature of this space, the public it sought to cultivate, and the forms of speech that would animate it are evident in the differences between Dancer's vision of Bath and previous versions of it. In his *History of Jamaica*, Long described Bath's hot springs, which had become a fashionable place of resort in the mid-eighteenth century for people from all over the island to take the waters, listen to music, and play cards and billiards, so that "from a dreary desert, it grew into a scene of polite and social amusements." It was a place of assembly where families "had used to meet each other in friendship, and united their talents of pleasing." It was a place for both men and women to enact polite engagements with each other and with nature (see figure 13). Yet, with the political divisions of the 1760s (see chapter 2), Long argued, "party-rage succeeded; the partizans of the different factions could not endure the thought of mingling together under the same roof; and the more moderate persons grew indifferent to the place, where chearfulness, confidence, and mutual respect, no longer held sway" (*HJ*, 2:166). When Long visited in 1768, he found the buildings decayed and the fine green baize of the billiard tables smeared with goat droppings. Dancer's garden did not aim to reinvent a heterosocial public based on cultured conversation and polite diversion. Instead, his form of botanical oration, and the garden it grew from, invoked a public defined by a masculine imperialism rooted in notions of utility, improvement, and duty, stacking imperial heroes such as Rodney against men who acted in their own private interests. While this vision for the garden was as much defined by the privileges of whiteness and freedom as the Bath hot springs had been, there was certainly little room within the public garden or the claims of Dancer's botanical oratory for the polite chitchat of men and women at their amusements.

However, making the case for this sort of public botany did require working within private gentlemanly communication networks. The fortunes of the botanical garden were a matter of conversation and correspondence on

Figure 13. The Bath Hot Spring. [Edward Long], *The History of Jamaica* (London, 1774).
© The British Library Board. 146.c.8.

and off the island. Thistlewood reported to Long that, on the garden's first establishment, "the Rev^d: M^r: John Poole was down in this Parish, and told me the Place made Choice of for the Botanic Garden, is very bad land, and in his Opinion a very improper Spot." Poole also noted the "Continual ill State of Health" of Thomas Clarke, the island botanist, since his arrival (Wright, remember, was much less charitable).[111] Clarke's replacement, Dancer, made great efforts to meet Long in London before departing for Jamaica, well knowing that his support was vital to the success of "an Institution . . . which I believe was partly founded on the Ideas you suggested of its Public Utility." When in Jamaica, Dancer continued to seek Long's help to develop support and counter criticism. Long wrote to his friend Bryan Edwards to persuade him to shift from being a publicly vocal critic of the gardens to an enthusiastic cheerleader for its public benefits. Dancer also worried that he had fallen out of favor with Banks, and he sought Long's assistance in reestablishing the connection, for the good of the garden.[112] Private patronage, through conversation and correspondence, could work in the service of public botany.

Dancer seems, however, to have been ill suited for sensitively managing the appropriate communicative forms of friendship and patronage. Of his self-perceived failure to cultivate Banks, he told Long that "my restricted Situation & uniform bad Health puts it out of my Power, to make those Demonstrations of my Respect that my Inclination prompt me to."

When he tried doing so, Dancer mistakenly assumed an intimacy between Long and Banks for which the historian had to apologize to the baronet.[113] Dancer also had to apologize to Long for potentially showing around too widely the letters Long had sent to Edwards and copied to him, "which as I knew they would have great Weight in recommending me to Him [Edwards] & others I have taken the Liberty of showing." Dancer asked for forgiveness "if I have in any Measure violated the Laws of Correspondence" by what he had said.[114] Most strikingly, Dancer's communicative misman-agement was evident in a comment from Edwards to Banks noting the trouble and expense to which Dancer's letters were putting the baronet. As Edwards pithily stated, "I know the man, and well remember that his visits to me, became at length so many *visitations*." He offered to "give him a hint not to let his zeal prove too nimble for his judgement."[115] Detectable here is the sound of Dancer declaiming in the mode of botanical oration when the quiet to and fro of polite (if loaded) conversation between gentlemen, friends, patrons, and the patronized was in order.

Public botany's fortunes were similarly mixed. In 1789, Dancer had in-formed Long "that the Prejudices which had long prevail'd in this Island against the Bot. Garden, are beginning to subside." However, Dancer fell ill and had to attend to his own medical practice, leaving little time for natu-ral history or for lobbying the politicians. He could not get what he wanted from them. As he reported, "after attending upon the Gentlemen of the House of Assembly for a whole Month," he only achieved a £100 increase in his salary and nothing more for the garden. It was not enough. It meant that he could not buy plants from Europe, and was reliant on those he received as gifts.[116] In 1804, he made a final attempt. He submitted further printed observations and suggestions "to the consideration of the House of Assembly and the public." He reminded them that such gardens flourished all over the world and had been of utility to Jamaica in the introduction of, among other plants, cinnamon and new sugarcane varieties, but the assem-bly's policy of trying to support the Bath garden and Hinton East's garden at Liguanea had left them both neglected. Unfortunately, as a handwritten note at the end of Banks's own copy of the proposals says, "The H. of A. having not thought proper to make Reform in the Bot. Gardens Dr. D. de-clines acting any longer as Isl[d] Botanist."[117] Public botany in Jamaica would remain a challenge.

Both botanical conversation and botanical oration were strongly de-fined by their inclusions and exclusions: by who could be part of the con-versation or was imagined as speaker or audience. Both were fundamen-tally represented as the preserve of free white men, however much they depended upon the work and knowledge of others. While they can be examined as imperial and colonial oral cultures of botanical knowledge, shaped by class, gender, and race, unlike the discussions of medicinal plants (botanical prescription), they reveal little of the basis on which indigenous, transplanted, and enslaved men and women engaged in either public or private talk about plants. There is, for example, little in the record of the Jamaican botanical gardens which gives even the limited access to the sorts of interactions with enslaved gardeners and free people of color which allowed the St. Vincent garden to grow and images of its plants to be disseminated.[118] Yet it is clear from the previous discussion of the plant knowledge of the enslaved, and the ways it was fostered and materialized within their own provision grounds and gardens—what Judith Carney and Richard Rosomoff call "the botanical gardens of the dispossessed"— that exchanges certainly occurred.[119] While not celebrating it in the same manner as Admiral Rodney's transoceanic gift, Dancer's 1792 *Catalogue of Plants, Exotic and Indigenous, in the Botanical Garden, Jamaica* did include the Bichy (kola nut) tree as "An African Fruit, introduced by the Negroes before Sloane's time."[120] And Sloane himself had informed his readers that his *Natural History of Jamaica* sought to "teach the Inhabitants of the Parts where these Plants grow, their several Uses, which I have endeavour'd to do, by the best Informations that I could get from Books, and the Inhabitants, either *Europeans, Indians* or *Blacks.*" Indeed, in his second volume, he recorded of the kola nut that "it is called Bichy by the *Coromantin* Negro's, and is both eaten and used for Physick in Pains of the Belly" (figure 14).[121]

Conclusion

Speech played an important part in making knowledge of plants in the eighteenth-century Caribbean. It was vital in shaping gentlemanly cultures of specialized botanical knowledge that intersected with the "polite" horticultural practices of experimentation and sociability in the gardens of slaveholders and their managers. It was central to the arguments for a public botany rooted in Jamaica's botanical gardens that depended on making botany public in its aims and to its audience in the age of abolition and revolution. And it was crucial to the ways in which healers found out about plants and their uses on islands that had much sickness and many ways of healing. Talk may have been denigrated in favor of text or image in certain circumstances, but when combined in various ways with objects, growing plants, specimens, written words, and pictures, it formed the often unreflected-on medium of communication within which plant knowl-

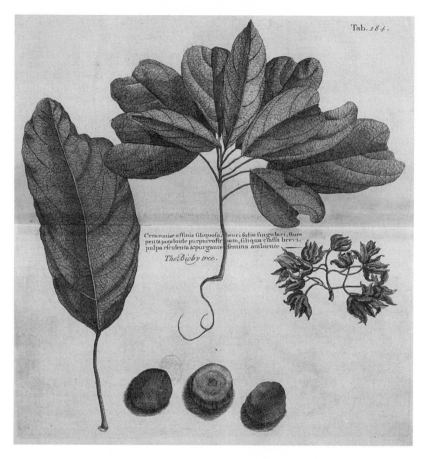

Figure 14. The Bichy Tree. Hans Sloane, *A Voyage to the Islands Madera, Barbadoes, Nieves, St Christophers, and Jamaica*, vol. 2 (London, 1725). © The British Library Board. 38.f.4.

edge was made.[122] Unlike writing or depicting, talking about plants was a practice in which diverse inhabitants of these islands, and those in other places concerned with their flora, could engage. Understanding knowledge making through speech requires and enables a broader view of who was involved in making that knowledge, through which practices, and on what basis.

Understanding that Europeans as well as the indigenous and the enslaved were deeply involved in oral cultures of botanical knowledge means emphasizing that speech is plural, differentiated, and power-laden as well as being a medium of exchange. This understanding involves disentangling and differentiating forms of talk and attending to their particular modes, locations, uses, and participants. Here distinctions have been drawn between botanical prescription, conversation, and oration, although it is clear that they also intersected in various ways. One of the central concerns has been to understand who could be involved in these speech situations and on what terms. The argument has been made that polite botanical conversation and the constitution of a public botanical audience and its orators were both gendered and rendered silent the majority of those who had something to say about the islands' plants—the enslaved population. It was the healer-patient dialogue, as constituted at the intersection of mobile, hybrid, and syncretic medical practices, that provided at least an opening, albeit one that could be used for appropriation as much as exchange, in the "biocontact zone." If we want to understand colonial botany as a matter of both exploitation and communication between Europeans and others, we need to attend to the inclusions and exclusions that different communicative practices made available.

It is also important to understand the relationships between these forms of talk. How might, in Latour's terms, chains of mediation and articulation be constructed to bring the identities and properties of Caribbean plant life across the Atlantic as matters of fact? Captain Thomas Walduck, in Barbados in the early eighteenth century, was certainly aware of the issues of trust and credibility, writing to the apothecary James Petiver that the "names and Virtues" in the "book of plants" he had sent were learned "from our Physicians (shall I call them) nurses, Old women and Negros," and he promised that "for the future I will take care by some Experiment or other not to be imposed upon but will serve you with truth to the utmost of my Capacity."[123] It is not clear how this was to be done, although Sloane, Barham, and others carefully triangulated what they had learned from speaking to people with what they observed (of both plants and their effects) and what they read in books. Others made the origins of this knowledge part of their credibility. John Southall of Southwark in London, advertising a commercial cure for bedbugs in his *Treatise of Buggs* (1730)—available at two shillings a bottle—told his readers that he had obtained the formula from an old, although uncommonly young-looking, freed "Negro" whom he had met while out riding in Jamaica. He had been given it after paying the former slave in food and drink and convincing him that he was not a hated "*Creolian*" but rather an Englishman. This man, Southall confirmed, had "great Knowledge . . . in the medicinal Virtues of Roots, Plants, &c." and tended a "piece of Ground" that "might be called a Physick-Garden, rather than a Provision-Plantation" where he produced medicines, the efficacy of which Southall could personally confirm. As Southall concluded, "'Twas owing to his Skill that he had thus preserv'd himself to so great an Age; and 'tis my Opinion, he had attain'd to a greater knowledge of the physical Use of the Vegetables of that Country, than any illiterate Person had ever done before him."[124] Yet the credibility of Southall's claims for his formula were not left to rest completely on what he had learned in the Jamaican woods. The book was dedicated to Hans Sloane, who, having responded to Southall's letter requesting him to "spare a few minutes to hear my observations on that Subject" and "have the first view of my performance," was thanked for reading it, "see[ing] the Experiments of my Liquor," and introducing him to the Royal Society. Southall clearly hoped to make his fortune through a transatlantic movement of botanical knowledge that conjoined the confidences of a former slave, the conversation of the man he judged "the most Judicious & Curious of mankind," and the authority of the Royal Society's public role in natural history.[125]

Examining the differences and connections between these forms of talk can also help explain change over time. In part, this was a matter of how face-to-face interactions, which may have then been recorded in manuscript as interpersonal transactions, were converted to more general, abstracted principles in printed works.[126] There were also differences across the long eighteenth century. The social and material infrastructure for botanical conversation was only effectively in place from the mid-eighteenth century onward, and the notions of the imperial public good that were developed within botanical oratory were shaped for an audience that had an ear to the late eighteenth-century debate over abolition. These developments reworked the speech situations within which botanical knowledge was formed, displacing the healer-patient dialogue as the primary locus of talk about plants and effecting the waning of respect for traditional knowledge noted by Drayton, Schiebinger, and Swan.[127] In the earlier period, doctors such as Barham and Sloane had attended to what the enslaved had to say to them about the islands' nature. Although they were often dismissive, they did admit the possibility of non-Europeans actively knowing and experimenting with nature.[128] For later commentators, the hardening racial categorizations that came with the development and defense of plantation slavery in the Caribbean, and which were worked out in practice through the conversational networks of white sociability in planters' gardens and the virtuous theatrical rhetoric of public botany in the botanical garden, however much they depended on the knowledge and skills of enslaved cultivators, meant that little experimental, philosophical, and communicative agency could be attributed to the enslaved. For example, while presenting extensive botanical information in his *History of Jamaica*, Long argued that "brutes are botanists by instinct," including "man in his rude state" along with the animals. For him, any cures that Africans effected were simply "random" (*HJ*, 2:380–81). Wright was less dogmatic, but he represented part of a more general move away from what was communicated in the field about "the medical powers of Plants" to what could be deduced, via other forms of observation and talk, from a plant's "sensible qualities."[129]

It was not that plant talk had stopped. Instead, the forms of talk that produced valued knowledge of plants had shifted. While there were still those calling for attention to the medical botanical knowledge of the enslaved in the 1780s, the center of gravity of European plant talk had moved from an asymmetrical dialogue about illness and cure in the islands' provision grounds, wooded tracts, and sickrooms to the conversational networks and gardens of botanical and horticultural gentlemen and the botanical gardens of an improving empire—and, ultimately, to the metropolitan laboratories that synthesized plant medicines.[130] At the same time, the natural knowledge of the enslaved was either reworked as folklore for a European audience or denigrated as superstition when obeah was understood through the religious ideas and speech practices explored in the next chapter.

FOUR

They Must Be Talked to One to One: Speaking with the Spirits

In the 1827 Jamaican gothic novel *Hamel, the Obeah Man*, the final speech by the eponymous "hero"—delivered to an audience of slaveholders in "the dialect of his country, the creole tongue"—diagnoses the political and spiritual ills of the island that have led to revolt:

> Your Missionaries have persuaded the Negroes that they are free; and they believe the king's proclamation telling them they are still slaves, to be a forgery! It will not be long, therefore, before they rise again; and they will take the country from you, except the king of England, and the governor here, keep these preaching men in better order. What do you want with them? You have a bishop and regular parsons; good men, who tell the Negroes their duty as slaves, and try to keep the poor ignorant things quiet and happy. If you let any other people turn their heads, believe me, they will twist off yours. I declare to my God, I never saw such trumpery, Your king, your governor, and all yourselves (forgive me, gentlemen) are afraid of these white Obeah men. What a fuss is made with them, and what strange nonsense they preach! . . . [If the enslaved] are not to be free by the law—forbid anyone to deceive them, on pain of death. I would hang or shoot the cunning, sneaking, fawning, fanatical, murderous villain, who tampered with the passions of my slaves, or dared to hint at such a circumstance as that of master Quashie holding a knife to his master's throat. But you are no worth— (forgive me, gentlemen—I spoke without caution)—I mean you are afraid of the white Obeah men.[1]

Pondering these questions of speech and belief, one planter, Mr. Guthrie, wondered at "that power which he had seen the Obeah man possess and exercise over the minds of his fellow Negroes," one, he was sure, derived more from the ignorance of the audience than "the talents of the conjuror, who was still a clever fellow, however, as he allowed more especially as he had heard his parting speech."[2] This, then, was a situation where words shaped minds in important, and potentially dangerous, ways; where speakers called on their gods to convince others, but where all might be conjurers: regular parsons keeping enslaved workers "quiet and happy"; obeah men controlling the minds of their "fellow Negroes"; and nonconformist missionaries turning the heads of the slaves. Indeed, describing Christian dissenters as "white Obeah men" suggests that distinctions between forms of speech and belief were unstable.

This chapter asks who could speak the word of the spirits in prayers, sermons, incantations, lamentations, and other oral forms and rituals. In doing so, it demonstrates what was at stake, particularly by the early nineteenth century, in the distinctions made between sorts of spirit talk, but also how unstable those distinctions were in the age of abolition. As in previous chapters, this brings together diverse forms of practice—in this case, spiritual practice—by attending to how they were spoken. Turning again to notions of speech and veridiction provides useful provocations, but ones that need reworking for the Caribbean context. Bruno Latour is keen to pay close attention to "religious speech" to reconfigure the long-standing debate about science and religion by conceiving them as distinct "regimes of utterance" with incommensurable felicity conditions. Concerned to revive a form of "religious utterance," and a mode of existence, that now "dries on his tongue" as "absolutely *meaningless*," Latour characterizes religious speech as "creating closeness" and convening communities in ways akin to the political speech discussed in chapter 2.[3] Yet, religious speech is also fundamentally different since there are other beings made real here. These utterances are, as he puts it, "very special words *that bear beings capable of renewing those to whom they are addressed.*" These beings are not "only words"; instead, "the words in question transport beings that convert, resuscitate, and save persons. . . . They come from outside, they grip us, dwell in us, talk to us, invite us, we address them, pray to them, beseech them." These are beings with agency. Where they appear and disappear is not under easy control, "the initiative *comes from them*, and *we* are indeed the ones they are targeting." That "these entities have the peculiar feature of being *ways of speaking*" means that this "regime of utterance" is very fragile: "If you fail to find the right manner of speaking them, of speaking well of them, if you do not express them in the right tone, the right tonality, you strip them of all content." The reality of these beings is that they exist through speech, and that makes access uncertain. As Latour puts it, "one lives constantly at risk and in fear of lying about them; and, in lying, *mistaking them for another*—for a demon, a sensory illusion, an emotion, a foundation."[4]

Latour's refusal to reduce such questions to sociological matters of belief is both challenging and instructive.[5] However, the terms of reference within his own work are determinedly Christian, even specifically Roman Catholic, such that his primary example of the beings of religious speech are "angels bearing salvation." Indeed, he makes a distinction between these beings and what he calls the "beings of metamorphosis," which are understood as the ways in which "the Moderns"—intent on "destroying fetishes, delivering people from their ancestral terrors, [and] putting an end to the power of sorcerers and charlatans"—strip "invisible beings" of their external existence and locate them "only in the twists and turns of the self."[6] However, this internalization is never complete, and there is constant reference to "the impression that one is being taken over, overcome by an uncontrollable force." As Latour's Moderns have it, "I am *alienated*, yes, *possessed*. I am an *alien*, one of the living dead, a zombie." This invocation of the Caribbean's best-known "being of metamorphosis" is a reminder of the limitations on his definition of "the Moderns." Along with his own insistence that the beings of metamorphosis are entities that grip, possess, appear, and disappear beyond human agency, that can be misrecognized as matters of the mind, and that have "a demanding form of veridiction, because one really can't say just anything at all about these beings: one ill-chosen word, one misunderstood gesture, one carelessly performed ritual, and it's all over: instead of freeing, they imprison; instead of taking care, they kill," means that there is little reason not to collapse the distinction between these beings and those of "religion." Indeed, doing so would seem more in tune with Latour's own injunction to close the gap between "the Moderns" and other "collectives" that have handled what we might see as spirits as "cosmological" entities.[7]

This chapter explores, on the asymmetrical common ground of speech under slavery, the forms of "spirit talk" invoked in Hamel's oration, whereby the ineffable was made present through words, and invisible beings and forces were called on to effect transformations. Such work with words and spirits always raised significant questions about the role of speech in understanding the beliefs of others and the power of talk to change people. What follows shows that various forms of speech were used to call upon spirits at large in the world and beyond but were always in danger of failing to effect the transformations desired of them and open to the accusation that they were the work of "sorcerers and charlatans." The chapter examines the forms of speech at work in three sorts of spiritual practice on these islands:

the sermons and catechetical teaching of the Church of England, including attempts to convert the enslaved to Christianity; the speech practices of what got called obeah, including ways of speaking with the dead, and how those were understood as powerful words by the slaveholders who sought to outlaw them; and the forms of preaching, praying, and oral instruction of the nonconformist missionaries, especially Methodists, who came to the islands to make converts in the late eighteenth century. Setting out these forms of speech, and the differentiations made between them, demonstrates how they each called on the ineffable to try to effect transformations shaped by the context of plantation slavery, and were therefore dependent, in various ways, on differentiating between the free and the unfree. Yet, as Hamel's oration shows, these forms of speech were in significant conflict with one another in the early nineteenth century as part of discussions over the future of slave societies in the Caribbean. The chapter examines these conflicts—over who could speak where and when—and demonstrates how, in figures such as the white obeah man, the distinctions between forms of spirit speech were inevitably exceeded.

How Shall They Hear without a Preacher?

Probably the most commonly noted characteristic of invocations of the divine among the white islanders of the Anglo-Caribbean was that they were irreligious. As Thomas Walduck wrote from Barbados in 1710, colonists were "Horrible profane and lude in their Discourse and Conversation" and "hardly think of God but in their Curses & Blasphemies."[8] Yet the Church of England was an active presence in the Caribbean and was, primarily, the planters' church. Barbados and Jamaica were divided into parishes with churches that were simple rectangular "preaching boxes" that "permitted worshippers to hear and see the actions of the liturgy without impediment. . . . . Their shallow chancels, simple communion tables, and towering pulpits topped with elaborate sounding boards articulated an Anglican vision in which word and sacrament were balanced in the worship of the church." Whom these words were for was clear from the memorials to prominent white families and pews ordered by social hierarchy and the ability to pay. Although churches admitted some of the islands' people of color, and more might hear the service through the buildings' open doors and windows, these were "liturgical spaces that created community for whites and offered only a low place in their hierarchy to free persons of colour and a few slaves" present as the domestic servants of slaveholders.[9] Indeed, as previous chapters have shown, the forms of speech heard in these places overtly bound together church and state and enforced divisions between black and white, free and unfree. They were where laws and proclamations were read out, elections to the assembly announced and conducted, and services of collective imperial thanksgiving undertaken.

Yet they were also venues for a variety of forms of religious speech that carried their own highly charged histories and geographies across an Atlantic world riven by religious controversies.[10] The words spoken in church, and at home, had different purposes and contested value as Anglicans opposed Catholicism and developing forms of dissent. What ministers said to their congregations was crucial. For example, Joseph Glanvill, a prominent seventeenth-century Anglican divine, argued that "Preaching is publishing God's Mind and Will, that the People may know his Truth, and obey his Laws." Although it was not a necessary part of divine service, there was an acknowledgment that "*Faith comes by hearing; and how shall they hear without a Preacher?*" For eighteenth-century Anglicans, the power of preaching meant that it was dangerous if not done by the ordained clergy of the Church of England. They had ample evidence from the previous century of the dangers of a world where everyone judged they had the right to speak.[11]

This meant attention was paid not just to who preached, but how it was done. Although there was no single approved style, Glanvill enjoined preachers to remember that they "speak as the *Oracles of God*, and as becomes his *Messengers* and *Ambassadors*." They should offer a "manly unaffectedness, and simplicity of Speech," rather than the "bastard kind of Eloquence that is crept into the Pulpit, which consists in affectations of wit and finery, flourishes, metaphors, and cadencies."[12] Plainness was also seen by John Jennings to be "*a Style becoming the Gospel of* Christ," but it should not shade into bluntness, sordidness, and rusticity with "the use of vulgar Proverbs, and homely Similitudes, and rude and clownish Phrases."[13] The whole should be presented with the "flower and vigour of our affections," not in a cold and heartless way. Indeed, preaching was performance since, as Glanvill argued, particularly for the "common People," who "have not Souls for much knowledge" but are moved by "the Voice, and Gesture, and Motion," "Religion is Zeal, and if Men are not earnest in it, if it hath not the warmth of their Affections, and the ardours of their Desire, and Love, it hath no considerable hold upon them." This meant that preaching was an embodied act. As they spoke, preachers should neither be "Mimical, Phantastical, and violent in their Motions" or stand stock still "like Images." Movement should be kept free and natural, "grave and decent," since employing "a rude loudness and apish gestures and boisterous violence" is "putting Religion in a Fool's Coat" and debases the Gospel, as much as the "jingles and quibbles . . . witticisms and flourishes" of "Foplins" and "wifflers."[14]

This was a technology of the self, using voice and texts, to hone a skilled accomplishment through self-examination and practice, as Jennings explains: "To arrive at any tolerable Perfection in Preaching *Christ*, is a work of Time, the Result of a careful perusal of the Scriptures, and studying the Hearts of Men: It requires the mortifying the Pride of carnal Reason, a great Concern for Souls, an humble Dependence on the Spirit of God, with the lively Exercise of Devotion in our Closets."[15] As such, church worship was not only reliant on the minister's words. The forms of worship set out in the Book of Common Prayer prescribed particular forms of words for both clergy and congregation. For Thomas Bray, the founder of the Society for the Propagation of the Gospel in Foreign Parts (SPG) and the Society for the Promotion of Christian Knowledge, the provision of these books in the colonies and instructions for their proper use were crucial "for all Ranks and degrees," "so as through the whole Service to read audibly, gravely, and devoutly, the parts allotted for them and all the Laity." If they did not, God would be affronted not glorified on the stage of "*publick Worship*."[16]

However, preaching and divine service could not be the vehicles for learning to be a Christian, particularly for children and servants, a category which included, while obscuring, the enslaved.[17] Although the holy scriptures were "the only infallible and certain Rule of all reveal'd Truth and christian Knowledge," the message of these complex and contested texts needed to be codified, reorganized, and appropriately delivered, particularly through catechisms—modes of question and answer—for those undergoing instruction. Accommodations needed to be struck between accessibility, on the one hand, and depth and theological precision, on the other. Such accommodations were particularly evident for what Isaac Watts called "the first Instruction of any of the more ignorant Parts of Mankind," which needed forms of speech that balanced learning texts by heart and a deeper understanding of their meaning:

> This Way of teaching hath something familiar and delightful in it, because it looks more like Conversation and Dialogue. It keeps the Attention fixed with Pleasure on the sacred Subject, and yet continually relieves the Attention by the alternate Returns of Question and Answer. . . . And thus the Principles of Religion will gradually slide into the Mind, and the whole Scheme of it be learnt without Fatigue and Tiresomeness.[18]

This approach might first involve memorizing required responses, but catechumens were subsequently to "be under a necessity of exercising in some measure their own Judgments, by returning, not to a certain Set of Words, a certain and determinate Answer, but appositely so much of the Answer, as belongs to some particular part of the Question." They would ideally move from collective class-based learning to individualized instruction. The catechist would recommend reading and "confer with them as it were, and that in a Familiar manner upon what they have last read," anchoring their questions and answers in scripture. This way of teaching would, for Watts, establish the difference between Protestantism—"That the Word of God alone is a sufficient Rule both of our Faith and Practice"—and Roman Catholicism, which he saw as a fog of incomprehensible mysteries, not least in the language through which it was conveyed. Catechizing would, Bray thought, be the means for reviving piety and the church in an impious age.[19] It might be combined with other oral forms, including singing psalms and prayer, in public, in families, and "in their closets." Indeed, what was required was not one form of religious speech but the proper coordination of several forms. As one catechetical question on the Sabbath put it:

Q. How should you keep this day holy?
A. By hearing God's Word read and taught, prayer private and public, often receiving the holy communion, reading good books, and using all sober and godly conversation.[20]

What were the implications of these speech practices for instructing enslaved Africans in Christianity? Most members of the established church thought Christianity and slavery were compatible.[21] For some, that meant not only that Christians could own slaves, but that the enslaved were not fit or able to become Christians themselves, while for others, it meant that enslavement by Christians brought—with the recognition that "Negroes and Indians" were "endued with a reasonable and immortal Soul, which alone constitutes him a *Man*"—the imperative of conversion.[22] Thus, they worked to counter the prevalent view that conversion to Christianity would mean freedom from bondage and to encourage planters to allow conversion via ordained ministers and the SPG.[23] Conversion was presented as an obligation due for the labor extracted from the enslaved and would, these divines proclaimed, ensure that slaves were more diligent and obedient. As Archbishop Gibson argued in 1727, the possibility of conversion accorded the enslaved a hyphenated humanity, treating them "not barely as Slaves, and upon the same Level with labouring-Beasts, but as *Men*-Slaves and *Women*-

Slaves, who have the same Frame and Faculties with your selves, and have Souls capable of being made eternally happy, and Reason and Understanding to receive Instruction in order to it."[24] Underpinning this argument in the early eighteenth century was the idea that if the enslaved were permitted by their masters to hear the word of God through the Anglican Church, then they would be converted.

This combination of textual and oral work designed to convince through reasoned dialogue was the basis of the Church of England's missionary work among the enslaved. SPG clergy were told "that in their Instructing *Heathens* and *Infidels*, they begin with the Principles of natural Religion; and thence proceed to shew them the Necessity of Revelation, and the Certainty of that contained in the Holy Scriptures, by the plain and most obvious Arguments."[25] Within this framework, preaching was understood as "an Evangelical Institution" that echoed the work of Christ's apostles, particularly Saint Paul, in spreading the Gospels to the ends of the earth.[26] This way of thinking about speech was written into some of the key texts used by the SPG.

Thomas Wilson's *Essay towards an Instruction for the Indians, Explaining the Most Essential Doctrines of Christianity* (1740)—better known in subsequent editions as *The Knowledge and Practice of Christianity Made Easy*—became what Travis Glasson calls "a global missionary classic."[27] The book's prefatory material established the unity of humanity, the biblical sanction of slavery, and its justification via conversion. The bulk of the work contained fifteen dialogues between an "Indian" and a missionary—although it was explicitly described as "for propagating the Gospel amongst the *Indians* and *Negroes*"—that worked through the basic doctrines of Christianity. Through these imagined catechetical conversations, Wilson's principle that conversion was a matter of "Patience in explaining the great Truths of the Gospel" was exemplified. The "Indian" interlocutor is an interested inquirer, always asking for further explanation, always seeking the truth. The missionary is willing and able to explain in full, avoiding theological controversies and setting the case out clearly, logically, and rationally. Language offers no barrier, and the concerns of the "Indian" or "Negro" do not spring from alternative beliefs or from them having been dispossessed or enslaved, but only from not having previously heard Christianity's truths. The imagined catechumen's final reassurance is that "I will most thankfully dedicate myself to [the Holy Ghost], that, by his Assistance, I may be able to please God, and perform what I have promised."[28] It really was conversion made easy.

However, the established church's missionary work had little effect on the sugar islands during the seventeenth and eighteenth centuries. The SPG had believed that the problem was the slaveholders not allowing the enslaved to hear the word of God. But the SPG's ever-closer accommodation of modes of Christian education to the views of the slaveholders—such as catechizing without teaching reading and writing—made little difference.[29] The planters' church offered little to the enslaved. Indeed, even the SPG's most substantial engagement in the Caribbean shows that the very forms of speech through which it worked served to further mark the differences between the free and the unfree and thus fatally undermined attempts at Christian conversion for those charged with doing it.

The largest SPG effort in the sugar islands was at Codrington in Barbados, where, in 1710, the society had been bequeathed a working plantation in order to build an educational establishment (figure 15). The bishops who ran the SPG had to be committed to both operating a full-scale sugar plantation based on enslaved labor and converting to Christianity those they held as slaves.[30] As well as exhorting their attorneys and managers to maintain sugar production levels, and the buying and selling of enslaved Africans, the bishops presided over a system of religious instruction. In 1740, following practice in New York and embedding the use of slave labor in the work of conversion, they resolved "to choose two of the most promising Negroe Boys now in the Societys Service, to be employed wholly in learning the Principles of Christianity, that when fully qualified, they may assist in the Instruction of their fellow Negroes." At the same time, the society's catechist reported that 65 of the 207 slaves on the estate were baptized, and he requested that he be sent "100 Bibles and Common Prayer Books, 100 Stiched Catechisms, and 100 Copies of the Whole Duty of Man" for use of those enslaved at Codrington and the poor people of Barbados. A few years later it was reported that it was the usher's duty to "Catechize the Negroes publickly in the Chapel every Sunday in the Afternoon."[31]

Figure 15. Codrington College, Barbados, 1720s. Detail from William Mayo, *A New and Exact Map of the Island of Barbadoes in America*, 1722. Courtesy of the John Carter Brown Library at Brown University.

The concurrence of the modes of religious speech deployed by the SPG and the need for the Codrington plantation to turn a profit generated particular sorts of oral encounter. As Glasson shows, Codrington's characteristic plantation demographics—of high death rates and low birth rates—meant that its labor force was predominantly male and African-born. While there were some positive reports on progress catechizing the enslaved, there were always more concerns that this effort was not working.[32] By the 1740s, it was reported that "the Negroes shew but little Regard to Religion" and that even those who agreed to be instructed "are very cool in the performance of what is taught." In 1745, Joseph Bewsher, the newly arrived usher, reported that few could repeat the catechism, and "as for the Old ones, some of whom have had the Priviledge of Instruction there for more than 20 years, they appeared as Ignorant of their Duty to God, as if they had arrived there from Guinea a Month before."[33] The blame was put on the unwillingness of the enslaved to adopt Christian marriage and to devote their Sundays to worship rather than growing food.

In summer 1759, Reverend John Hodgson, a young Oxford-educated clergyman, arrived in Barbados to take up the position of usher at Codrington. By July 1760, he was writing to the SPG in London to tell them that his and their hopes for enslaved converts were unfounded, wrecked on the shoals of speech. Hodgson wrote that he had only been able to instruct those who worked at the college rather than on its plantations. "These," he wrote, "being more civilized than the rest, are proportionately more capable of comprehending the plainer principles of our Religion." His method, on Sunday afternoons after prayers in the chapel, had been "to explain the Apostles Creed, the summary of our Duty to God and our Neighbour at the end of the Commandments in the Church Catechism, and the Lord's Prayer," all of which they were expected to learn by heart. Yet there had only been slow progress. More damningly, he informed the bishops that almost nothing could be expected of the majority of those working the plantation lands:

> They must be talked to in a language which they do not understand, they must be talked to upon Subjects which from their being void of all the principles to which application should be made, or filled with such as render their minds still more inaccessible, cannot take the least hold of them, and they must be talked to, not in a collective capacity but one to one, it being absolutely necessary to inculcate upon each individual, even when they are the most civilized, the several particulars that are taught them to a degree of repetition which can hardly be conceived by those who have had no direct knowledge or experience of them.[34]

The Church of England's disciplined control of texts, speech, and religious knowledge, worked out within hierarchical catechetical forms of reasoned question and answer in a face-to-face oral engagement, had reached its limit—a mutual incomprehension, repeated again and again to no effect, at least in Hodgson's terms—in the power-laden and sweat-soaked cane pieces of the Caribbean.[35]

Other clergymen had been here before. Indeed, it was a long-standing position of Anglican ministers in the Caribbean who felt that metropolitan desires for conversion were unrealistic. In 1707, Richard Tabor and Thomas Lloyd informed the bishop of London from Jamaica that they knew that many "warm Gentlemen" blamed them for not converting the enslaved, and they emphasized the difficulties in doing so. The enslaved, they wrote, were of "a falce, base, & slavish temper in whom Barbarism is so radicated: y^t tho born here, they will still retain their Heathenish coustoms." Moreover, enslaved Africans spoke ten languages, none of which had been "brought under ye Regulation of letters," and, as James White of Kingston argued in the 1720s, mixed with "other infidels like themselves, who are almost all Liars, cheats, diabolical Cursers & swearers, & lewd talkers, without the least shame." The clergy had, they argued, enough to do "as to ye lost shell of ye Church of England in these parts, y^t we have not Space to make a progress in matters of so difficult a nature." This was, these ministers noted, very different from the work of the early Christians, since they lacked "ye happy gift of languages, [with] w^h ye Apostles were blest in primitive times." As White argued, since they could not undertake the three or four years of catechetical instruction that would bring the enslaved to a proper knowledge of scripture, none were proper practicing Christians who could "give a satisfactory reason of the obvious attributes of God . . . [or] why our Saviour is the Son of God."[36]

Despite these racialized accusations of ignorance and barbarism, and the Anglican missionaries' adoption of the slaveholders' view of the enslaved, the promise of universal Christianity depended upon the assumption that if the social and political conditions for speech were different, then instruction, and even conversion, would be possible based on a common humanity (see, for example, HJ, 2:429). This social theory of language and apprehension is clear in Hodgson's letter. He argued that the principles of Christianity would not be received through the forms of instruction proper to the Church of England "'till there is in some sort a general change in their condition by introducing among them the regulations, & advantages of civil Life," particularly monogamous marriage.[37] He could not, however, see that happening in the near future.

In London, the SPG's response to Hodgson's letter was sympathetic. They read it in the contradictory context of their members' declining commitment to the conversion project because of such difficulties, and the bishops' own direct, and at that point successful, experience of examining black people on the Church catechism. Philip Quaque and William Cudjo had appeared before them in late 1758 expressing their desire to be baptized and had been examined as to their knowledge. In 1765, following further instruction, Quaque became the first black SPG missionary to Africa, although Cudjo ended his days in a madhouse.[38] The bishops certainly recognized the limits of what missionaries could do but noted in reply to Hodgson that "as we are fully satisfied of the ability and attention of these gentlemen in this pious work, we trust in God that their perseverance will be rewarded with better success."[39] In Barbados, however, this was taken to be an acknowledgment of the impossibility of converting all but a very few of the enslaved. Hodgson having fallen seriously ill in late 1760, and dying aboard a ship to England in January 1761, his successors effectively stopped trying to achieve conversions.[40]

Some years later, when the SPG expressed surprise that its response had been interpreted that way, the reverends James Butcher and Thomas Wharton reiterated Hodgson's arguments from Barbados. For Wharton, fifteen years in the Caribbean had convinced him that the lack of religious instruction was not the planters' fault: "after longer Experience I was daily more and more convinced of their [the enslaved Africans] untractable & perverse Dispositions; dispositions utterly averse to imbibe any Impressions of either religious or moral Duties." While the society's demand for a fuller program of religious instruction would be met, a situation in which children were put to work "as soon as they are able to speak," and where the enslaved refused to make Sundays a day of rest and worship, offered little hope, Wharton believed: "The Time I trust will come when Persons of all Complexions will embrace one Hope, and one Faith, but I am persuaded that that Period is not yet arrived, and in vain will it be to attempt inculcating the purest & most perfect System of Religion, without first implementing amongst them the Principles of civil Government and social Life: And such a Government, I may venture to affirm, the Policy of this Colony will never admit of."[41] Indeed, it was not until these ideas of Christian education intersected with the movement for the abolition of the slave trade and the imperial state's concern for what "civil Government and social Life" should look like in these colonies that the social and political conditions for speaking and listening would change.

In all this sermonizing, and through these unequal catechetical dialogues, there was only the most partial attempt to understand what the enslaved already believed. Their existing forms of spiritual practice—however well they had survived the middle passage, or however they were being remade in the Caribbean—were seen as an absence or a barrier, like the languages they spoke among themselves. Tabor and Lloyd argued not only that "they seem to have no notion of Spiritual things," but that "neither have they (as we have been informed) any words in y^r languages to express y^m by."[42] Yet there were signs of something else. In Barbados in the 1720s, Reverend Arthur Holt reported to the SPG that those enslaved on its plantation needed to be stopped from performing their "Plays," often on a Sunday, where, with "horrid music, howling & dancing about ye Graves of the Dead, they [offer] victuals & strong Liquor to ye Souls of the deceas'd." In particular, he noted that the "Oby Negroes or Conjurors are the Leaders to whom the others are in slavery for fear of being bewitch'd." These obeah practitioners would intercede with the spirits to do good or ill and, if allowed to do so, Holt argued, Christianity would "make but slow advances here."[43] There were even some hints of attempts at translation between spiritual vocabularies for invisible things, albeit mixed with planters' "common knowledge" about the spirit worlds of the enslaved. As Tabor and Lloyd scornfully noted, "ye best apprehension most of y^m have of a Soul is y^t is like y^r Shaddow, & all think w^n they die y^t they go to their own Country."[44] What modes of belief, and what ways of speaking of the world of spirits, could understand the Christian soul as a shadow and speak of dying as returning home?

The Professors of *Obi*

The soul as a shadow may be the translation of Christian conceptions of what moves between life and death within an animated world of spirits where objects, places, people, and animals were not necessarily as they seemed. In the Caribbean, a person's "shadow" could, as in some African and Amerindian cosmologies, take animal form, wander, get lost, or be stolen, particularly at the point of burial. As previous chapters have shown, in such a world, good or bad fortune might be attracted, deflected, or passed on to others. It was a world of divination and prophecies, but also of action at a distance, where powerful forces were at work, animated by some and affecting others. This was a spirited world that was made in the black Atlantic. The huge silk cotton, or kapok, trees—common to both West Africa and the tropical Americas—were the place where shadows lodged, and Edward Long told his readers that enslaved Africans in Jamaica "firmly believe in the apparition of spectres. Those of deceased friends are *duppies*; others, of more hostile and tremendous aspect, like our raw-head-and-bloody-bones, are called *bugaboos*" (*HJ*, 2:416). Indeed, duppies were also known as shadows.[45]

As with all the spiritual worlds discussed in this chapter, this one, and its beings and forces, needed interpreters: practitioners who could speak in appropriate ways and settings of what the connections between humans, plants, animals, spirits, and even gods, might be, and what those connections might mean for their interlocutors. This interpretation could be more or less everyday, or a matter of life and death. There were the "Nancy" stories of the trickster spider, spun daily on the plantations, along with tales of duppies and bugaboos. Beyond this, obeah men and women administered oaths of various kinds—invoking gods and ancestors to bind people to their words in law or in revolt—and sought prescriptions for the sick which merged the ills of the body and those of the spirits under the social relations of slavery. Those who have attempted to trace the term *obeah* back to transformed West African spiritual practices—Igbo-speakers' *dibia* or the *bayi* of the Gold Coast—have stressed the malleable, localized, and noninstitutionalized forms that it took in Africa and the Americas. Without shrines, fixed rituals, or large groups of adherents, and only invoked when need arose, these practices were most present in the words spoken by ritual specialists as they engaged the spirits "through the use of material objects and the recitation of spells," to effect divination, healing, and protection.[46] Indeed, the use by obeah men and women in Barbados and Jamaica of everyday, if liminal, materials—feathers, bones, bottles, eggshells, grave dirt, beads—to manage the power of the spirits suggests the transformation of these objects into something more than material through the performance of ritualized forms of speech and action, which would have been gendered in ways now difficult to uncover and may have included unintelligible words and other beings speaking through human mouths. Obeah was, in this sense, a speech act.[47]

Yet it is also clear, as Diana Paton argues, that "obeah is a creation of colonialism as much as it is a construction of Africans in the Caribbean."[48] For white planters and colonial officials, the part obeah was understood to have played in Jamaica's 1760 revolt made it something singular, enduring, dangerous, and powerfully real in its effects (see chapter 2). They made it present in the world in their own powerful stories, told in other settings, of spirits, words, and objects. Thus, in 1788, Stephen Fuller, the agent for Jamaica, informed a Lords of Trade committee on the slave trade of an enslaved boy who would not pass a glass bottle hung on a stick by the side of the road. Neither "Threats nor persuasion" from his master would move him, until the man "dismounted, and by breaking the Bottle, destroyed the *Obi*."[49] Here, obeah was more powerful than the master's commands—words that should have been obeyed. It could not simply be dismissed as a figment of the imagination. Its power had to be engaged with and countered, even if that gave it legitimacy as a force in the world. Indeed, the same committee heard that obeah men captured in 1760 who were shown apparitions from magic lanterns and given "some very severe Shocks" with "Electrical Machines" were not immediately converted to European modernity's versions of phantasmagoric specters or action at a distance. Instead, these demonstrations either had "very little Effect" on them, or, in one case, the *dibia* "acknowledged that 'his Master's *Obiah* exceeded his own.'"[50] This was a matter of concern. As the Barbados council reported, "Of their Arts we know nothing; but of the Effects produced by them, on those on whom they are exercised, are a Dejection of Spirits, and a gradual Decay." Indeed, Fuller told of a planter who had lost a hundred slaves to obeah in fifteen years, although his evidence was designed to suggest that it was the spiritual practices of the enslaved rather than the spirit of capitalism that brought death to the islands and thus justify the continuation of the slave trade.[51]

Obeah's work at the boundary between life and death was also evident in the final proclamation on the limits of white power made by the Jamaican ritual specialist executed in 1760 (see chapter 2). As such, it was part of a range of forms of communication with the dead noted by white observers of the spirit worlds of the enslaved. Bryan Edwards, for example, described the graveside sacrifices that families of "Koromantyn Negroes" in West Africa made to their "tutelar deity" and the dancing and singing at funerals of the enslaved in Jamaica (*HCC*, 73 and 87). Edward Long called these rituals "the very reverse of our English ceremony," observing that "every funeral is a kind of festival; at which the greater part of the company assume an air of joy and unconcern; and, together with their singing, dancing, and musical instruments, conspire to drown all sense of affliction in the minds of the real mourners. The burthen of this merry dirge is filled with encomiums on the deceased, with hopes and wishes for his happiness in his new state." Yet, for Long, these sung "encomiums" did bear comparison with the oral rituals of the British past and its margins: highland Scots' *coranich*, and its songs "in praise of the deceased, or a recital of the valiant deeds of him or his ancestors," and William Camden's account of the ancient Irish "custom of using earnest reproaches and expostulations with the corpse, for quitting this world, where he (or she) enjoyed so many good things, so kind a husband, such fine children, &c." (*HJ*, 2:421–22). Indeed, the point of these practices for the enslaved was to speak with the dead. As John Shipman, a Methodist missionary to Jamaica, wrote in 1820, "putting the corps into the grave, they dance, beat their drums, and make a feast about it, offering a part of what they have to the dead with some of those liquors the person loved most during his life, at the same time speaking to the deceased as though he were still alive and present with them."[52] There were other ways of talking across time and space here too. As the funerals took place, channels of communication were opened "to all friends t'other side of the sea," and messages for those elsewhere were spoken into the open graves.[53]

Since spirit forces were at work in the world, all deaths needed to be interpreted. As Edwards put it from the planters' point of view, the enslaved "never fail to impute it to the malicious contrivances and diabolical arts of some practitioners in *Obeah*" (*HCC*, 90), and John Stewart described in 1808 how "previous to the interment of the corpse, it is pretended that it is endowed with the gift of speech, and the friends and relatives alternately place their ears to the lid of the coffin to hear what the deceased has to say."[54] Consequently, funerals became what Vincent Brown calls a "spirited inquest," where the deceased's relationships to his or her community, living and dead, might be interrogated.[55] As Long skeptically described it:

> Sometimes the coffin-bearers, especially if they carry it on their heads, pretend that the corpse will not proceed to the grave, notwithstanding the exertion of their utmost strength to urge it forwards. They then move to different huts, till they come to one, the owner of which, they know, has done some injury to, or been much disliked by, [t]he deceased in his life-time. Here they express some words of indignation on behalf of the dead man; then knock at the coffin, and try to sooth and pacify the corpse: at length, after much persuasion, it begins to grow more passive, and suffers them to carry it on, without further struggle, to the place of repose. At other times, the corpse takes a sudden and obstinate aversion to be supported on the head, preferring the arms; nor does it peaceably give up the dispute, until the bearers think proper to comply with its humour. (*HJ*, 2:421)[56]

Such rites (figure 16) were part of a variety of questions posed and answers given at the boundary between life and death, where fortune and misfortune, ill will and protection, were brought into play by talking through the relationships between the living and the dead, the human and the nonhuman.

Invisible forces and their effects were made present through these spoken consultations. Through these shared, and often communal, moments

Figure 16. Funeral in Jamaica. James Phillippo, *Jamaica: Its Past and Present State* (London, 1843). © The British Library Board. 1304.h.4.

of harming, healing, and protection, spiritual practitioners wove accounts of what was wrong and prescribed their remedies (see chapter 3).[57] Illness, pain, or misfortune were signs of malevolent acts attributable to others. Collective and open-ended speech practices, such as the forms of gossip that animated the vampire stories that Luise White uses to tell the history of twentieth-century East Africa, would have helped explore ideas of cause and effect, addressed questions of power and uncertainty, evaluated reputations and relationships, and "assert[ed] social values, not as static traditions but as learned and lived practices."[58] Where conclusions could be reached, the malevolence might be countered, by setting another "obi." For the planters—seeing only the loss of laboring bodies—this was simply a matter of what the enslaved believed, a sign of their barbarity, a trick with words and poisons:

A Negro, who is taken ill, enquires of the *Obiah-Man* the Cause of his Sickness, whether it will prove mortal or not, and within what Time he shall die or recover? The Oracle generally ascribes the Distemper to the Malice of some particular Person by Name, and advises to set *Obi* for that Person; but if no Hopes are given of Recovery, immediate Despair takes place, which no Medicine can remove, and Death is the certain Consequence. Those anomalous Symptoms, which originate from Causes deeply rooted in the Mind, such as the Terrours of *Obi*, or from Poisons, whose Operation is slow and intricate, will baffle the Skill of the ablest Physician.[59]

That all this work with people, plants, objects, and spirits was tied together with words—something, for the Enlightened, that was only of this world since they would admit no magic spells, no spirit speech—made obeah's orality both everything and nothing. They termed those doing this work "the Professors of *Obi*."[60] On the one hand, this made obeah little more than "only words"—they were merely "professing" to have such powers, or, as the post-1760 Jamaican antiobeah legislation had it, "pretending to have Communication with the Devil and other evil spirits" to delude the weak and superstitious.[61] On the other hand, it expressed the power of spiritual exposition and the hold it might have over a susceptible, rather than skeptical, audience. However, such "authorities" had to admit that they knew nothing of what was actually said, since "A Veil of Mystery is studiously thrown over their Incantations, to which the Midnight Hours are allotted, and every Precaution is taken to conceal them from the Knowledge and Discovery of the White People."[62] Yet this veil of mystery only helped obeah to become, as dramatized in Britain and the Caribbean, a theater of spirited talk. James Grainger's poem *Sugar Cane* (1764) had obeah practitioners who "mutter strange jargon," and the Jamaican doctor Benjamin Moseley imagined for his readers the making of an "OBI" from animal parts, grave dirt, "potent roots, weeds, and bushes, of which Europeans are at this time ignorant," and then buried or hung up "with incantation songs, or curses, or ceremonies necromantically performed in planetary hours, or at midnight . . . [with] the person who wants to do the mischief . . . sent to the burying-grounds, or some secret place, where the spirits are supposed to frequent, to invoke his, or her dead parents, or some dead friend, to assist in the curse."[63] For Moseley and his readers, the words of witchcraft echoed in the Caribbean night.

Spiritual practitioners could also speak the future. They were, as Moseley put it, the "oracles of the woods," and Shipman, while acknowledging that he "never saw anything of it," reported the forms of speech of early nineteenth-century "Myal-Men" who "work themselves up into a state of perfect phrensy, until they foam at the mouth and the body becomes violently agitated and the countenance distorted in the most frightful manner and they are ready to die away, and the first words they utter are considered as a sure prognostic of what is to happen."[64] Yet the authority of such oracles was by no means absolute. Working between humans and spirits; between past, present, and future; and between life and death was a fraught business, and those who made claims could be denied by others.[65] The report of the leeward Maroon obeah man who had once been "Consulted on every Occasion" but was now "disregarded," or the account of a "counsell" interpreting patterns of fallen water that might point the way to a successful revolt (see chapter 2), demonstrate that divination was always subject to discussion. What did the signs mean? Had, or would, that future come to pass? Where the credibility of practitioners was primarily located in their way with words and objects, and with localized and immediate evaluations of their claims, those who paid or trusted them to work with the spirits were always on their guard. Thus, Thomas Thistlewood noted in his diary in 1754 that "a Negro man belonging to old Tom Williams, named Jinney Quashe (a noted Obia man)" had come to the estate "pretending to pull bones, &c. out of several of our Negroes for which they was to give him money." However, he "was discovered by them to be a cheat" and they chased him away, reminding Thistlewood of when he had seen a Yorkshire crowd turn on the "noted conjurer" Black Lambert.[66]

Obeah-talk was, therefore, an extended discussion of the power of words. Evidence given to the Lords of Trade included a case where a husband, attempting to discover his wife's infidelity, had forced her "'to take a Swear' (as she called it)" by drinking grave dirt and water "accompanied with the usual Imprecation (that her Belly might swell and burst, and her Bones rot, if she was guilty)" (see chapter 1). She, being guilty, had fallen so ill that the doctor, unaware of the "Swear" and unable to find a physical cause, had pronounced her incurable. The "Swear" being discovered, however, the man's master told him that he must either face execution for practicing obeah or "go directly to his Wife, and endeavour to cajole her into a Belief, that it was nothing more than a sly Contrivance of his own to get at the Truth." He was now to say "that the Mixture she had swallowed was not made with *Obiah Dirt*, but only with a little common Earth, which he had picked up on the Road, and therefore it was very harmless." He was, to save himself, to demonstrate that it was all just talk, and, indeed, "he acted his Part so well, that the poor Woman was completely duped a second Time, and recovering her Spirits, was soon restored to perfect Health." The words had done their work and restored her, and everyone else, to the land of the living. Indeed, the story was told to make the point that the only evidence on which a conviction for practicing obeah could be really securely made was that which came directly or indirectly from the mouths of the practitioners themselves: evidence from others that someone had sworn to put obeah (and had the materials to do so); threats made or overheard (such as the butler "overheard . . . threatening *Revenge*, and vowing that he would *buy some Obi to put for his Master*" just before the well water went bad); and, most secure of all, the accused's own "free Confession of the fact."[67] In recognizing the power of belief, but needing to deny any supernatural cause, prosecutions could only be based on what someone said they were doing, not proof that what they did was magic. Obeah was a speech act in legal terms too.

However, despite obeah being increasingly defined through the law, it could not be neatly separated from questions of religious belief.[68] Nor could different forms of belief and different stories of spirits, gods, and their modes of speech be easily kept apart. In endeavoring to explain obeah to the Lords of Trade, Fuller and Long suggested that there were continuities with ancient forms of snake worship that might have originated in the seductive words spoken by the devil to Eve. They then used linguistic evidence to find potential "old world" derivations which mixed the biblical and the nonbiblical: the Egyptian word for serpent ("*Obion*"); the biblical "Dæmon *Ob*" which Moses warned against as a "Charmer, or Wizard, Divinator, aut Sorcilegus"; the witch of Endor ("*Oub* or *Ob*") who summoned up the dead prophet Samuel's spirit; and "*Oubaios*," the "Basilisk or Royal Serpent, Emblem of the Sun, and an ancient oracular Deity of Africa."[69]

Indeed, the spirit-animated world that obeah opened was that of white people in the Caribbean too. Thomas Walduck, one of the first English commentators on obeah, wrote that, as well as being impious, "most people in the West Indies are given to the observation of Dreams & omens, by their conversation wth Negroes or the Indians, and some of the Negros are a sort of Magicians and I have seen Surprizing things done by them." His own unsuccessful attempts to account for these surprising things combined natural philosophy and religion. He sought rational explanations and told his correspondent, the apothecary and botanical collector James Petiver, that he himself was "very free from Superstition or predestination." Yet, noting his own extreme "vicissitudes of fortune," he resolved to "never despond but hope to live wth comfort to myself and reputation to my family, yet Gods will be done, tho: it be to my undoing." Such a world, where Manichean forces worked themselves out in everyday life, also offered him an explanation for the "Magick and ways of Divination" of "Indians" and "Negroes." As he wrote to Petiver, "One would wonder how these Simple Creatures, in every thing else should come by the Knowledge of Destroying att a Distance without any application but when we Consider how Ancient Idolotry is, and what a busie Agent the Divell is to doe and propagate misschiff our Admiration will Sease." Since he said he was "forced to doe Violence to my Reason to believe any thing of it," Walduck distanced himself from accepting the same supernatural forces at work in the world as those he called "Witch negro[es]," but he had brought close other beings—God and the devil—to explain what he saw them do.[70]

If the devil at work in the Caribbean might explain obeah for Thomas Walduck, it was also the case that obeah could incorporate Christianity. As Katharine Gerbner shows for Moravian missionary work on a Jamaican plantation in the 1750s, the enslaved understood Zacharias George Caries and the version of Christianity he presented within their own cosmology. As he put it, "they call me obea"; and, as he was more reluctant to admit, they prevailed upon him to baptize them as quickly as possible. Conversion ran both ways.[71] Thus, in 1788 a Mr. Rheder of Jamaica indicated to the Lords of Trade committee that obeah and Christianity occupied the same ground: "On searching one of the Obeah-men's Houses, was found many Bags filled with Parts of Animals, Vegetables, and Earth, which the Negroes who attended at the Sight of, were struck with Terror, and begged that they might be christened, which was done, and the Impression was done away."[72] Baptism might not be a rejection of other spirit worlds but an active incorporation of the power of Christianity and its spirits, rituals, and words. What the enslaved said as part of these rituals might not mean what Christians thought it did. As Caries complained—again mishearing a form of spiritual practice as the words of those who merely profess or pretend—they can "imitate our language without feeling it in their heart."[73] But maybe they meant exactly what they said.

This concern over the power of words, and what it meant for the enslaved to say they were now Christian, when they were really making another form of obeah—what Moseley called "*white* OBI"—haunts Long's discussion of conversions to Christianity in Jamaica.[74] Like some other Anglicans, Long argued that the enslaved rejected baptism, since it meant giving up their vices. For him, the only motive of "Creole slaves" to convert was "to be protected from the witchcraft of obeiah-men, or pretended sorcerers; which affords a plain proof of the influence which superstition holds over their minds." Under these circumstances, he argued, "the mere ceremony of baptism would no more make Christians of the Negroes, in the just sense of the word, than a sound drubbing would convert an illiterate faggot-maker into a regular physician." This was particularly true, he argued, for the established church, "it being founded on the principles of reason, and therefore adapted only to rational minds; which, by their own natural strength, are capable to judge of its rectitude, and embrace it on account of its purity and refinement from that very grossness which pleases, while it enslaves, other minds, that are clouded with ignorance." His targets here—forms of Christianity that enslaved the mind with superstition or enthusiasm, and might therefore attract those he thought born to slavery— were both Catholicism and dissent, and their ways with words. Indeed, he made them sound a lot like obeah. There were "the awful ceremonies, the indulgencies, injunctions, mummery, and legerdemain, of the Romish church and its ministers," with their magical objects: "crosses, relicks, and consecrated amulets." There was also the "abundance of enthusiastic rants and gesticulation" of "Quakerism, Methodism, and the Moravian rites" that "would operate most powerfully on the Negroes" (*HJ*, 2:428 and 430). Indeed, it was nonconformity and its modes of speech that Long found most threatening, and which he believed would significantly change the relationship between religion and slavery.

To Hab de Tongue to Tell of Him Goodness

In 1824 the Wesleyan Methodist missionary John Crofts reported on his activities in Jamaica:

> Sunday 6th [June] Whitsunday. At 6 O'C[lock] this morning I preached to a good congregation from Acts 2, 38–9. And afterwards administered the Sacrament to about 110 or 120 persons, during which time the deepest solemnity and devotion reigned. Before I had time to take a little breakfast one of the leaders came to say that his Class was waiting for Tickets, and as several other Classes were appointed to meet this forenoon it was after twelve O'C[lock] before I could dismiss them, by which time my own Sunday noon Class were assembled so that after having attended to them and other occasional applicants such as I have almost every Sabbath, I had but little time left before the people assembled for Evening worship. I preached at 4 O'C[lock] from Matt^w 12, 31–2 On the blasphemy against the Holy Ghost, deep attention was paid . . . and I hope the word will not be without effect as I endeavoured to make it a practical discourse. I was much fatigued after the labours of the day, but whilst my dear wife was reading to me the Memoirs of several eminent Christians and ministers from the magazines which we lately received, my spirit was revived and I felt a fresh resolve that as long as I live my time and strength shall be fully devoted to the work of my Lord & Master.[75]

While such accounts were inevitably shaped by missionaries' knowledge that their letters might be recirculated in the very periodicals that they read, or had read to them, it is also clear that what they expected and the expectations of them were that they should spend all day—certainly every Sunday—talking about God.

There had been religious dissenters of various kinds in Barbados and

Jamaica from the late seventeenth century onward, with Quakers involved in both developing Barbados's slave economy and provoking other slave-holders' concerns with their commitment to the religious education of the enslaved. In addition, Moravian missionaries were active in Jamaica start-ing in the 1750s, and black Baptists, such as George Liele, had moved there following the American Revolution to establish their own churches.[76] How-ever, it was only at the end of the eighteenth century that a concerted effort was made to send nonconformist missionaries from Britain to educate and convert the enslaved population to Christianity, and to bring an "energy" to the battle against "sin of any kind" that the established church lacked.[77] These missionaries, and the mission societies who sent them, were keen to avoid political arguments about slavery in favor of individualized argu-ments about salvation, and to demonstrate, particularly for the Methodists (who remained part of the Church of England until the 1790s), that their "Missions happily tend to the stability of internal peace, and good order, the due discharge of the duties of the lower classes, and the total destruc-tion of every principle of discontent and insubordination."[78] Yet, as will be shown, the changing relationships between the enslaved, the planters, the imperial government, the dissenting missionaries, and the established church meant that both islands experienced a shift in the conditions of religious speech in the early nineteenth century which made the freedom of speech central to the debate about the future of slavery. Close attention to the speech practices of the Methodist missionaries and those they spoke with can demonstrate the part they played in that process of change.

First and foremost, the white nonconformist missionaries who came to the Caribbean from Britain understood themselves as "preachers," but, unlike most of their Anglican counterparts, they saw the enslaved black population as their primary audience in "this land of sin, sun and death."[79] Missionaries understood enslaved Africans as fully human and therefore both born in sin and having the capacity for conversion and the need for salvation. They took the view that "they are placed under our care and gov-ernment consequently we shall have to give an account to God for them" and considered that the major barrier to their work was that the slavehold-ers wanted to prevent the enslaved from hearing the word of the Lord.[80]

Preaching, then, was the great purpose. John Shipman described him-self as "one, called by the providence of God, to this Colony for the pur-pose of preaching the gospel," and he argued that "we are commanded to go into all the world and to preach the gospel to every creature."[81] Ideally, therefore, there was to be as much preaching as possible, in as many places as possible, and missionaries despaired when they were prevented from speaking. Preaching was a vocation closely tied to personal histories of conversion and self-knowledge. William White, an aspiring Methodist missionary in Jamaica, told of "God's dealings with me, in reference to my conversion and call to preach the Gospel; both of which are much interwoven." As he put it, "here I am a monument of divine forbearance and love, persuaded that my happiness in life depends upon my preaching the Gospel, resolved by God's grace to live and die in this service." Faith, identity, morality, and speech were tightly bound together for these men, so that, as the missionaries in Kingston wrote, "The character of a Preacher is a sacred thing."[82]

Preaching was certainly understood as something learned: by example, reading, and practice. George Johnston, in Kingston, remembered the help others had given when he was just an "arrant *babler*," and John Wiggins, also in Kingston, asked to be sent six copies of the "Pulpit Assistant" and "Simon's Skeletons" to help him prepare his sermons.[83] Much of the advice to preachers echoed that for Anglican sermons.[84] Isaac Bradnock, in Jamaica, appreciated those in "Sound plain pointed Gospel language," and Methodists were urged to follow Saint Paul's advice: "Except ye utter words easy to be understood ye shall speak into the air."[85] But there was less emphasis on reason, and more on the soul, as the source of the words spoken. As Bradnock wrote, "With respect to my own Soul, I bless God I feel it prosper, I feel the power of that blessed Gospel to be the life of my own Soul, *I can, and Do, preach a full and free Salvation*, because I have received it *freely*, and blessed be God I enjoy it fully."[86] Done right, there was little that preaching could not achieve. As Shipman argued, "The word preached [in] its purity has . . . conquered the most inveterate evil[,] subdued the most depraved passions . . . [and] introduced order, and morality into all ranks of society, and diffused happiness through all stations, and classes of human beings."[87]

In turn, what came from the soul—assisted, Crofts wrote, by "The Lord . . . so that the word was powerful" —was directed at the heart, not the head.[88] This was particularly true, missionaries argued, for black listeners:

Though our Congregation *here* chiefly consists of Blacks and coloured people, yet we have frequently a number of White inhabitants, and occasionally a few respectable Magistrates and Merchants, who hear the word, with attention, and obvious approbation. Many of the people of Colour have rec[d] liberal education and profess enlarged and scriptural views of divine truth. The negroes in general understand what we say in the pulpit and treasure up select sentences in their minds, on which they *feed* during the

Week whilst labouring in the *field*, or at the *Mill*. It is however very requisite that in the application of our public discourses we sh[d] be plain pointed and lively, for I have often found it difficult to raise a suitable interest in their minds whilst explaining—than impress their consciences by feeling application.[89]

This assessment of the audience meant preaching in a style that had been criticized, including by Long, for "enthusiasm" but which its proponents sought to defend. Shipman argued that preaching was part of the "momentous" work of salvation and required the preacher to "use every argument in his power," and the most impressive language, against the hardness of the human heart. As he asked, "can he be cool and dry in his discourses, if duly impressed with such an idea, will he not rather do as St Paul did among the Epesians, 'warm them with tears'[?]" If this was "enthusiasm," Shipman continued, then it came from the minister's "awful responsibility to his God," and the apostles had been guilty of it too. Moreover, it was, he argued, even more necessary "in preaching to negroes," since they "require stronger language" and "greater warmth of expression to make them believe that you are in good earnest, that you mean what you say, and render them from a natural stupor which seems to hang about them." Such racialized generalizations also led to calls to introduce more music and singing into Christian services. It would, Shipman suggested, "be better for them to sing praises to God in the field than to hear them vociferating their heathenish songs, in which they are sometimes inveighing against their Masters, and at other times against each other."[90]

These ideas of the spiritual power of speech inevitably sowed doubts for those who felt their performances did not live up to either the fluency of divinely assisted exposition or the requisite emotional intensity. William Horne complained, following one sermon in 1822, that he "had a degree of freedom in Speaking but not much of that feeling which makes preaching . . . a triumph & delight." Peter Duncan, of Bath in Jamaica, judged a couple of his sermons as "rougher . . . than I am accustomed" and "lifeless," although on another occasion he said he "felt more freedom of speech as well as more enlargement of spirit than usual." William Gilgrass complained that his Bridgetown congregation was not satisfied with him since "God has not formed my voice for Singing. I am told they want a singing Preacher."[91]

Receptivity to the message was, however, not simply a matter of the preacher's performance. The quality of the event also depended on the size and engagement of the audience and the setting, and so the missionar-

ies built chapels. By 1817, the Methodists were proud to have "the largest Place in Kingston for the worship of God, and probably in the whole
Island." Their chapel held two thousand people, more than the parish
church, the Baptist chapel, the Jewish synagogue, and the French church,
and William Ratcliffe noted that "the deep seriousness which pervades our
Assemblies would do credit to the most order[ly] and enlightened European congregation."[92] Yet the demand for such spaces also came from the
congregations themselves, as did the building funds. Robert Young wrote
from Stoney Hill, Jamaica, in 1825 that "the Negroes" regarded "'Buck-
rah house,' the name they give to the house of their master," as "improper
places for Gods worship, [since] they think that good cannot come out
of Nazareth." Rather than be "boiling house Christians," he reported that
they were willing to walk long distances to worship in chapels their own
contributions had built.[93]

Missionaries sought to preach to all comers but would let only those
who were serious in their faith become members of the church. These converts were predominantly women, who underwent a rigorous program of
Christian education via more intimate forms of talk.[94] This education included Sunday schools, where young children and older girls might be instructed by the minister's wife, and where the forms of catechizing were familiar from the established church.[95] They might, however, be undertaken
in ways judged more appropriate to the enslaved:

> Let a question be proposed to them out of a catechism as they stand round a
> Room, if none of them can give a rational answer, let the answer be given by
> the Catechist repeating it a few times. Then let the first person able to give a
> correct answer stand at the head of the Class, after repeating both the ques-
> tion and answer again, let the second person giving a correct answer stand
> next, till they have all answered the question. And in like manner let the
> other questions be proposed and answered. This mode of catechising may be
> adopted without teaching them to read, and will inspire them with emula-
> tion, and gradually open their minds to discover the nature of the christian
> religion.[96]

Indeed, Shipman wrote a catechism "particularly suited to their capacities, customs and condition"—based on John Wesley's *Instructions for Children*—and including specific attention to the duties relative to masters and
"servants."[97]

Those who wanted to progress further formed classes, divided by gender, each with a "Leader" chosen from among themselves. The extent of

this system set the nonconformists apart from the established church. The leaders—who were men and women; black, white, and mixed race; free and enslaved; and literate or not—played a crucial role.[98] In part this was a matter of numbers, since there were too few missionaries to undertake intensive oral work with so many "hearers." In part, and when done by a "Black Exhorter," it involved ways of speaking the missionaries knew they could not perform:

> Put twenty heathen Country Negroes together as Catechumens and they will not know as much either of the Doctrines of Religion or of its power in three years tho in the hands of a Missionary as they will learn in twelve months when mixed with a Class of holy zealous Christians. In the former state the teacher must begin with the head in the other he will begin with the heart and kindle a flame that will throw light on the mind—If a Miss[y] preach and pray like an Angel the Negroes are not soon affected or moved. They think he has learned his trade better than some of the Church Ministers. But when met in Class by an able Pious Black Leader and hearing other Blacks speak of the love of God with so much life and pray with liberty and power they are overcome at once and both ignorance and unbelief soon give way.[99]

Thus, while leaders (or deacons) and ministers' wives were not permitted to preach from the pulpit, at least the former could demonstrate their faith through "extemporaneous prayer"—a public performance of scriptural knowledge and eloquence prized by nonconformists along with preaching but differentiated from it. As the Baptist minister William Knibb put it in 1832, "Prayer is an Address to the Divine Being for Mercy, Preaching is the Communication of Knowledge."[100] While he was keen to stress that only white, male ministers preached, it was also clear to both black nonconformists and their political and religious critics that forms of talk such as group prayer offered modes of public speech to black people, both free and enslaved, unavailable almost anywhere else.[101]

The significance of this distinction was evident in attempts to keep "leaders" and "deacons" under strict subordination to the white ministers. The leaders had no formal teaching role but only, as Shipman put it, "to meet the people for the purpose of praying with them, and conversing with them respecting their views of the gospel, and their habits and practices." Reporting to the minister on who advanced toward baptism and who were "backsliders" would allow him to commend or expel members from "Christian conversation" as appropriate. This judgment was based primarily on oral evidence: How morally were they reported to live and what could they communicate of what they knew of the gospel? Did they pray regularly, or indulge in idle talk, lies, or stories "about *jumbees* and such nonsense"?[102]

Yet what was most striking in the Caribbean context were the forms of equality that characterized dissenters' speech. They were the only white people on the islands who called black people "Brother" and "Sister"— and were addressed that way in return. They stopped to speak to enslaved people in the street, shook hands, and sometimes invited them into their homes.[103] Baptist and Methodist ministers were to be as open to their flocks as possible. They needed to talk face-to-face to congregants about their spiritual lives to listen for the words that were true signs of the "inward and . . . spiritual grace" which was part of being *regenerated, born again, or made new creatures in Christ Jesus.*[104] Missionaries therefore imagined, and presented in script and print, idealized conversations with the enslaved that enacted the process of spiritual transformation. One series of dialogues, produced by a Jamaican Methodist, involved missionary explanations of "divine Truth" via questions and answers reinforced by quotations from scripture, collective prayers, and direct addresses to God. They included "Quashee," the imagined plantation driver, saying to the minister, "Me begin to feel different about it since you talk with me so much."[105]

Nonconformist missionaries certainly knew what they wanted to hear back. Isaac Bradnock, after spreading the gospel in small Jamaican villages, informed the society in London that "it would do you good to hear them relate their Artless Experience." In contrast, Shipman praised the verbal skills of those he had heard "pray extempore in such a manner as would make many of our own complexion blush for themselves." What was crucial for both was that what they heard came from the heart and was often accompanied by tears shed by both speakers and audience. Shipman wrote of chapels where "the negroes may be seen weeping under the word."[106] The manner of speech was also important for missionaries in what it conveyed about faith and feeling. For example, as John Crofts told of visiting one older woman:

> On asking her a few simple questions I was struck with the propriety of her answers but when I asked her what Jesus Christ had done that all sinners might be saved she hesitated a little. Henry said to her, "me sister Mary what you say bout Jesus Christ precious blood when we upon prayers just now? Hab Jesus [Chris]t blood done any ting for *you*. What you say him hab done for all we sinners?["] This was as it were touching the Springs, for she instantly exclaimed "O my sweete massa! me hab not de sinful tongue and de polluted lip to praise him for all de goodness him do me. Ah! me no know

what would have come of me by dis time if it no be fo him mercy. Him fader say sinner shall die, but him say he will come in dis world and dead for we, so him say we must truss in him and den him forget we sin and suppose we lab him him will take we to dwell wid him. Ah! my dear massa, I neber full wid de good word I no should care if me yere it all day me lab it so much." This and much more she said whilst the tears rolled down her sable Cheeks and the one eye she had left sparkled with lively joy.[107]

Here, and from deathbeds and funerals, as well as catechetical instruction, creolized forms of speech—often spoken by women and children—were heard as markers of the authenticity of missionary reports, but even more as signs of the depth of acceptance of a religion of the heart and the power of the preached word rather than "mere notions of the head, expressions of the lips, self righteousness and an attention to some of the forms of religion." As Crofts later reported, Old Mary Wilson had said to those she prayed with the night before she died, "My heart quite full, minister to night him fill me wid good word."[108]

Dissenters held specific gatherings in which these forms of speech were encouraged. Methodist "love feasts" gave church members an opportunity for communion by sharing food, drink, prayer, and testimony in imitation of Christ's apostles. Transplanted to the Caribbean, they provided a space for the enslaved to speak about God in their own words. As Isaac Bradnock reported from Jamaica in 1806, "My Soul has been truly blest with the Simplicity of our Lovefeasts both in Town and Country. Many give proof of the power of saving Grace. Saying when they first heard the Sound of the Sweet Gospel, It *overhall'd* their hearts and they could never rest until *Massa bring* them to hear the word again, *Den* him make peace with their Immortal Soul, and more they feel very happy, and more *me Beg sweet* Massa, to make *me* hold on and hold out to the End." As Bradnock added, "methinks He that will stoop to hear the simple language of Children, will not despise their broken Accents."[109] Yet, relying on people, as one black Spanish Town congregant reportedly put it, to "hab de tongue to tell of him goodness," meant such events did not always succeed. Peter Duncan, who had been concerned about his own preaching, stated that a love feast at Manchioneal was "though not the liveliest I have witnessed was not entirely dead." Yet one he organized at Montego Bay in 1829 was much worse. He complained, "It was any thing but a lively one, I therefore at the close spoke sharply about it, and then mentioned the remainder of the names, who had grown weary in working out their salvation."[110] Transformation would not come if speakers did not engage in the proper forms of talk.

These nonconformist missionaries saw themselves as spreading the word and "enlarging [the] . . . empire" of "divine Truth" by responding to "the cry come over and help us."[111] The process of bringing people to a personal knowledge of God involved all these forms of speech and an intimacy of encounter through words that were understood as freely spoken and heard, particularly in their "broken Accents," as expressions of a deep faith that came from the heart. Yet there were many tensions here, both within the dissenting churches and between them, the established church, the enslaved, and the slaveholders. Missionaries' forms of speech led to accusations of "propagating the seed of equality," which not only threatened more "rudeness & contempt" to "white people" but also provoked Jamaican planters' fears that they would all be "massacred under the Sanction of Religion."[112] Through these ways of speaking, congregants had opened up spaces where the enslaved could articulate a critique of the wealth and power that planters' materialistic greed had built on their backs. It was this, as much as a notion of waiting for the rewards of salvation, that was spoken at Crofts's Spanish Town love feast when one man said of Jesus Christ that "him pardon all my sin, him clode me and gibe me joy an non of all de riches in de world set before me I take Jesus [Chris]t an him Cross before dem all for it better to take up de Cross in dis world den for enjoy pledgsher of sine which come to an end."[113] Crofts may have spoken about God all day every day, but others wanted their say too, and in their own accents.

The Freedom of Religious Speech

It is impossible to understand the contours of dissenting speech, indeed any talk about god(s), spirits, and belief on these islands, without recognizing the contentious and violent context within which it occurred. This means examining the relationship between the many forms of Christianity, abolitionist movements (about which we will see more in chapter 5), and debates over the nature of slave society after the end of the slave trade in 1807. In particular, the so-called amelioration of slavery in the early nineteenth-century Caribbean involved arguments over whether slavery was to end and what sort of subjects slaves or free black people should be. Evangelical movements within the Church of England meant that the Christian contribution to this debate was no longer restricted to an opposition between dissenters speaking and hearing a religion of the heart and the more precise cadences of Anglican rational religion. In many ways the early nineteenth-century context for talk about belief was utterly transformed. The question of the conversion of the enslaved to Christianity, and how that might be effected, shifted from being a minority pursuit to becoming the fundamental condition of any sort of social transformation in the Caribbean as viewed from the perspective of the imperial power.[114]

In 1783, in the SPG annual sermon, the bishop of Chester, Beilby Porteus, took up the cause of abolition and raised the question of responsibility for the *"one class* of our fellow-creatures which has such distinguished pre-eminence in misery of almost every kind . . . the AFRICAN SLAVES in our West Indian Colonies?"[115] Here, and in the plan he subsequently submitted to the SPG, Porteus returned to Hodgson and Wharton's complaints about the impossibility of speaking with the enslaved about the Christian God in the cane fields of Barbados. As the bishop noted, the SPG had appointed catechists for its enslaved workers and had exhorted them to make converts. But, it was, he argued, "too evident, both from the letters of our catechists, as well as from other undoubted testimony, that the endeavours used to civilize and to christianize our Negroes have not been attended with the desired success." While baptisms were still taking place, it was not clear that the enslaved were "any thing more than mere nominal Christians" because "no effectual impressions of religion seem to have been made on their minds, nor any material change produced in their principles, dispositions, and habits of life."[116]

This failure was not, as "some of our catechists have alleged," Porteus argued, "an impossibility in the nature of the thing itself, an absolute incapacity in the minds of the Africans to receive, or comprehend, or retain religious truths."[117] It was a matter of both oral technique and context. First, it was the way the catechists put it:

> Although several of our ministers and catechists in the college of Barbadoes have been men of great worth and piety, and good intentions, yet in general they do not appear (if we may judge from their letters to the Board) to have possessed that peculiar sort of talents and qualification, that facility and address in conveying religious truths, that unconquerable activity, patience, and perseverance, which the instruction of dull and uncultivated minds requires, and which we sometimes see so eminently and so successfully displayed in the missionaries of other churches.[118]

As well as talking more like Moravians, Porteus suggested that the existing catechists and the "assistant women" employed by the society be overseen by a new, and zealous, missionary—to be called "The Guardian of the Negroes"—who would select, perhaps even write, appropriate new texts and devise a mode of instruction, including sermons, lectures, lessons, and prayers, suitable to "their tempers, capacities, dispositions, manners, customs, peculiar idioms, and modes of thinking." These would be new forms of Anglican speech shaped by an evangelical mission. Missionaries, managers, and the enslaved would pray together in the fields at the beginning and end of the day; the enslaved would be encouraged to sing short hymns that adopted "their own simple melody"; and the Guardian and the catechists would engage in "constant intercourse and conversation" with the enslaved, "mingling even in their entertainments, their festivities, and amusements," using a "kind of familiar and friendly intercourse with their slaves" to turn "every little incident into an instrument of moral and religious improvement."[119]

Yet none of this effort would work on its own; it needed the right context. Here Porteus agreed with Hodgson and Wharton that there could be no real conversion to Christianity—no grounds for the formation of Christian speaking subjects—without attention to their "civil life," what Porteus called "some of the benefits and blessings of society and of civil government": protection by fixed laws; legal marriage and stable domestic life; attachment to the soil and "a little interest" in its profitable working; and the prospect for "the most deserving to work out their freedom by degrees." As he put it, "They must in short be considered as men, and as moral agents, before they can be made christians." This was not a vision of the end of slavery—the bishop was sure to reiterate that Christian slaves were still slaves, and better slaves at that—but it did imagine an end to the slave trade. Porteus was writing immediately after the American Revolutionary War, and he argued that disruption to the slave trade had required planters to treat their enslaved workers better, particularly women, to ensure the reproduction of the labor force. No more "fresh slaves" also meant an end to the constant influx of "heathenish practices and foreign manners" able to "obliterate in a few weeks all those sentiments of morality and religion" which might now be retained by those "well acquainted with the English language [and] . . . familiarized to the English customs." His was an abolitionist vision of an improved system of slavery based on Christianity, patriarchal families, education, and a slow transition, for some, and maybe for all, toward free wage labor.[120] It had implications for the moral and political action of slaveholders (most immediately the SPG), colonial and imperial legislators, and the enslaved themselves.

As far as this vision might effect immediate changes on the SPG's plantation, it fell on deaf ears. Porteus was thanked for his trouble but told that the society's circumstances meant it could not adopt the plan. As Glasson argues, not only was there strong resistance to abolition within the SPG, but the financial crisis at Codrington, and its resolution by leasing out the plantation for a secure return, gave the society little control over its management.[121] However, by the early nineteenth century, and through Porteus's continued work and that of many others, it was the bishop's agenda that shaped the prospects for Christianity in the sugar islands. The central issue was the role that religion would play in a slave society no longer reliant on the slave trade, and where social and sexual reproduction was as important a question as the future of slavery itself.[122] At the heart of this issue was what sorts of subjects the enslaved might become. Thus, in 1808, a year after the abolition of the slave trade, and a year before his death, Porteus outlined a plan for the education and conversion of the enslaved that, he argued, would transform them as people. Once more he accommodated this Anglican program to the slaveholders' interests, insisting that "the religion of the meek and humble Jesus" would not "produce ambition, pride, discontent, and resistance to lawful authority." Most important, since he was suggesting something he knew would be considered dangerous—teaching enslaved children to read—he reassured his readers that not teaching them to write as well "will always be a strong mark of discrimination, a wall of partition, between them and the white inhabitants; [it] will always preserve a proper distinction and subordination between them and their superiors, and present an insurmountable barrier against their approaching anything like an equality with their masters."[123] As imagined within Anglican abolitionism, conversion based on universal humanity, and reading but not writing, could reproduce a fundamental distinction between the free and the unfree. Porteus's work, from the 1780s onward, is a reminder of Anglican abolitionism's contingencies, and it highlights the contested terrain of religious speech in the Caribbean in the early nineteenth century.

The established church—in Britain and the Caribbean—was divided on how, or how far, to pursue the conversion and education of the enslaved, and what that meant for the system of slavery. Where they did pursue it, that was in direct competition with the nonconformist missionary societies whose work was conducted in the face of trenchant opposition from the majority of the planters, magistrates, and members of colonial assemblies, who were often the same people.[124] Older conflicts between the imperial government and the colonial legislatures (described in chapter 2) thus got a new lease of life in relation to legislation on the treatment of the enslaved and questions of religious freedom. One of the most important areas of conflict as far as the missionaries were concerned was attempts to control who could preach and where they could preach.

As shown in the previous section, nonconformist missionaries' forms of talk, with their presumptions of "equality," were a direct threat to the slave system. Their thundering sermons might, it was argued, drive the enslaved out of their minds "by threatening them with the terrors of the future state" or could accidently incite them to revolt by using biblical texts which were "good and proper for enlightened Christians" but which "have been eagerly laid hold of in a literal sense" by "ignorant slaves" who could be "led to very mistaken and dangerous views of the subject they embrace." While they were instructed to avoid politics, the missionaries certainly did not align themselves with the planters and, as one critic put it, "apply themselves direct to the slave & coloured population, as if their business was alone for their ear." This "familiar intercourse & freedom" or even "Dreadful familiarity with Slaves" was seen to be coupled with "a degree of suspicion and ungracious stiffness in their manners and intercourse" with the white population, which spoke of differences in both status and cultural politics.[125]

As well as following the fault lines of class and race, this was a matter of gender. The enthusiastic religion of the heart put men and women into intimate contact with one another, provoking new emotions and raising heightened feelings through novel forms of talk and action.[126] For the missionaries, this was the path to spiritual awakening. In 1806, Bradnock reported from Morant Bay on the conversion of nine young women of color who had initially come to chapel only to hear the singing but had become aware of their sinful state by talking to other women and hearing him preach of the Day of Judgment. The next morning, one of them, Elizabeth Ann Barnard, began to fast and pray. She fainted, "overcome with grief," saw herself suspended by the neck over a dark doorless prison filled with miserable creatures, and "cry'd out loud enough to be heard a great way off." At that moment, "suddenly the Lord spoke peace to her Soul," and her family sent for Bradnock. He was pleased to say that weeks of intensive admonishments, praying, instruction, preaching, and singing had brought them to a "good confession . . . That Christ dwells in their hearts."[127] Yet, for others, even within his own church, Bradnock's actions were evidence of "An Unchristian familiarity with Coloured and Black Women." He was said to have installed the women in the chapel's front pews, where they were administered the sacraments "to the great mirth of ungodly men, and sorrow of the aged and pious part of society," and to have given them "liberty from the Pulpit to visit him at any time" while closing his door to others. This Methodist missionary was, some thought, no better than the white men who had previously sexually exploited these women: "The

Chapel house was more like a tavern, than a preachers abode, [and] among the Gentlemen in town it went by the name of Mr Bradnock's Seraglio."[128]

These conjoined threats and disorders were the basis for attempts to silence or regulate the speech of these Christians. In the 1790s, a white man rode a horse into George Liele's Jamaican Baptist chapel and demanded it be given the sacrament.[129] In 1804, when he was in Barbados, Bradnock reported having to "have four young Men (Fops of the Town) bound to the Grand Sessions for a Riot in our Chapel," and in 1810, William Gilgrass also wrote from Bridgetown that "the young men of the Town" had stopped him four times while delivering one short sermon and forced him to leave the pulpit. Gilgrass would not, however, be silenced by "the devil and his young emissaries," promising that "the glorious gospel shall be sounded, in sinners ears, whether they be finily saved or damned, they shall know a prophet has been among them."[130] In turn, in 1813, James Whitworth gave the governor of Barbados the names of several young men who had attended his chapel "for the sole purpose of disturbing the Congregation & sporting with religion." Again, whiteness, class, and masculinity intersected in their words and actions as they followed women home "with the most abusive language & awful imprecations." Even where the magistrates might be sympathetic, going to law was hampered by the rules and expectations surrounding giving evidence (as explained in chapter 1). As Richard Pattison put it, facing a similar case, "we have not a White Man in Town in Society, and such are the Laws of this Island that a Black or Coloured Man's Evidence is not admitted in Law against a White Man and the White women we have in Society are poor, and of little weight."[131]

Indeed, the law was increasingly used to silence nonconformist preachers. The Jamaican assembly passed legislation in 1802 and in 1807, the year slave trade was abolished by Parliament, that enjoined slaveholders to "as much as in them lies, endeavour to instruct their slaves in the principles of the Christian religion" but restricted that instruction to the established church.[132] Although this prohibition on nonconformists was disallowed by the imperial government as an attack on religious freedom, Kingston's corporation also passed an 1807 local ordinance that included the following injunction:

> No person not being duly authorized, qualified, and permitted, as is directed by the Laws of this Island, and of Great Britain, and in the place mentioned in such licence, shall under pretence of being Minister of Religion of any Sect or denomination, or of being a Preacher or Expounder of the Gospel or other parts of the Holy Scripture, shall presume to Preach or Teach, or . . .

[lead] Public Prayer or sing Psalms, in any meeting or assembly of Negroes or persons of colour within this City or Parish.[133]

Nothing, the press argued, did more to bring the true practice of religion into disrepute "than the pretended Preaching, Teaching, and Expound[ing] the word of God as contained in the Holy Scriptures, by uneducated, illiterate, and ignorant persons, and false enthusiasts," especially if that was "to large numbers of persons of colour and negroes, both of free condition and slaves, assembled together in houses, negro houses, huts, and the yards thereunto appertaining, and also in divers lanes and bye-places within this City and Parish."[134] The magistrates closed the chapels down and sought to end preaching through this system of licensing.

There were objections, from both the missionaries and the missionary societies, on the basis that Britain's Toleration Act should hold across the empire. There were also imperial interventions to disallow legislation, or have it subject to suspending clauses, that provoked more legislation and claims from both missionaries and legislators that the process of making and unmaking these laws had violated "the liberty of the British subjects."[135] In Jamaica, legislation in 1810, 1816, and 1826 set strict rules on who could preach, where and when, with magistrates licensing where religious talk could happen and who could do it, and having any individual who was judged to be "a fit and proper person to perform the office of preacher or teacher, at a meeting, or assembly of persons of colour or negroes" swear in court the oaths of allegiance and supremacy, and a declaration against popery.[136] This constrained geography of religious speech was a clear threat to those wanting to "preach zealously against sin." As one Methodist missionary wrote of the 1807 Kingston ordinance, "It looks like Satan's last shift here, but I hope he will be disappointed."[137]

On the ground, missionaries faced the difficult task of working out how they could speak within this changing landscape of legal control: determining what laws were in force and how they were actually being enforced.[138] Could they, for example, be imprisoned for singing God's praises in their own homes, as Gilgrass and one of the "Black Exhorters" had been? What would happen if they tested gaps in the law by preaching in previously closed chapels, as John Wiggins did in Kingston in 1812? In this instance, Wiggins had said to the lieutenant of the city guards, "Sir, tell the police Officer, that the people will come in the afternoon, and that I shall preach if spared, because there is no Law of the imperial Parliament, or of the Assembly of this Island, to the contrary." Although, he reported, the "people's prayers, praises and floods of tears" which accompanied his sermon had "sufficiently proved that the Lord was with us," he was imprisoned for a month as an unlicensed preacher.[139]

Such defiant opposition was uneasily mixed with attempts at accommodation and negotiation, especially by the Methodists, who stressed their orthodoxy as a basis from which to speak.[140] Aware that Kingston's corporation was considering its ordinance in 1807, Bradnock, Knowlan, and Gilgrass—presenting themselves as ministers licensed under the Toleration Act—drew up an address which expressed their appreciation of the need "to prevent disorderly and illegal conduct" and their own desire to "be distinguished from every other Sect [or] Body of professing Dissenters in the Island, as we are not and cannot be accountable for the misconduct of those who are not in our Society—being altogether a distinct Church from every other denomination of Dissenters whether Baptists or others." They were, they stressed, noted for their loyalty to the Crown and that "the purity of our Doctrine and the Utility of our labours are acknowledged by the first characters in all the British Dominions."[141]

However, the missionaries were also aware that such legislation could, as it racialized the freedom of speech and its geography, threaten the basis of their mission in the sugar islands. This threat was particularly clear in relation to the 1816 law. Commenting on it, John Wiggins in Morant Bay pointed out its potential drastic consequences: "We will not on any acc^t obtain License to Preach in any Brown or Black person House tho Free[.] Indeed this Clause may be employ'd to silence all our Local Preachers[,] Exhorters and even Class Leaders[.] Any assembly of Slaves for the purpose of Religious instruction will be deemed unlawful unless the person who officiates be regularly Licensed. But no Brown or Black Man will obtain License therefore none of our Church Officers of either shade of Complexion can conduct another meeting with safety." As he continued, "I immagine that *our* missionaries will meet with great encouragement in future in this Island if they will decline employing Brown or Black persons as Leaders or other Officers in the Society. That however we cannot do." Yet the law itself was not clear; everything depended on enforcement. Wiggins was optimistic that the magistrates "are *generally* very indulgent to *our* people" and thought it likely that the law would only be applied to "persons not actually attached to any Religious party governed by an Authorised Minister." The difficulty was that "the Clause is apparently from design worded so loosely as to throw us completely within the grasp of prejudice" and, crucially, that the missionaries could not be certain of their clear separation from such ungoverned persons—both within their churches and outside— and their ways of speaking.[142]

As Wiggins's letter suggests, it was the leaders, deacons, and exhorters within the nonconformist churches that concerned the magistrates and others, especially whether what these leaders said to congregations of the enslaved and free(d) people of color was sufficiently controlled by white ministers. This was also a concern for the ministers themselves as conflicts emerged within the churches. For example, the questions over Bradnock's relationships with women of color were pursued not only by "ungodly men" but also by the church's male leaders, who saw their position eroded by his new congregation and their church threatened by the behavior of the ministers.

After the 1802 Jamaican legislation was disallowed, the chapels re-opened and, as Mary Smith, an established Kingston Methodist, put it, "sinners flock'd to hear that Word which has been, to many, the power of God unto Salvation." Many were young women of color whose salvation required "more frequent Application too, & intercourse with the Preachers, than would in different Circumstances be allowed."[143] For others, this extra attention, as well as Bradnock's "art of flattering and insinuating himself" and his "levity & ambition"—which had meant building a more ostentatious pulpit to address, in the front row of pews, "women dressed in all the enticing & alluring fashions of the world, lightly cloathed with naked arms up to their shoulders, their Breasts some entirely & some almost exposed," while relegating "persons of a dark complexion" to less prominent places—had questioned his moral authority. It was even alleged that the minister's housekeeper had been seen by a female leader rubbing Bradnock's naked body as he lay in bed, although he claimed she was nursing him, and she said that she was "catching his graces."[144]

Consequently, the leaders' meetings—where the minister's authority should have been exercised—became "a continued Scene of Riot & Contention" with "loud words . . . used on both sides" and, as Mary Smith wrote, "the most Rude & unmannerly behaviour, & grossest expressions, I ever heard from Men professing godliness."[145] Matters came to a head when the ministers discovered that the leaders had been holding secret meetings of their followers and had been administering the sacraments to them, a rite reserved for the ministers. These were, Bradnock wrote to London, the actions of men—led by a white leader, William Carver (who aspired to be a preacher)—"that have long been striving to take the government of the Church into their hands, giving much abuse and annoyance to their Ministers, Whome they strove to trample under their feet." It was, he argued, a product of Satan's displeasure at the ministers' success in making converts. The dark lord had been "calling up all his friends to his assistance" and led Carver to claim that "we are all holy as you and have a just right to the Priests office." The stage was set for a confrontation between what Mary Smith saw as "Ambitious, & designing Men, who have for some Time been endeavouring to Rule the Ministers & cannot be satisfied to keep their own place & Station" and what the leaders saw as a "wicked man" who should not be permitted to preach in their chapel or administer the holy sacraments to their classes. Members of each side claimed the other had excluded them from access to the hearers in the classes, and each side was keen to know whether "women leaders" had an equal vote with men or not, as had been the custom, as that would tip the balance of power one way or the other.[146]

At the next leaders' meeting, one of them began to "read to Mr Bradnock the rule or paragraph 4th page 77 & 78 Section 34," governing the conduct of their meetings. In response, "Bradnock turned away for a moment . . . [and] then took out a paper from his pocket" and "attempted to Read out these violent Men," expelling them from the church by reciting their names. William Carver "snatch'd [the leaders said "took"] the Paper out of his Hand, tore it in Pieces, & scatter'd the Pieces on the Floor," while Bradnock "threatened loudly to call for the Guard to take [them] . . . into custody &c." Meanwhile, Gilgrass and Knowlan ushered in what the leaders called "the partial young male & female friends of Mr Bradnock." They rushed into the hall, and "a loud clamour of different voices now commenced," with some of the young women abusing the leaders and Bradnock talking "most vehemently." Carver, the leaders said, tried to hold Bradnock's arm to speak to him. In contrast, Mary Smith reported that he "took Mr. B by the Right Arm & Twirl'd him Round." Bradnock later accused Carver of hitting him. This "unexpected riot" continued for an hour until the leaders— unable, they said, to reason with the ministers—decided to leave.[147]

Opinions differed on what happened next. The ministers sent London an account of an orderly and consensual process whereby the offending leaders were expelled by Bradnock reading out their names at a meeting. They admitted that "one or two attempts were made to interrupt him but the whole Society almost to a man stood up and commanded silence which had the desired effect."[148] In contrast, the leaders wrote that they heard "our names publickly read out to the world, as men not fit to be looked at." For them it was a show trial, an illegitimate performance orchestrated by the ministers. Bradnock had even packed the congregation with "the many back sliders who had aught against their leaders for having complained against them." When one of the leaders attempted to speak, "immediately a loud & sudden voice or rather voices Issuing from the Company of Backsliders and their Gang demanded 'hold your Tongues' which they repeated several times, lifting up their hands some of whom had provided cudgels for the better enforcing their threatenings." They had, they said, been "Publickly Scandaliz[ed] . . . before the Community of Kingston." Tellingly, it was decided in Britain that Bradnock should not be removed, with Thomas Coke concluding, "We must not strengthen the hand of those boisterous (I was going to say wicked[)] oppressors of the Preachers."[149]

This was a dramatic instance of a broader tension. On the basis of their intensive face-to-face engagement with "hearers," the leaders wished to maintain authority over their classes, particularly who might come forward for baptism or be excluded from the church as a "backslider." Missionaries, in turn, wanted overall control of the process, believing that a church—or holy community—was constituted as much by those it excluded as by those it included.[150] However, they had to balance their desire to oversee the work of leaders, and "displace" them if necessary, against the mission's dependence on their "Zeal, Piety and usefulness" in delivering converts and funds to support preachers and chapels.[151] Such a situation, which always had to be worked out locally in the end, could lead to difficult confrontations at the leaders' meetings or quarterly meetings where the work of the church was discussed. Here the leaders were not afraid to express their opinions, and might not do so in "a very pleasant manner." The missionaries could not simply impose their will. They needed to negotiate. As Peter Duncan put it, having held a "face to face" meeting to resolve "a disagreeable dispute," "O! What wisdom what Patience what Piety does this work require!"[152]

The difficulty of making distinctions between who could speak and act in particular ways was also evident in the division between preaching and praying. The Baptist minister William Knibb, discussing the case of Samuel Swiney, who had been arrested for illegal preaching, told a House of Lords Select Committee that prayer was "an extempore Effusion of the Person who holds forth" and that it might be spoken by "a Negro" deacon of the church even in the absence of a minister. He reassured them that although Swiney had not undergone any formal examination, he was "chosen from the most respectable and most consistent of the Members" and that Knibb had frequently heard him offer prayers of his own composition to the congregation and discussed matters of faith with him. Swiney, he said, had spoken for about five minutes without notes to a congregation of around twenty at Knibb's house and had stuck to the key distinction that "Prayer is an Address to the Divine Being for Mercy, Preaching is the Communication of Knowledge."[153] As has been shown, it was crucial to dissenting ministers that preaching be restricted to those properly qualified. Yet, the certainty of this distinction between modes of speech was not evident to the Jamaican magistrates who prosecuted Swiney, or to those who complained that the Methodists had repeatedly "Brought Coloured and black men to the Quarter Sessions to be qualified as Ministers of the Gospell," "some of which were very forward and bold."[154] Nor was it evident to the member of the Lords committee who asked Knibb if, holding as he did to an understanding of "Preaching to be an Exhortation addressed to others to a certain Course of Duty" or conduct and "Praying to be individually addressed to an Almighty Power, in which the other Individuals congregated unite," he had "any Doubt that you could so arrange a Prayer that, though in point of fact it was addressed to the Divine Being, it might point out to the Individuals joining in it the Mode of Conduct you wished them to follow?" Knibb had to admit that it was "rather a difficult Question."[155] This difficulty was compounded by the fact that there were always other Methodists and Baptists on the islands who were not under the discipline of these ministers and who spoke the word of God in ways that both they and the planters saw as dangerous.

White Baptist missionaries, like Knibb, presented their mission to Jamaica as a response to the heterodox beliefs and practices of the black Baptists who had come to the island in the 1780s and had built significant congregations.[156] In turn, the Methodists faced being confused with the Baptists (of all sorts) and being undermined by the proliferation of preachers who called themselves Methodists. Since politicians and magistrates were keen to use evidence of dangerous preaching by any nonconformists to restrict them all, the white missionaries wanted to establish the differences between these voices. The problem was who was preaching, where they were preaching, and how they were preaching. For slaveholders and magistrates, this was simply a matter of deceit that justified legislative regulation: "There are frequent instances of Slaves possessed of cunning and plausibility, who under the cloak of religion, and by preaching and teaching have acquired a surprising ascendency over the minds of their fellows which they have used for the most wicked and mischievous purposes, bewildering their minds, robbing them of their little property, and leaving them a prey to poverty, discontent and despondency." They tricked not only the enslaved but the missionaries too. Calling themselves Methodists, "like all others who make any pretension to religion," or Baptists, they had become class leaders, "overrunning the country, preaching nonsense, and sedition, in their nocturnal meetings; . . . [so that] there was a black preacher on every estate in this neighbourhood, before it was known to the white people."[157]

This array of "persons of all colours and descriptions and of every vari-

ety of creeds (and some perhaps of no creed at all but mere political inno-
vators)" calling themselves Methodists, when they were really "Antinomi-
ans," inevitably threatened the reputation of the Wesleyan Methodists.[158]
Shipman, writing in 1820 on the "State of Religion among the Negroes"
in Jamaica, recalled that when he arrived five years earlier, "the Baptists
had no regular Missionary in the island, but every thing among them was
under the management of a few poor black men,—the consequence was
that many extravagances were introduced among them, and frequently
these were palmed upon the Methodists."[159] Yet, for him, the problem was
more general:

> Perhaps there is no part of the world where there are so many religious In-
> structors of this description as in Jamaica. Every man who is able to use a
> form of prayer, however imperfectly, immediately thinks that he has the right
> to set up as a *Teacher* or *Preacher*, and endeavours to turn it to a good worldly
> advantage, and his life being irregular is not considered any obstacle to his
> becoming a public teacher by those who are led away by his cunning. Hence
> the people instructed by them are often grossly ignorant, very superstitious,
> and incorrect in practice. So true is the old adage "As priest as people."[160]

He then detailed the character, practices, and "baneful effects" of these
"low, illiterate, and immoral men, who have set themselves up as Preachers, and taking upon them the name of baptist or annabaptist (although
generally by other people called Methodists) have travelled all over the
country, pretending to preach, and extorting money from the poor people,
who are ignorant of these things." These men were fornicators, whoremon-
gers, and adulterers, Shipman asserted. They preached salvation through
sin, mixed Christianity with Judaism, and thought that communication
with the divine came not through reading and prayer, but through fast-
ing, and time in the "wilderness": "believing that they will meet with Jesus
Christ, or extraordinary blessing under some bush during the night."[161]

For such worshippers, religious speech was both less and more important than for Methodists like Shipman. In relating what he knew of
Mr. Magee, "a poor Man in Falmouth, Trelawney," Shipman showed how
orality took curious forms within their services, often subordinated to
other sounds and movements:

> His mode of worship is very extravagant it consists chiefly in queer gesticula-
> tions and painful operations on his body. . . . [H]is few followers will belch

up wind apparently from the bottom of their stomachs, in the most painful manner to an observer, and this is attended with a strange apparent commotion of mind which seems to thrill them. . . . In their acts of worship they scratch their heads till they have made the back part sore. In their worship they have some tunes, which are not unpleasant to the ear, and on some occasions pronounce some words about Peter going down to the water at which time they move towards a post and kick it with their feet.[162]

Yet speech was also absolutely central because this group, like others, rejected the written word of God. As Shipman put it, "He imagines that the Almighty gives him that immediate intuitive knowledge, which must be received as infallibly correct, so much so that the Scriptures can be of no service to him, because he receives his revelation more *directly* from the Divine Being." One worshipper, with whom Shipman had a "long conversation," told him when the missionary "introduced the word of God . . . that they had a surer guide in Mr Magee." The words Magee spoke were how the divine made itself known in that part of the island.[163]

The Baptists faced similar problems. Knibb told the Lords committee that there were "a great Number" of people in Jamaica who called themselves Baptists but "who are not in the least connected with us."[164] They were, he said, in serious error:

I am sorry to speak against any Body of Men, especially because they may be considered inferior on account of their Colour; but it is a lamentable Fact, that both the Doctrine and their Example are very bad; we could hold no Connexion with them. When I state that they deride the Bible, they call themselves Spirit Christians; they say they are taught the Spirit by Dreams, and they have this Passage of Scripture to confirm what they consider to be the Truth: "The Letter killeth, but the Spirit giveth Life:" [2 Corinthians 3:6] that before White Ministers came, they used to have the Mind of God revealed to them by Dreams, and that guides them.[165]

It was not just that they derided reading and attended to dreams rather than to ministers preaching the gospel, but that "they had their own Ceremonies and their own Principles," and that they "walk about the Country" at night preaching "on their own Responsibility" and going "to the Slave Houses when nobody knows it." Indeed, everything about how they spoke was anathema to the Baptist Missionary Society, as well as the established church:

Q. Did you ever hear of any unknown Tongues there?

KNIBB. They have had them a long Time; they speak with unknown Tongues; some Persons stand in a hollow Tree, and speak with unknown Tongues; they call themselves Anabaptists.

Q. They have no Connexion with you?

KNIBB. No, they hate us.

Knibb, fending off accusations of the involvement of Baptist missionaries in the massive uprising of the enslaved in Jamaica in 1831–32 (as we will see in chapter 5), sought to distance his church from these practices. He connected this version of the Pentecost—where the Holy Spirit's gift of tongues to Christ's apostles was manifest in a glossolalia rather than a disciplined transcendence of language's divisive geography to make it possible to preach the gospel to the farthest ends of the earth—to pagan tree worship, including that of the silk cotton tree. And, knowing the effect it would have, and that English politicians would understand what he meant, he spoke of Spirit Baptists "administering Oaths mixed with Blood, and Things of that Kind," and called what they did "a Kind of Obeahism," understanding it as a "Superstition" or "Species of Paganism" with "the same Influence over the Minds of the Negroes the Obeah had before." Yet it was also something new, an unholy mixture whose devotees might be called "christianized Obeahs."[166]

Conclusion

These conflicts return to where this chapter started: the working out in *Hamel, the Obeah Man* in the 1820s of the troubled relationships between those who invoked invisible and transformative beings and forces, pro- and antislavery politics, and speech. As in the novel, the cast of characters involves Church of England clergymen "who tell the Negroes their duty as slaves"; practitioners of obeah whose words exercise power over their minds; and nonconformist ministers who preach rebellion. Each attempted to communicate with others—gods, spirits, and the dead—to effect transformations in the world. By detailing their speech practices—always both real and imagined—and the highly charged interpretations of them, the intention has been to demonstrate how the definition and differentiation of forms of spiritual speech—sermons, prayers, catechetical questions, incantations, lamentations, and messages for the dead—sought to establish the modes of speech and belief proper to the free and the unfree, black and white, and how spirit speech signaled legitimate and illegitimate modes of freedom.

What Bruno Latour signals as the demanding "felicity conditions" of these spiritual speech acts are evident throughout. In all cases, these regimes of utterance were uncertain in their invocation of other beings, whose presence and transformative power could not be taken for granted. Spiritual speakers—whether what they were doing was characterized as "religion" or not—all had to endeavor to ensure that who spoke, the words used, the ways of speaking, the places of speaking, the relationships between speakers and listeners, and the use of other objects and texts were all aligned to render possible the closeness of other beings and the transformations they could effect. Accusations that this was all "only words"—that it was conjuring or trickery with no access to anything else—were continually present, both in the self-examination of those who spoke and in the interpretations of what they said and how they said it by believers and skeptics of all sorts.

In the Anglo-Caribbean world, every element of these contested forms of communication—the attempts to provide the right conditions for spiritual speech and the accusations of deception—was made through the social relations of slavery, even where that was under question. The Anglican bishops and clergy of the SPG sought transformations through hierarchical and rationalized forms of speech by trained men who aimed to deliver the truth to the ignorant and used and reinforced the status of the enslaved to do so. This work brought clergymen such as John Hodgson to the necessity to speak "one to one" to those enslaved on the Codrington plantation, but these face-to-face oral encounters only produced an endless repetition of catechetical questions as the enslaved found no use or meaning in his words and the contingent transformations they promised. Significantly, the diagnosis of that failure, by Hodgson and others, combined blaming the capacities of the enslaved and their languages and a political critique of slavery for not providing the forms of civil life that could create Christian speaking subjects who would still be enslaved. For their part, early nineteenth-century nonconformist missionaries attempted to maintain strong distinctions of who—ministers and leaders, black and white, men and women—could speak in specific ways and challenged the conditions of speech under slavery along with their enslaved "brothers" and "sisters." Their attempts to deny the spiritual speech practices of other Baptists and Methodists—who spoke in tongues or took the word of those who interpreted their dreams or night visions before the scriptures—and the conflicts between the leaders and the ministers within the churches demonstrate the ways in which race and slavery shaped these missionaries' struggles for recognition, a struggle that was, within the attempts to control religious practice on these islands, always a struggle to speak and be heard.

Finally, while there was little attempt on the part of any of these Christian missionaries to really understand what enslaved Africans believed, there was, particularly after 1760, a set of ideas about "obeah" among supporters of slavery which emphasized its power over the enslaved and at the same time dismissed it as just talk. This understanding of obeah as a speech act held some truth in the sense that, as a fugitive practice, obeah could have little institutionalization beyond the words and actions of its practitioners, both men and women, and their transformation of everyday materials, and that its effects were always subject to collective affirmation or denial. The difference lay in how the power of those words was understood. Thus, within slavery's geography of dislocation and death, gravesites became places for collective spirit speech, for communication with other worlds beyond the here and now, and for the interrogation of collectives which included the living and the dead. This speech had a freedom that could not be contained. Its powers of transformation were real on the plantations, and to the planters, and it could not easily be separated from Christianity. That it might be spoken and made within practices that called on the Christian God, Jesus, and the Holy Ghost was there in the characterization of dissenting missionaries as "white obeah men" by the enslaved and the slaveholders and of black Baptists as "christianized Obeahs" by white missionaries such as William Knibb. This characterization, along with Thomas Walduck's invocation of obeah as the devil's work and Stephen Fuller and Edward Long's etymology of "obiah" within diverse religious traditions, once again signals the impossibility of making and enforcing clear distinctions in the flow of spoken words.

Throughout this chapter, the discussion of the enslaved as believing and speaking subjects who might undergo significant transformations of mind and status—as Christian "servants," educated and converted free waged labor, or rebels inspired by their beliefs to seize their own freedom—was a vital part of the contests over the abolition of the slave trade, and then of slavery itself, from the early 1770s onward. That the politics of abolition transformed the conditions for the performance and evaluation of the spiritual speech of the enslaved—as a support or challenge to the system of slavery—can be seen, for example, in the discussion of obeah in 1788, in Bishop Porteus's response to Hodgson's letter, and in the missionaries' attempts to negotiate the contested geography of religious speech on these islands. The next chapter more directly examines the forms of speech against slavery within those movements for abolition and emancipation.

FIVE

They Talk about Free: Abolition, Freedom, and the Politics of Speech

The Jamaican slaveholder and novelist Matthew "Monk" Lewis recorded in his *Journal of a West Indian Proprietor* (1834) for March 1816 that a conspiracy had been discovered at Black River when a funeral conversation was overheard by an overseer hidden behind a hedge. The plot involved 250 people, a "*black* ascertained to have stolen over into the island from St. Domingo, and a *brown* Anabaptist missionary." They had elected a "King of the Eboes" and intended to massacre all the whites on the island at Christmas. The plot discovered, evidence was given at the king's trial that the words to a song had been found on his person, and that he had sung it at the funeral feast, with others chanting the chorus:

> Oh me good friend, Mr. Wilberforce, make we free!
> God Almighty thank ye! God Almighty thank ye!
>   God Almighty, make we free!
> Buckra in this country no make we free:
> What Negro for to do? What Negro for to do?
>   Take force by force! Take force by force!
>
> CHORUS:
> To be sure! to be sure! to be sure![1]

Lewis mocked the "King of the Eboes" for the failure of the revolt and hinted at the dangers posed by dissenting missionaries and the interisland movements of Haitian revolutionaries. His story invoked the political and religious voices raised against the Jamaican plantocracy and tied them, through the song's lyrics and the name of William Wilberforce, to those who spoke out in Britain for the abolition of slavery and the emancipation of the enslaved.

Although the relationship between speech and antislavery is the focus of this chapter, these concerns—within both the movements to abolish the slave trade and to emancipate the enslaved, and in the ways in which the enslaved sought to free themselves—have been foreshadowed from the beginning of this book. They are there in the "insolent words" heard by Madam Sharp, which led to the execution of the "old Negro man" in Barbados in 1683, and in the knowledge that Edward Long's *History of Jamaica*, and its considerations of who could speak, was a response to the antislavery activity of the early 1770s. Indeed, these matters are an important part of all the intervening chapters. The consideration of slave evidence as part of the reform of slavery, as suggested by Joshua Steele in 1789, was a crucial question throughout the late eighteenth and early nineteenth centuries as both abolitionists and imperial reformers sought to bring slavery under the rule of law. Indeed, political battles over this issue, and any other reforms that were interpreted by the islands' legislative assemblies as limits on their deliberative power, were understood within the framing of political talk hammered out over the previous century of debates on the freedom of speech.[2] Indeed, as this chapter will also show, the nature of political talk among the enslaved remained a crucial matter in the early nineteenth century as it became entwined with abolitionist discourse. More broadly, the questioning of the future of the "civil government" of the sugar islands, a questioning of the organization, if not always the existence, of slavery, was played out through forms of speech in natural history, as Thomas Dancer's oratorical work in the botanic gardens in 1790 attests, and in religion, as the battles over speech between clergymen, missionaries, planters, and the enslaved testify.[3]

However, instead of knitting all these threads together, this chapter engages directly with the ways in which it was possible to speak in favor of what Manisha Sinha calls "the slave's cause": abolition of the slave trade and freedom for the enslaved.[4] How were abolition and freedom talked about and what difference did that make? While there is important previous work among the vast literature on abolition concerning how those who spoke out against slavery and the slave trade spoke, it has tended to concentrate on set-piece presentations—parliamentary speeches and abolitionist sermons, for example—and on the rhetoric of abolition as a component of a wider discourse of antislavery.[5] However, taking seriously the range of voices raised against slavery, the modes and contexts in which they spoke, and the variable evidence we have of those speech acts—that is, dealing with abolition on the asymmetrical common ground of speech set out in this book's introduction—requires more attention to speech practices, and oral cultures broadly conceived, than to discursive structures and rhetorical work, although questions of practice, form, and content can never be completely separated. Pursuing this emphasis on speech practices, the chapter works through three distinct, but related, inquiries into orality and antislavery: first, how the voice of the enslaved was represented in the abolitionist and proslavery literature of the late eighteenth century; second, how questions of class, race, and gender shaped the speech practices of abolitionist debate and activism during the attempt to build a mass movement in the late eighteenth century; and third, how the abolition and antislavery movements in Britain intersected with the political talk of the enslaved in the Caribbean during the conspiracies and revolts of the early nineteenth century. Taken together, these inquiries show that examining the oral culture of antislavery can add significantly to the rethinking of abolitionist activism that has already encompassed print culture, visual culture, and material culture. Such an examination also opens the possibility to understand abolitionist politics on the same ground as the antislavery activities of the enslaved themselves.

The Slave's Two Voices

The questions of racial difference and the unity, or otherwise, of humanity at the heart of the debate over abolition made speech as important an issue in the 1780s and 1790s as it had been for Long in the early 1770s. In his 1784 abolitionist *Essay on the Treatment and Conversion of African Slaves in the British Sugar Colonies*, James Ramsay argued that "before we proceed to claim the rights of society, and of a common religion for Africans we must first put them in possession of that humanity, which is pertinaciously disputed with them." He questioned David Hume's assertion of innate racial difference and argued that "as far as I can judge, there is no difference between the intellects of whites and blacks, but such as circumstances and education naturally produce." He admitted that there might be some superficial physical differences—in hair texture, nose shapes, and skin color—"But their tongues are as musical, their hands as elegant and apt, their limbs as neatly turned, and their bodies as well formed for strength and activity as those of the white race."[6] One abolitionist clergyman even argued, drawing on Volney's reading of Herodotus, that since the Egyptians were "actually Negroes," then we "owe [them] our Arts, Sciences, and even the very use of Speech."[7]

Instead, for Ramsay, differences between people were a product of "Sacred history" and contingent circumstances. He argued that humanity was originally unified, but that God, at "the confusion of Babel," then "divided them into families and languages, giving to each distinctive features, and a separate speech." This divine action set social groups on divergent paths but also promised "the ultimate re-union of mankind." Differences in capacity could not simply be lined up along racial lines. So Ramsay argued, against Long, that although Francis Williams's "verses bear no great marks of genius," an argument based on innate intellectual inferiority would have to show that "every white man bred [at the same university] . . . has outstripped him" (see chapter 1). In turn, Ramsay heard forms of speech among the enslaved as evidence of both their intellect and the repressive influence of slavery on them. Drawing on his experience in the Caribbean, he argued that "Negroes are capable of learning any thing that requires attention and correctness of manner. They have powers of description and mimickry that would not have disgraced the talent of our modern Aristophanes." Yet slavery suppressed these talents so that "a depth of cunning that enables them to over-reach, conceal, deceive, is the only province of the mind left for them, as slaves, to occupy." As a magistrate he had "heard examinations and defences of culprits, that for quibbling, subterfuges, and subtilty, would have done credit to the abilities of an attorney, most notoriously conversant in the villainous tricks of his profession."[8]

Ramsay's view supported a vision of slavery's future similar to that of Beilby Porteus, the bishop of London (chapter 4). Enslaved men should be "objects of civil government," and as such, Ramsay contended, they should be protected by the law and encouraged to marry, have families, and tend their own small properties.[9] Such "civil privileges" would, Ramsay argued, "go hand in hand" with religious instruction, with "the union of liberty and religion both advancing slowly together, without any abrupt or violent change in the condition of the slaves themselves."[10] Like Porteus, Ramsay admired what he saw as the French practice in relation to "new Negroes" whereby "with the first rudiments of a new language, they draw in the precepts of a religion that mixes itself with every mode of common life," as opposed to the situation where "foreigners are said to learn English, by the oaths and imprecations with which our tongue abounds."[11] However, such visions of slavery's future, and the place of speech within it, could leave slavery and the slave trade in place. For example, an anonymously authored pamphlet from a member of the "Society of Universal Goodwill" in Norwich argued, in 1788, that "this race of men, only wants a proper education and instruction, to answer every good purpose in society." The place to start was the way they spoke. If "new Negroes and children" were "taught to speak the English language, with propriety, fluency, and correctness" rather than the "almost unintelligible jargon, of native and half English words, placed and used without regard to grammar or pronunciation," and which was dangerously shared by slave and slaveholder alike, then "every future proceeding would be rendered easy & pleasant."[12]

It is unsurprising that when Ramsay represented the words of the enslaved in his *Essay*, describing moments of fine sentiment, honor, and self-sacrifice, he portrayed them as men speaking standard English.[13] This portrayal was common to most of the sentimental literature of abolition, in prose and poetry. For example, William Cowper's "The Negro's Complaint," first published in 1788, was, when republished to encourage readers to boycott slave-grown produce, accompanied with the instruction that readers should "place themselves in the same position" through a poem that offered "a simple and pathetic delineation of what may naturally be supposed to pass, at times, through the mind of the enslaved negro." This was acknowledged as an imagined moment of speech, but one that, in the act of identification, carried the truth with it: "However incapable he may be just in such a manner to speak the sentiments of his mind, yet, from his condition and circumstances, we may easily imagine that similar with the following he, as a mere percipient being, must frequently feel."[14] Cowper's "Negro" then speaks a clear poetic English, evoking the sympathy of the reader via sentimental identification uninterrupted by barriers of language and difference. The same effect might be achieved by evoking the sighs, groans, and cries of the enslaved, along with their tears.[15]

However, questions of race, language, and difference could not be so simply dealt with in abolitionist attempts to ensure that the "cruelties of the Slave Trade" were no longer "talked of . . . with an indifference, common to other commercial considerations."[16] If, as well as sharing a common humanity that should invoke sympathy, enslaved Africans also needed to be represented as requiring moral guidance, education, and conversion to Christianity, then that needed representing too. This tension is well captured by an abolitionist artifact that combined speech, script, and print into something that, like Cowper's poem, might be circulated to engage those who read, saw, and discussed it in relation to the boycott of slave-grown produce (figure 17). Probably created by the Quaker printer James Phillips, whose presses produced vast numbers of abolitionist tracts, including those incorporating Josiah Wedgewood's famous "Am I Not a Man and a Brother" emblem, the printed side echoes that emblem through a simulacrum of a trade card on which the enslaved subject speaks

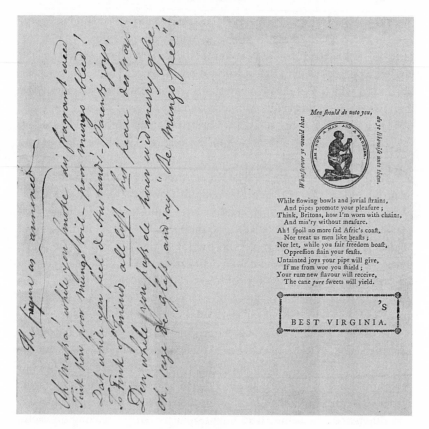

Figure 17. Card with abolitionist poem, probably printed by James Phillips.
Library of the Society of Friends, London, Thompson-Clarkson Illustrations, II f.35.
© Religious Society of Friends (Quakers) in Britain.

a poetic language of "mis'ry," sentiment, and sympathy in a supplicant appeal based on a common brotherhood with those who could provide relief from "Oppression."[17] Yet, in manuscript, and probably in Phillips's hand, there is an alternative text for "the figure as annexed" that used the voice of "Mungo" to appeal to the reader or listener. Here, the authenticating and heartfelt accents of an imagined creole speaker calls on these male smokers and drinkers to "Tink how poor Mungo toil—poor Mungo bleed!" They are asked, in the accents of a racialized difference, to pledge themselves to the cause of this victimized husband and father, and by implication his family, and to "seize de glass, and say 'Be Mungo free!'"[18]

This form of speech was also deployed in the 1792 pamphlet *No Rum! No Sugar!* that presented "Half an Hour's Conversation, Between a Negro and an Englishman, shewing The Horrible Nature of the Slave Trade" and encouraged abstention from the products of slavery. To bring the message home, the pamphlet described "the Voice of Blood," combining as its epigraphs Genesis 4:10 ("What hast thou done? The *Voice of they Brother's Blood*, crieth unto me from the ground") and 1 Chronicles 11:19 ("My God forbid it!—shall I drink the blood of these men?") to argue that using slave-grown produce was to consume the blood and bodies of the reader's enslaved brothers.[19] The text then worked abolitionist arguments into a dialogue between "Cushoo" and "Mr. English," which eventually persuades the latter to join the boycott through a combination of economic argument and sympathetic identification: "*E*: I thought you blacks had no such fine feelings. *C*: Me beg you pardon, Massa, dat be da whites—de blacks all feel."[20]

This way of representing the speech of the enslaved carried risks for abolitionists. Enslaved Africans speaking in imagined creole tongues were also extensively deployed in proslavery visual and textual materials, along with representations of grotesque black bodies, to challenge claims to humanity, intellectual and moral capacity, and, ultimately, freedom.[21] These representations circulated in private as well as public spheres, connecting them just as abolitionist representations did. For example, a manuscript shorthand book, given by Theodore Barrell to his Bostonian relatives in 1790 "as a small testimony of esteem," contained "A Negroe Song" by Samuel Matthews—"as humerous a character as every inhabited this Island of St Christophers"—that mocked, in an imagined creole, the words and feelings of an enslaved man "Buddy Quou" on finding that the baby born to his beloved "Quasheba" was not his but that of the "Buckra man."[22] It was not, therefore, immediately evident what such speech signified in the public sphere or whether it was a suitable vehicle for abolitionist argument. So, when emphasizing "the natural taste of the Africans for music" and their ability to learn European forms, Ramsay noted that "instruction and assiduity might change mungo's silly stage gibberish into the soft thrills and quavers of Italian eunuchs."[23] For him, the slave subject worthy of sympathy and assistance spoke standard English and might sing like a European.

Ramsay's reference, and also that of Phillips's abolitionist artifact, is to Charles Dibden's play *The Padlock* (1768) and to Mungo, the most famous contemporary black character on the eighteenth-century Anglophone stage and one that signals the contingency of this cultural politics of speech. Dorothy Couchman has argued that Mungo—an enslaved servant in a comedy of domestic sexual intrigue—was a novelty, as the first character to speak a version of Caribbean creole in the British theater. In eighteenth-century London, he was played by white actors in blackface, "making grammatical, phonological and lexical changes to their speech" to produce a form of language that demonstrated racial difference. Indeed, even though Scottish in its origins, "Mungo became a name for those who spoke in the manner of enslaved West Indians." Yet, as Couchman shows, in this opera buffa, Mungo sings not in creole but in standard English, so that "the opera's score downplays the libretto's construction of racial difference."[24] While *The Padlock* was not an abolitionist play, it was turned to that end. In 1787, Reverend Samuel Disney wrote an "Epitaph to Mungo" which gave the black character the last word in a new ending. This switched after the first two lines from a creole to standard English, and from comedy to sentimental, patriotic, and patriarchal abolitionism, with Mungo saying, "Let me speak" and declaring that he should be "like the Briton, free," "For though no Briton, Mungo is a man!"[25] Indeed, this can be paired with the explicit call for freedom in the voice of Mungo in Phillips' manuscript text that is not there in its printed counterpart.

Yet such a voice could not simply be equated with calls to end slavery. For example, *No Rum! No Sugar!* concludes with a reassuring message, also in creole, for slaveowners about the productive and reproductive capacity of slavery after the end of the slave trade: "Massa, Negro more care taken, better fed, better cloath'd, live longer, able do more work, have more pickaninny—Den pickaninny more take care of, dey no want slaves."[26] The question of the slave's two voices, and how they might be deployed, demonstrates above all else the complex and contested cultural politics of speech in the debate over abolition and freedom, showing that matters of race and gender were crucial in shaping the meanings of voice within both pro- and antislavery arguments.[27] These cultural politics were also evident when the debate came off the page and was engaged face-to-face.

Debating Slavery

Attention to the importance of print culture, visual culture, and material culture for abolitionist strategies, especially in Britain, has been crucial in showing how these media contributed to making a mass movement and generating political change.[28] However, less attention has been paid to the details of how print, images, objects, petitions, and so on were actually put to work and how they were meant to effect change by changing minds. What follows argues that the politics of speech, and the politics that was made through speech, played a key role at crucial moments in the formation of the abolition movement in Britain. This close attention to abolitionist speech practices begins by considering the discussion of slavery within the London debating societies of the late eighteenth century, in order to provide the basis for closer attention to who spoke about abolition, where and how they spoke, and how that speech was shaped not only by religion and politics but also by class, gender, and race.

London's debating societies demonstrated that "the passion for public speaking had become epidemical" among the capital's men and women of the middling sort in the 1780s and 1790s.[29] Societies such as the School of Eloquence, the Oratorical Academy, and the University for Rational Amusements met several times a week to debate social, political, religious, and moral issues. Competing as commercialized evening entertainments as well as venues for political deliberation and the demonstration of oratorical skill, the debates were orchestrated by a "Moderator" and advertised on the basis of the quality of the rooms, their acoustics, and the refreshments available, as well as the matters under discussion.[30] Promoted by their proprietors as schools for the development of the eloquence necessary for the bar, the pulpit, and Parliament, and as a necessary counterpart to the more restricted deliberations of the House of Commons, they were also derided as scenes of "riot" and "harangue." Involving both men and women as orators and audience, they raised the question of the power of speech to change minds, of who could speak in public, and how they should speak.[31] First flowering in the early 1780s and taking on all the major questions of the day, the debating societies were effectively removed from the political sphere after November 1795 when the "gagging acts"—William Pitt's reaction to the Age of Revolution—forced them to advertise that "All political remarks or allusions are utterly inadmissible."[32]

Inevitably, these forums addressed slavery and the slave trade, and their discussions mirrored the shape of the national debate, within Parliament and outside, in terms of both timing and content. Those meeting at Coachmakers' Hall on October 28, 1779, debated the question "Is the slave trade justifiable?" as did those at the Westminster Forum on April 24, 1780, deciding at least in the latter case that it was not.[33] Such debates were occasional topics in the early 1780s, taking their place alongside associated questions of imperial warfare and policy, and of religion and morality, some of which addressed allied issues. For example, on March 22, 1785, the Ciceronian Society asked, "Can the eternal Salvation of every Individual of the human Race be proved from Scripture and Reason?" and others that took place on dates ranging from 1779 to 1786 debated the changing fortunes of war in the Caribbean.[34] However, from early 1788, as Parliament met to discuss the slave trade, and into 1789, when abolitionist campaigning was in full swing, the debating societies took up the issue in earnest, questioning whether the slave trade could be justified on the basis of "Justice, Christianity, Policy or Humanity" and whether its abolition was "consistent with the political and commercial interests of Great Britain."[35] Although arguments came from both sides, where outcomes are recorded, abolitionists prevailed. The question of the slave trade was also bound up with other issues: compared to East India Company corruption, addressed through the influence of women on their husbands, and tied to contested notions of English constitutional liberties. It continued to be debated in 1791, turning in November to the causes of the slave revolt in St. Domingue, and in 1792, again coinciding with parliamentary debates, and addressing the "moral obligation" of abstention from slave-grown sugar and rum. After that it was addressed again only in 1795.[36]

Who took part and how they spoke is harder to determine. The majority must have been the usual suspects: men with experience or ambitions in the law and politics, or "learned Divines," all expounding theories and evidence of what they saw as sound policy, humanity, and morality. They did include women as both listeners and speakers, but their involvement was always qualified in a context where women's public speech and the relations between men and women were themselves important debating points.[37] In February 1788, for example, the *Morning Chronicle* commended "the fair sex" for preferring to hear a debate on the slave trade than attending "places of trifling and uninstructive amusement," and, most tellingly, those at the School of Eloquence in the same month were "honoured by a circumstance never before witnessed in a Debating Society," when "a lady spoke to the subject with that dignity, energy, and information, which astonished every one present, and justly merited what she obtained, repeated and uncommon bursts of applause from an intelligent and enraptured auditory."[38] Identifying this performance as exceptional, and the speaker remaining unnamed, is of a piece with the contested position of women speakers in mixed debating societies, where they might speak wearing masks, and the contention over women's debating societies such as the Belle Assemblee, established in 1780, whose participants were described both as speaking with "uncommon propriety, elegance and dignity" and as "absolutely disgustful."[39] In April 1788, they debated whether politicians' wives should "exert their influence over them for the abolition of the Slave Trade," again demonstrating women's contingent voice. Yet it is also evident that abolition was a subject on which women could have a public say. The possible participation in future debates of this unnamed "LADY, whose intellectual accomplishments, and wonderful powers of eloquence . . .

[had] delighted a public audience," was certainly used to draw interested listeners to experience her "amazing oratorical excellence."[40]

Indeed, the organizers of the Westminster Forum hoped that she would participate in a debate on the political and commercial justifications for abolition in February 1788 that would also include "Several popular Gentlemen, who have interested themselves in the petitions presented to Parliament" and "A NATIVE OF AFRICA, many years a slave in the West-Indies" who would "communicate to the audience a number of very remarkable circumstances respecting the treatment of the Negroe Slaves, and particularly of his being forcibly taken from his family and friends, on the coast of Africa, and sold as a Slave." Perhaps she would offer her opinions on his recounting of "several interesting circumstances relative to the conduct of the Slave-holders towards the African women," although these would certainly be countered "by the speakers on the other side," and no black women were recorded as speaking in this or other debates.[41] This "African" might have been Olaudah Equiano (better known then as Gustavus Vassa), a black activist, who spoke in the same debate as the "Lady" in May 1789, the year that he published the widely read *Interesting Narrative* of his life.[42] Equiano might also have been the "African Prince" or the "ingenious African" who took part in debates that year, since the *Times* reported that the prince was told that "reading his Narrative from the Chair would be unprecedented."[43] But he was not the only "African" in London debating slavery. Indeed, there was something of a collective voice, certainly in writing, as a group identifying themselves in 1787 as the "Sons of Africa" and consistently including Vassa, Ottobah Cugoano, and George Robert Mandeville, published letters in support of abolition.[44] As speakers, these men might draw on their own experience, but they also matched their opponents in eloquence. The *Times* reported that, in a 1789 debate at Coachmakers' Hall, "One Gentleman only opposed the Abolition, which he did in a Speech of great Fluency and Strength of Reasoning. He was replied to by an African (not Gustavus Vassa) who discovered much strong natural Sense and spoke with wonderful Felicity." Such speakers were noteworthy, if not entirely exceptional. For some, their performances would have confirmed abolitionist sentiments. For others, their words would have raised the question, debated in 1791, whether "the insurrection at Saint Domingo [was] to be attributed to the exertions and writings of those who have interested themselves to procure an abolition of the Slave Trade."[45] Taken together, then, these debates raise questions about the ways in which men and women, black and white, could and could not speak about slavery and the slave trade, both in these public forums and in the other spaces where abolitionist mobilization took place, and which involved other forms of talk.

Some Stir and Noise: Early Abolitionist Activism

As the London debating societies show, the discussion over slavery, as conducted in Britain and the British Parliament, became focused on the combination of political, commercial, moral, and religious questions raised by the slave trade. This focus, as others have argued, can be explained through the prominent role of Quaker activists in the early history of abolition, especially in the founding of the Society for Effecting the Abolition of the Slave Trade (SEAST) in 1787, and, as what follows emphasizes, the role of particular speech practices in shaping both their political strategies and the content of the debate.[46] In this, 1783 was a crucial year. It was not a year in which the London debating societies considered slavery or the slave trade, but there was plenty of discussion going on elsewhere.

An important starting point is a single conversation. On March 19, 1783, Olaudah Equiano "called on" Granville Sharp—widely known for his antislavery views since the Somerset case (see introduction)—at his house in London "with an account of one hundred and thirty Negroes being thrown alive into the sea, from on board an English slave ship." This was the *Zong* case, as reported in the previous day's newspaper.[47] The slave ship's insurers had questioned in court their liability to pay out for the loss—as insured property—of 132 African men and women thrown overboard on the orders of the captain. Equiano recognized the importance of the case for those who opposed Atlantic slavery, and the potential for further legal action.[48] He had met Sharp in 1774 while unsuccessfully trying, despite the Somerset judgment, to use the law to prevent John Annis from being returned to the West Indies.[49] Sharp certainly understood the power of speech in changing the law on slavery, at least certain sorts of legal speech. As well as getting the briefs written for Somerset's lawyers, he had sent one of them "an iron Gag-Muzzle" from one of the many London ironmongers who supplied the slavers: a metal mask which, with its "Flat Piece of Iron upon the Tonge" and "suffocating Plate before the Mouth," "must altogether be esteemed a diabolical Invention." This was exactly the same sort of "iron muzzle" that had so shocked Equiano when he saw one silencing an enslaved women in Virginia, "which locked her mouth so fast she could scarcely speak." Sharp aimed to turn the gag into speech that would free the enslaved, calling it *"an Iron Argument,* which must at once convince all those whose Hearts are not of a *harder metal,* that Men are not to be entrusted with an absolute

Authority over their Brethren." Indeed, the very same mask was deployed by an abolitionist speaker in a debate on abstention from sugar at Coach-makers' Hall in January 1792.[50]

Equiano and Sharp were unsuccessful in their attempts to turn the *Zong* case into a murder trial. Yet, accounting this a failure of the abolitionist movement, at least in the short term, because of Sharp being "silenced" by the Admiralty's lack of response to his calls for justice and by his not using print "to publish evidence of the villainy of British officials at the highest level," would be to underestimate the significant effects of the many con-versations that spiraled out from Sharp's meeting with Equiano.[51] As James Walvin shows, Sharp called on a series of acquaintances and contacts to build the basis for intervention. He met the Oxford legal scholar Dr. Bever; Beilby Porteus, then the bishop of Chester—who had delivered his abo-litionist sermon to the Society for the Propagation of the Gospel (SPG) only the previous month (see chapter 4); the bishop of Peterborough; the Unitarian reformer Dr. John Jebb; and the member of Parliament (MP) General James Oglethorpe, with whom Sharp had previously shared his long-standing view that Britain's support of slavery could only lead to di-vine retribution.[52]

As well as trying to mobilize these influential figures, Sharp talked to other interested parties. He and the Quaker William Dillwyn—who first came to Britain from Philadelphia in 1774 and who lived in London be-ginning in 1777—went to Westminster Hall on May 19 and 20, 1783, to try to attend the second *Zong* hearing but were disappointed when it was not scheduled. The day after the case concluded, May 23, Sharp, who had been present (along with "Several Negroes"), spent two hours with Dillwyn, who had not.[53] Quakers, particularly in North America, had a long history of antislavery writing and activity and had, more than twenty years before, outlawed slavery within their own communities.[54] Significantly, since the early 1770s, Philadelphia's Quakers had repeatedly requested those in Lon-don to take the matter of the slave trade up with the political authorities in Britain who could do something about it. However, their appeals had been deflected by the English Quakers' unwillingness to become embroiled in political activity.[55] Under this pressure, by April 1783, Dillwyn had become part of a committee appointed by the London Meeting of Sufferings to consider the slave trade and was, with others, keen to bring further pressure to bear on the yearly meeting in London that summer to get more done. In early May he wrote to John Pemberton, another Philadelphia Quaker who had crossed the Atlantic since the end of the American Revolutionary War, giving him the details of this "shocking case" and, more important, telling him that people were talking about it. As Pemberton subsequently reported in a letter to his brother James, back in Philadelphia, the case "makes some stir & noise, & may tend to open the Eyes of some, it seems that few know the iniquity of the trade indeed numbers among friends, that more general knowledge is necessary to be spread."[56]

The opportunity was evidently there to get more people talking about the *Zong*, particularly other members of the Society of Friends. More specifically, the slave trade would be discussed at the London Yearly Meeting in June with the aim of prompting a petition to the king and Parliament. Indeed, another prominent Philadelphia Quaker had the same idea. In late May, Anthony Benezet, who had campaigned against slavery since the 1740s, wrote to John Pemberton that he hoped that Pemberton's presence in England would "be instrumental in putting friends there upon a weighty Consideration whether it is not high time for them Individually and as a Religious Society, to lay this important Concern before the King & Parliament, the great Senate of the Nation." In support of that hope, Benezet informed him that he had read in the Pennsylvania news an account of the first *Zong* hearing excerpted from the *London Advertiser*, which had reported that the evidence given had "made every person shudder," especially in realization that permitting "such a flagrant act of Villany with impunity . . . makes the Crime general & provokes divine Wrath against the Nation." As Benezet added, pointing the way forward, those who "will not comply with the injunction 'To open their Mouths for the Dumb, in the Cause of such as are appointed for Destruction' Prov. 31, 8" could only expect to suffer divine punishment themselves. He suggested that Pemberton and others "confer with Granville Sharp upon this weighty subject" and told him that Dillwyn had already written that Sharp had "attended" almost all the bishops and found them "disposed to discourage the African Trade."[57]

This was not heard as idle talk. As the abolitionist activist Thomas Clarkson put it in his 1808 history of the movement, "These subjects occupied at this time the attention of many Quaker families, and among others, that of a few individuals who were in close intimacy with each other. These when they met together, frequently conversed upon them. They perceived, as facts came out in conversation, that there was a growing knowledge and hatred of the Slave Trade, and that the temper of the times was ripening towards its abolition."[58] More directly, Quakers such as Dillwyn and Pemberton turned a moral and religious concern into political action in 1783. As Christopher Brown has shown, what the London Yearly Meeting would decide hung in the balance, with influential English Quakers including David Barclay opposed to petitioning the king.[59] Much depended on what was said and heard in a context where political speech was inseparable from forms of religious speech in which the speakers and listeners, and also the silences through which Quakers communicated, brought close more powerful spiritual beings and forces than simply their own opinions.[60] As one diarist noted of the discussion of the slave trade at the meeting:

> Many words were gone into on this head all seeming to be well intended but full of the Arts & contrivances of the Creature[.] [A] degree of awful solemnity gradually spread itself over the Meeting and as this prevailed the contrary nature fell under it[.] I saw this coming as an overshadowing from the highest in which my soul rejoiced, this increased more & more over the meeting so that the glorious all powerful dominion of the blessed truth was witness'd . . . to the praise of the most high for which my heart was filled with thankfulness, blessed be his holy name for ever.[61]

Thus concluded, it was agreed to petition Parliament on the matter, and a committee, including Dillwyn, immediately set about inserting antislavery material in newspapers and, from 1784, distributing large numbers of *The Case of Our Fellow Creatures the Oppressed Africans*, a pamphlet written by Dillwyn and John Lloyd, to politicians, judges, clergymen, West India and Africa merchants, and ships' masters.[62] Again, as Brown argues, this was not just about the pamphlets. As far as possible, pairs of Quakers would hand deliver and explain these texts directly to the recipients.[63] Their strategy combined extensive printing with intensive talking.

Indeed, these face-to-face encounters were crucial to developing the arguments and strategies for abolition. By the summer of 1785, two years after the Quaker petitions, there was still no action by the legislature. Dillwyn's committee judged that further extraparliamentary attempts were needed "to inspire the publick with such Sentiments of Justice" that would produce a call on government "so general & so loud" that it could not be ignored.[64] After debating how that should be done—and perhaps kept to it by an October 1785 letter to them signed by Gustavus Vassa and seven other "Poor Oppress'd needy & much degraded Africans," which praised their "Benevolence, unwearied Labour, kind interposition & laudable Attempts . . . towards breaking the Yoke of Slavery"—it was decided that the best avenue was to get Benezet's *A Caution and Warning to Great Britain and Her Colonies* read by the pupils at the major public schools.[65] However, addressing "the rising generation" meant asking their teachers. This proved to be a decisive moment. As was later reported, "This last Service, more particularly requiring personal application, & from its confining Nature more easily admitting of it, produced diverse opportunities of free Conference with Men of liberal minds, equally perhaps desirous for the removal of oppression as ourselves, although (from not having so long looked at the religious side of the Question) not equally convinced that all sinister advantages must be waved [sic] to accomplish that end."[66]

These conversations had shown that "Men of liberal minds"—exactly those who should support the cause—required not the religious and moral message of pamphlets like the *Case of Our Fellow Creatures* and the "righteous judgments of the Lord" it contained, nor notification of the "load of guilt that lies upon our Nation generally and individually" set out in Benezet's *Caution and Warning*, but something more practical on British trade and the good conduct of emancipated Africans, what the London Quakers called "the temporal good that arises from the practice of Piety."[67] No longer a matter of discussion just among the faithful, the committee had heard it from the horse's mouth.

The Philadelphia Quakers, to whom the London meeting appealed for useful information on the behavior of emancipated slaves, told them in no uncertain terms in 1786 that they should avoid "being diverted from a patient & faithfull Adherence to the sure Foundation of christian Equity, Truth & Righteousness [by] be[ing] drawn into a vain Work of beating the Air or pursuing the uncertain Windings of a Labyrinth of carnal Reasoning."[68] Despite this warning, the broadening of the argument beyond purely religious considerations which emerged from these "free Conference[s]" decisively shaped the activities of the Society for Effecting the Abolition of the Slave Trade (SEAST), the most significant early abolitionist committee. This committee was formed in May 1787 and brought a number of Quakers, including William Dillwyn and James Phillips, together with Granville Sharp, Thomas Clarkson, and, eventually, William Wilberforce. SEAST certainly effected the production and distribution of a great variety of printed materials addressing all aspects of the iniquities of the slave trade, focusing as much attention on commercial and political questions as on moral and religious ones.[69] Significantly, this work was also done through talk.

The abolitionist activism of Clarkson in England, Scotland, and Wales starting in 1787, and William Dickson's later work in Scotland, represented a key part of SEAST's work to gather evidence against the slave trade, to disseminate it, and to encourage petitions to Parliament so the matter would be debated there through a coalition built by Wilberforce.[70] Clarkson's and Dickson's work intimately entwined speech and print at all stages. For example, as Dickson, a former secretary to the governor of Barbados and therefore well versed in the iniquities of slavery, recorded in his diary for February 4, 1792:

> After calling on several gentlemen [in Perth] whom I found rather unin-
> formed, at 12 met seven gentlemen who had been requested to come to
> Mrs Marshall's.
>
> Could scarcely get them to separate emancipation from abolition—
> Stated facts & reasoned & satisfied most of them—they had not digested the
> Abstract [of evidence to a House of Commons select committee in 1790–
> 1791], wh[ich] is scarce; but Mr Mellows to write to Edin[burgh] for more—
> They are disposed to do what is right but think the mode should be left
> to the legislature who are better informed—Are determ[ined] However to
> petition on the ground of the trade being radically cruel and unjust. . . . At 9,
> met 10 gentlemen who had been asked to sup and converse on the subject—
> Convinced them all—Mr Young confessed I had put the subject in so many
> and so satisfactory lights that he was satisfied and wished to hear nothing
> more on the subject, till Monday. . . .
>
> Rec'd Letter fro[m Clarkson] of the 25th Jan. saying our adversaries en-
> deavour to avail themselves much of Domingo affair & suggest the dan-
> ger of stirring the people at present—Caution not to touch on Fr[ench]
> Revol[ution].[71]

While print was crucial, and Dickson—the author of *Letters on Slavery* (1789)—worked very hard to distribute the *Abstract of the Evidence . . . for the Abolition of the Slave Trade*, it could never be flexible enough for all the necessary arguments and counterarguments, or, in the case of the revolutions in St. Domingue and France, what had to be avoided.[72] Neither could its effects be counted on. That took talk. As Dickson wrote to Phillips, "had the Abst[ract] been spread here & Clarkson or myself had gone round to explain it, I think we might have carried all Scotl[and] hollow. . . . I have met with no man that has read it without conviction. But the evidence *must* have time to operate. A Scotchman will not join any cause unless you explain to him the *cur* and the *quare*. But *then* he perseveres."[73]

The importance of these face-to-face interactions again raises the question of who could perform such modes of speech, where, and when. Clarkson's and Dickson's oral engagements were both enabled and constrained by a speech politics that combined whiteness, masculinity, and class. They were in many ways speaking a new sort of public politics and had to make decisions on how to do it as they went. For example, Clarkson was very uncertain about an invitation to give the Sunday "discourse" in a Manchester church, doubting "whether the pulpit ought to be made an engine for political purposes."[74] These uncertainties were multiplied in more private settings such as those Dickson entered in Perth, but activists could also be more flexible in their responses there too. This might mean using "spirit," "mildness," or even "ridicule" as the case demanded. As Clarkson advised Dickson—suggesting that such debates were conducted among relative equals, and in shared idioms of race, class, and gender—it was "necessary to know your man and to know how to address him, whether by working on his vanity, interest etc."[75] Yet these extraparliamentary activists might be silenced by class and patriarchy too. As Wilberforce noted in a private letter to Lord Grenville, a powerful member of Pitt's government, "Clarkson is a very modest man, & you must encourage him, pat him on the head or he will never venture & come near you or speak his mind with the requisite freedom."[76] However, if there were constraints on male activists' freedom of speech, then that was even more the case for women who wanted their voices heard.

Changing the Conversation: Women and Abolitionist Speech

As a political intervention, women's speech about abolition was always shaped by gendered differentiations of the public and private, even when their talk blurred such distinctions. The women who participated in the London societies' debates either had their influence on politics marked as emanating from the private sphere, as the wives of politicians, or were seen as exceptional when contributing eloquently to public debate. In addition, Thomas Clarkson and William Dickson negotiated a strongly gendered terrain as they encouraged petitioning (an almost exclusively male preserve) and generated popular support though discussion of abolition.[77] For example, as well as noting which ways of speaking might work best when confronting men, Dickson attuned his speech to how he thought women felt hearing him. In mixed company at Whitburn, arguing with a male "apologist" who had been in the West Indies and with whom Dickson said he would have been glad to continue to debate over a bottle, he noted that "I defeated him on every tack as . . . my two companions especially were pleased to say. Neg[roe]s universally lab[ourin]g under the whip like cattle had great effect. The poor ladies one of them pretty, absolutely looked pale. I begged pardon, but s[ai]d [it was] necessary to speak out the truth."[78] Feeling was, of course, as powerful for the abolition movement as reasoned argument, and women articulated a strong political voice—and produced important early works of abolitionist thought—through sentimental literature that aimed to change minds and shape the political process. They also mobilized a gendered politics of domestic consumption in the early 1790s sugar boycott based on a combination of economics and sympathy.[79] In the metropolitan elite circles involved in the parliamentary campaign and in the provincial abolitionist networks that generated popular support, women's talk was a vital part of the politics of abolition.

For some observers, women's speech was central to the origins of the movement. The Moravian minister Christian Benjamin Latrobe argued that "the abolition of the slave trade was, under God, and when the time was come, the work of a *woman.*" He meant Lady Margaret Middleton, whose circle, at the house of her friend Elizabeth Bouverie—Barham Court in Teston, Kent—included James Ramsay, Hannah More, Clarkson, and Wilberforce, as well as her husband, Sir Charles Middleton, the comptroller of the navy. Latrobe even identified the key moment as a "remarkable conversation" in summer 1786 when he, the Middletons, and Ramsay were at breakfast, and Margaret told her husband that he should bring the matter of the slave trade before the House and demand a parliamentary inquiry. While he deferred, having not yet given his maiden speech, "an interchange of opinions" lighted on Wilberforce, as an MP "who had lately come out, and not only displayed very superior talents, and great eloquence, but was a decided and powerful advocate of the cause of truth and virtue, and a friend of the [prime] minister." This particular exchange was, Latrobe noted, itself a product of the "frequent conversations" among Ramsay, Bouverie, and Margaret Middleton on "the state of the slaves in the West Indies, and the abominable traffic in human flesh and blood." These conversations had led Lady Middleton to encourage Ramsay to publish his *Essay on the Treatment and Conversion of African Slaves in the British Sugar Colonies* and prompted her to "urge the necessity of bringing the proposed abolition of the Slave Trade before parliament, as a measure in which the whole nation was concerned." Such speech was part of the Teston circle's concern for a national reformation of manners and encouragement of the role of evangelical Anglicanism in politics.[80]

Finding the origins of abolition in what Lady Middleton said over breakfast was both contested and belies the multiple and collective nature of the movement from its inception, even at Teston.[81] Clarkson, Wilberforce, and More each pledged their allegiance to the cause there, though they deployed their voices in different ways. For More's part, she wrote poetry to coincide with the parliamentary debate in 1788 and suggested to Lady Middleton in September of that year that "opening the eyes, and diffusing humanity in to the hearts of people" might be done by persuading Richard Brinsley Sheridan, the manager of the Theatre Royal, Drury Lane, and a supporter of abolition, to put on "Oroonoko, or the Royal Slave." More argued that "many people go to a play who never go to a church"—she judged three thousand a night—and only Porteus and Ramsay were preaching abolition from the pulpit anyway. The play would, however, need adjusting. The "indecent and disgusting" comedy should be removed, and "some good poet" would have to "write an affecting prologue, descriptive of the miseries of these wretched negroes, and what a glory would accrue to this land from their redemption," More explained. She suggested that Sheridan himself do it, since someone who, as an MP, "could talk so eloquently for five hours and a half on the oppression of two Begums," as he had recently done at the trial of Warren Hastings, "could not refuse pleading the cause of millions of greater sufferers." Yet, while expressing the necessary modesty, she did have an eye toward writing it herself. She could, it seems, imagine her own words ringing out across the stage, even though spoken by someone else.[82]

This theatrical proposal was part of More's engagement with Clarkson and SEAST's attempts to generate popular support. Indeed, she worked hard to make these arguments in the slaving port of Bristol, a tough town for abolitionists, telling Margaret Middleton that "Clarkson desires us to canvass for him from the Member of Parliament down to the common seamen; he wishes to turn the tide of popular affections in favour of this slave business."[83] This effort continued what the Quakers had started, with evangelical religious commitment giving women a basis for political action.[84] More could, therefore, be critical of Quaker commitment to the cause, accusing the Quakers of "*Talking* humanely, perhaps, but when anything is to be *done*, or any assistance given, they are as cold and prudent as the most discreet and money-loving churchman." Whether Quaker reticence was rooted in continued concern over political involvement, More's critique sprang from her concern with the place of religion in politics and the desire to speak a new evangelical language in the public sphere. Reflecting in 1786 on the question, "Why one never hears any religious conversation unless it is tinctured with Methodism or hypocrisy?" she robustly argued that the language of "*regeneration, grace, new birth, Divine influence, the aids of the Spirit*" should not be mocked when introduced to "common conversation" because since it is "very near the heart, and filled its secret affections, it would now and then break out into words." Concerning religion, "should we not," she asked, "talk of it as often as we do of money, business, beauty, pleasure, or wit?"[85] Coming from the heart, it was a political language of religion that women could and should speak in public.

Whatever the tensions between evangelicals and Quakers, much of their discussion of abolition, and allied causes such as political reform, occurred within what Kathryn Gleadle calls "intermediate spaces of sociability," where the medium was conversation rather than formal debating or the more directed speech of political activism designed to convince skeptics or defeat opponents. This focus on conversation is signaled by the ways in which abolitionists sought to mobilize women through the sugar boycott of the early 1790s. For example, the reprinting of Cowper's "The Negro's Complaint" (discussed earlier in this chapter) came with an abstentionist message on a small folded sheet entitled, "A Subject for Conversation and Reflection at the Tea Table." Yet the alignments of gender, the private sphere, and abolitionist table talk were not as simple as this message implies. As Gleadle shows, using the extensive personal notebooks of Katherine Plymley of Shropshire, going beyond abolitionist tracts, with their appeal to the public sphere of print, and detailing the conversations, and ways of understanding conversation, of middle-class women involved in the abolition movement "can help to reveal an extremely diffuse and complex narrative of the relationships between domesticity, politics, and gender."[86]

Katherine Plymley lived with her brother Joseph, the archdeacon of Salop and a pivotal figure in the provincial abolitionist movement. Their home was a hub of abolitionist activity—including frequent visits from Clarkson as he toured the country drumming up support—and of political discussion. Gleadle argues that for people like the Plymleys, conversation was understood as "the currency of civil society," a form of politics in itself which helped "define the relationship between the individual and society." As she shows, Katherine was involved in political conversation but occupied a liminal position in some of those discussions. Indeed, a domestic microgeography is traced within which Katherine performed "a finely constructed set of gendered behaviours" whereby she could be, as she put it herself, "more unrestrained in conversation" with Clarkson in the interstices of her household's formal entertaining—playing with the children, or waiting for breakfast—than with anyone else (although she noted the idiosyncracies of his manner and conversation). However, she was also made marginal to overtly political conversations between men, particularly when women were expected to withdraw after dinner and she had to rely on her brother's reports of what was said. Indeed, even when she was present, it was more as audience than equal participant. For example, she reported that "Mr C[larkson] and my Br[other] now proceeded to converse in a strain of calm wisdom & true goodness that was highly gratifying & edifying to be present at. I am most thankful to have sat in such company." As Gleadle puts it, "women could inhabit the same space as men and yet remain excluded from the civic meanings perpetuated" there.[87]

Yet, as Gleadle also shows, this was not simply the exclusion of women from discussion of abolition, or of politics itself in the 1790s. Katherine Plymley frequently debated such matters with her female acquaintances, such as Mrs. Vaughn of Otley Park, a wealthy local widow, whom she engaged on the "liberty of the subject," and although she did not meet Clarkson's wife Catherine until 1805, she was told that she was "'a great politician' and gifted in political argument." Indeed, Catherine Clarkson and her friend Sarah Jane Maling, a disciple of Mary Wollstonecraft, were, as Gleadle argues, dedicated to the idea that the radical democratic political project of the 1790s, which included abolition, depended on restructuring personal interaction—particularly conversation, and especially within domestic settings—via "alternative customs of egalitarian converse." In this vein, Katherine Plymley viewed the 1795 gagging acts that had silenced the London debating societies with great concern as an attempt to "repress liberty of speech," "even in private conversation." As Jon Mee notes, Plymley "also reported with concern Wilberforce's revelation that [as a result of the acts] . . . he would no longer entertain [Thomas] Clarkson in his home because 'his conversation was so very unguarded in politics.'"[88]

Indeed, from the earliest stirrings of the popular and parliamentary abolitionist movement, women's friendship networks had operated—through script, print, and speech—as modes of political participation that were both separate from and connected to the male-dominated domains of raising petitions, orchestrating parliamentary debate, and working through SEAST and the Quaker committees that preceded it. As Judith Jennings shows for the cross-denominational friendship of Mary Knowles, Jenny Henry, and Anne Seward, women's networks produced modes of political participation—such as consumer awareness of slave labor, or moves to liberate the enslaved—in advance of their more general adoption. The conversations that were the lifeblood of these networks had significant effects. Jennings argues that Henry's determination to liberate her mother's Jamaican slaves, something well known to her friend Knowles, must have shaped the work of the crucial Quaker committee, including William Dillwyn and Mary's husband, Thomas, which met at the Knowles's house beginning in 1783. These friendships, like all abolitionism's coalitions, were formed and fell apart on the basis of political commitments. So Anne Seward broke with Mary Knowles over the French Revolution when, on a visit to Seward's house in Lichfield, Knowles "made flaming eulogisms on French anarchy, which she calls freedom and uttered no less vehement philippics against everything that pertains to monarchy." As with Clarkson and Wilberforce, these women could no longer make conversation with each other.[89]

There were other limits to these spaces of sociability. In June 1793, Katherine Plymley noted that her brother—who had bought Equiano's *Interesting Narrative* and later assisted the author in promoting it and the cause—expressed his concerns at Equiano "going through the country for this purpose, as he feared it would only tend to increase the difficulty of getting subscriptions when wanted, for carrying on the business of the abolition. The luke-warm would be too apt to think if, this be the case, & we are to have Negroes coming about in this way, it will be very troublesome."[90] How, then, did race shape abolitionist speech? How, in particular, did Equiano speak, and how might he have been heard?

Hearing Equiano: Race and Abolitionist Speech

As has been shown, Olaudah Equiano and other "Africans" played an important part in the London societies' debates on abolition in 1788 and 1789. Indeed, he was already well known in the capital—as Gustavus Vassa—before the publication of his *Interesting Narrative* in 1789, having publicly engaged, in person and in the newspapers, with abolition and the cause of London's black poor. Many people felt the strength of his voice. The "address of thanks" to the Quakers, as they renewed their activism in 1785, was personally presented to them at their Gracechurch Street meetinghouse by a group of eight "distressed Africans," led by Equiano, and reported in the newspapers.[91] James Ramsay, who knew him from St. Kitts, noted both his "truly surprising" knowledge and understanding of the scriptures and that those in charge of the scheme to resettle London's black poor in Sierra Leone, on which Equiano had been a "commissioner," "dreaded so much his influence over his countrymen, that they contrived to procure an order for his being sent ashore." Although Equiano attended the parliamentary debates on the slave trade in 1788, had his views on them published in the newspapers, and offered to give his own testimony, no formal evidence was taken from those who were or had been enslaved, and they played no direct part in the parliamentary process.[92] Ramsay, however, thought to use Equiano's presence, perhaps even his voice, to influence Parliament when he suggested in September 1788 that an address he had written, "purposed for the two houses" and of which Bishop

Porteus approved, should "be distributed by Vasa in person at the opening of the Session."[93]

More broadly, while Equiano is best known as a writer, he should also be understood as a speaker. As John Bugg argues, the *Interesting Narrative* worked through a concern for the voice of enslaved and free(d) black people, from silencing by iron masks to the continuing inability to be heard in court, due to the laws on evidence, and in Parliament. Countering this concern, the book then offers extended testimony on the slave trade as the "oath of a free African" and demonstrates through autobiography how Equiano found the power of his own voice. As Equiano noted of the name he used as its author, Olaudah "in our language, signifies vicissitude, or fortune also; one favoured, and having a loud voice and well spoken."[94] The book's frontispiece shows that author (figure 18). A poised, polite, pious, and literate Equiano directly engages the reader's gaze, and the illustration shows his Bible open at the book of Acts, chapter 4, verse 12.[95] This classic evangelical text—"Neither is there salvation in any other: for there is none other name under heaven given among men, whereby we must be saved"—signals his own moment of spiritual rebirth when not only did "the Scriptures become an unsealed book," but he found a new voice: "It pleased God to pour out on me the spirit of prayer and the grace of supplication, so that in loud acclamations I was enabled to praise and glorify his most holy name."[96] Indeed, Acts 4 tells of the "unlearned and ignorant" apostles Peter and John being imprisoned by the rulers of Israel and "commanded . . . not to speak at all nor teach in the name of Jesus," to which they had replied that "we cannot but speak the things which we have seen and heard."[97] And that is what Equiano proceeded to do, both in his book and his extensive tours around the British Isles from 1789 to 1795, bringing its message, and that of abolition, directly to a broad audience.

Like Clarkson and Dickson, Equiano spent much of his time on the road speaking to people. His itinerary and contacts were shaped by the same Quaker, evangelical, and radical networks, and his encounters occurred within the same sorts of "intermediate spaces of sociability," often negotiated by women abolitionists.[98] There is little evidence that Equiano gave, rather than attended, set-piece speeches. Instead, he and his book were present at events in private houses opened for the occasion or semipublic places, such as the houses of clergymen or Quakers, taverns, or the premises of hairdressers or booksellers, and the interaction was described, in the messages of thanks he published in newspapers and gathered in later editions of his book, as an engagement with "friends," or those who

Figure 18. Frontispiece of Olaudah Equiano's
*The Interesting Narrative of the Life of Olaudah Equiano, or
Gustavus Vassa, the African* (London, 1789). Courtesy of
the John Carter Brown Library at Brown University.

might become so.[99] These modes of conversation paralleled the practices of white abolitionists, both men and women, but, as Joseph Plymley's concerns show, there were other matters at stake when Equiano spoke.

In August 1790, Equiano's tour took him to Sheffield, where he sold—from the house of Reverend Thomas Bryant—the recently published second edition of his book. In notices in the *Sheffield Register* he thanked "his numerous Friends for the Reception this work has met with," but the same newspaper also had to defend him against attack. It noted that "the most *pitiful* thing we have lately seen, appeared in the London papers of last week, in the form of a petition from the 'Ourant Outangs, Jackoos, and other *next of* kin of the African Negroes,' attempting to prove them the same species; and under the appearance of admiration, ridiculing the favourers of the abolition."[100] This item, in the *Oracle* on June 15, 1790, and the *General Evening Post* on August 10, purported to be a petition from the "OURANG OUTANGS, JOCKOS, *and others next of kin to the* AFRICAN NEGROES *resident in* BATAVIA." These apes, citing the authority of Linnaeus, Buffon, and "other eminent Naturalists and Anatomists," and noting their own ability walk upright, behave politely (including drinking tea), display the "powers and passions of mankind," and "convers[e] together in their proper dialect, and communicat[e] their ideas, wants and desires to each other," claimed their place in "the same chain with the inhabitants of Europe, and only one degree farther removed than the African Negroes . . . and of the very same species." These pretend petitioners then praised the British Parliament for having "warmly espoused" the cause of "their distressed brethren the African Negroes," hoping that the "same refined tenderness, sublime humanity, and sentimental pity" would be extended to them.[101]

This crude satire was an attempt to cast abolition as a threat to Britain's supposed racial purity via migration and miscegenation. The "petition" noted that the while "all Blacks were a few years since driven out of France by a Royal Decree," their own intention was to come to Britain in large numbers since it was evidently so welcoming to their kind. They had decided to hold a "Congress" and appoint a "Committee," and they had procured European schoolmasters—"HUMANITAS" and "DEMOCRITAS"—who had written the petition for them, "to teach them the English and other modern languages and accomplishments, the only circumstance of consequence in which they are at present deficient." Thus equipped, they looked forward to "forming connexions with, and becoming allied to the descendants of many great and genteel families who so warmly espouse their cause, . . . endeavouring one day to become members of other Congresses besides their own, and by commixing with Europeans of both sexes . . . to become good European subjects in a few generations." All that was needed was for the governor to make "suitable provision" for "a large importation into Europe, of their Committee and Deputies, by the next fleet from this country."[102]

As this "petition" indicates, the talking apes central to Long and Monboddo's accounts of race, speech, and slavery nearly twenty years earlier re-emerged in the late 1780s abolition debate as part of what one newspaper called the "African-humanity mania" on which, as another newspaper put it, "so much appears to depend for the decision of this great question."[103] The earlier ideas were, therefore, recycled and embellished in newspapers and by proslavery writers such as John Gardner Kemeys.[104] "Civis," writing to the *Morning Chronicle* in February 1788—when the slave trade was before Parliament—cited Hume, Buffon, Voltaire, Monboddo, and, especially, Gibbon to argue that, when it came to speech, "the choicest production of an African would not be superior to the jabbering of Monkeys, as it would be inferior to the polished eloquence of the modern historian." These ideas were also reproduced for all those interested in the debate, as in Reverend John Adams's *Curious Thoughts on the History of Man* (1789), which presented Monboddo's work on animal and human speech to "the youth of both sexes."[105]

Abolitionists, of course, issued strenuous refutations. Countering ideas of "the stupidity of negroes," Reverend Robert Boucher Nickolls, dean of Middleham, in a letter printed by Phillips in 1788 and distributed by SEAST, argued that "nothing has been written by the late defenders of slavery, that discovers half the literary merit of" Phillis Wheatley and Francis Williams. As he concluded, "I never heard of poems by a monkey, or of Latin odes by an oran-outang."[106] This ongoing debate also helped fill the newspapers. "Oroonoko," replying to "Civis," argued that "if regular speech, which manifests itself by verbal discourse, be the result, as philosophers inform us it is, of a regular train of thought," and that the "oran-outang" is able to speak but does not "from a want of thought, properly digested," then "can there be a stronger reason than this to believe . . . that [Negroes] are rational beings, endued with good sense, and capable of every degree of civilization"? "Civis junior" countered a month later to again attempt to blur the distinction, arguing that "speech is merely a mode of communicating ideas, the brute possesses it in some degree, and having been in the West-Indies, you probably know that the language of the Negroes very much resembles the jabbering of monkies." "Oroonoko" replied in turn that the well-known fact that "some few animals are known to use words" was not enough to "overthrow a general rule" on the difference between the human and the brute. The debate continued.[107]

As boundary creatures, talking apes were as useful to satire and entertainment as to philosophy and natural history, domains that were by no means separate. Newspaper accounts from the 1770s of "the Man-Monkey" might clearly be social or political satire or, as in the account of a "Lady" from Sheerness who had an "*Ourang Outang*" that was "possessed of the powers of articulation" and could "for a few moments, hold a *tolerable* conversation," but which had so misbehaved at "a Barbicue" that it had to be muzzled, might be a factual account of poor pet keeping, or something else entirely.[108] Some of this political satire had obvious, if now obscure, West Indian targets, such as the squib on City of London electoral politics that presented one candidate as "a species of the human kind, called the *Oran Outang*, taken in the bottom of a sugar cask in the island of Barbadoes, when a *whelp*, from whence it is conjectured by the naturalists it sucked its *nutriment*." Other pieces satirized the American colonists.[109]

Such apes were a real presence in London too, and their relationship with the human was what was always on show.[110] From April 1785 until at least May 1788, and perhaps as late as 1796, "The surprising MONKEY GENERAL JACKOO" performed at Astley's Amphitheatre by Westminster Bridge, initially on the same bill as "A new Dance called The MERRY NEGROES" and later alongside the exhibition of "wild born human beings" the Three Monstrous Craws "whose country, language, and native customs . . . are yet unknown to all mankind."[111] Jackoo, declared the most popular monkey in the metropolis in 1786, excelled at rope dancing and, while not speaking, imitated human activities, dressed as a Russian hussar, and, when featuring on the same bill as "the Pig of Knowledge" and "The Learned Dog," provoked discussions on the boundaries of humanity.[112] Londoners could also visit an "Orang Outang" in the Tower of London menagerie from at least 1777 and another at Exeter Exchange on the Strand from the late 1770s until the early 1790s. These creatures were also ripe for satire in the age of abolition. On the death of the Exeter Exchange's ape in 1792, the *Morning Chronicle* reported that James Boswell, having visited it often in its last months, and having previously read the account of the creature by his friend Lord Monboddo, had collected enough material for a biography. Indeed, Boswell had found the animal both talkative and quick-witted, willing to joke about those who did or did not have tails. Perhaps surprisingly, the Johnsonian "O.O." had also read Monboddo's *Origin and Progress of Language*, which he judged "a pretty Essay" but "chimerical." It was, he said, perhaps unnecessarily, not true "that those of our nation cannot speak," but he saw how Monboddo and the European travelers he relied on "might be easily led into that mistake" since, like the African apes who stayed silent to stop themselves being enslaved, "we have a rule to prevent any intercourse with strangers."[113] Boswell, who had written crudely comic antiabolitionist poetry and mocked Monboddo's view of the "Orang outang," was getting a taste of his own medicine.[114]

The vicious whipping up of fears about the black presence in Britain that the petition played on, and which can also be heard in Joseph Plymley's concerns over Equiano's provincial travels, were part of this discussion too.[115] A letter entitled "BLACK MEN," printed in the St James's Chronicle in February 1791, proclaimed an abhorrence of the slave trade, but a greater abhorrence of the threat posed by "the incredible number of negroes which, since the American war; have got a footing in this metropolis and kingdom," a population the author estimated at twenty thousand. Repeating arguments aired around the Somerset case over "why these people should not mix with us"—including Long's report of the infertility of mules and "mulattoes" and the idea that while "a black woman and an *Ouran Outang* might produce a human being . . . that would not pass unnoticed or unpunished" by God—the author noted that "they are men; but very different from Europeans," calling them "a distinct race, and a lower link of human being" whose blackness was not a matter of "climate or heat of the sun" since "were they to live and cohabit only with their own complexion, in this, or any other more Northern climate, they would for ever remain *black, and all black.*" Ideally, the writer argued, they "ought to be confined to their own soil, and their own complexion," but at the very least there needed to be laws passed "to compel them from all connections, *lawful,* or *unlawful,* with white women," or Britain would go the way, in both complexion and vice, of "Portuguese Brazil."[116]

This, then, was the context in which Equiano—married to a white woman, Susan Cullen of Soham, from 1792, and with two daughters—spoke and was heard.[117] As the *Sheffield Register* put it in 1790, countering the pretend petition, "Surely this unfortunate race is sufficiently degraded by being the objects of an iniquitous traffic, without being in *every* degree leveled with the beasts that perish." It then paraphrased Shakespeare's *Merchant of Venice*—"Hath not an African eyes, hands, organs and dimensions, senses, affections, passions?"—and invited "any within the circle of readers" to decide for themselves: "let them see GUSTAVUS VASA, the free African, now in Sheffield—his manners polished, his mind enlightened, and in every respect on a par with Europeans."[118] While they did not directly discuss his speech, an enlightened mind was not something that could simply be seen; neither, in many ways, were polished manners, although dress and bodily comportment were vital signs. They were, however, something that might be heard at the house of Reverend Thomas Bryant, along with evidence that the former slave was indeed the author of his own book. Audiences could hear both at the same time as he spoke, and even more so in the back-and-forth of polite conversation. Yet, as Plymley's concerns show, white abolitionists were evidently concerned that "the luke warm" might see and hear something else.

Race also mattered in the relationship between radical politics and abolition, which Equiano had played an important role in forging in the revolutionary decade of the 1790s. The London Corresponding Society (LCS), the largest and most influential of the popular reform societies, led by the shoemaker (and Equiano's landlord and friend) Thomas Hardy, began when Equiano put Hardy in touch with the very same Reverend Bryant who had hosted Equiano's visit to Sheffield and who shared Hardy's political views.[119] Indeed, Hardy's wife Lydia wrote to him from Chesham in Buckinghamshire, where her father lived, to say of the slave trade that "the people here are as much against it as enny ware" and to ask after both the LCS and "Vassa," who was then touring Scotland.[120]

Yet there were tensions too, particularly following the crackdown on radicals like Hardy who were tried for treason in the mid-1790s. For example, at "a numerous meeting of the Friends of Freedom" at the Crown and Anchor Tavern in the Strand on November 5, 1796, called to celebrate the anniversary of Hardy's acquittal, the more than seven hundred participants dined, sang songs, drank toasts (to "The Deliverance of the Nation on this day from Arbitrary Power," "The Rights of Man," and "A fair, free, and full Representation of the People") and heard speeches from John Thelwall, Earl Stanhope, and General Banastre Tarleton, who, after a toast of "Liberty all over the World," noted that it was his first appearance before such a gathering and decried the government for having "violated that liberty which is the birth-right of Englishmen." When he had finished, however, "Gustavus Vasa, the celebrated African," stood up and reminded the audience that Tarleton, as MP for Liverpool, had been one of the strongest opponents of abolition in Parliament.[121] Taking the chair, Equiano "proposed a toast relative to the slave trade."[122] While the newspapers derided Tarleton for his new alliances and what they saw as his hypocrisy, it was also evident that the cause of liberty was not necessarily universal.[123] A satirical reworking of the meeting in the avowedly reactionary Anti-Jacobin of December 4, 1797—one which has mistakenly been taken as wholly fictional since it seems to report events that happened after Equiano's death in March of that year and certainly contains errors—notes, with a telling pun, that Tarleton's words "unfortunately gave rise to an altercation which threatened to disturb the harmony of the evening: OLAUDAH EQUIANO, the African, and HENRY YORKE, the Mulatto"—who had also been tried for treason in 1795 and was not actually at the November 1796 meeting— "insisted on being heard; but as it appeared that they were entering upon a subject which would have entirely altered the complexion of the Meeting, they were, though not without some difficulty, with-held from proceeding further."[124] While Equiano had finally been able to confront face-to-face one of the leading parliamentary supporters of the slave trade, his raising of the question of "complexion," of racial slavery, had done more to reveal disjunctures between radical politics and abolition than it did to cement them.

Speech was an essential part of the wiring of the abolitionist movement, connecting people together and engaging hearts and minds. Abolitionist speech took many different forms, even when the content stayed the same. For example, Cowper's poem provided an imagined intimate connection to the voice of the enslaved within a politics of sympathy; animated tea-table conversations led by women in the 1790s sugar boycott; and, in the 1820s, was sung by ultraradicals including Robert Wedderburn, author of *The Horrors of Slavery* (1824), in the coffeehouses and taverns of Spitalfields and Bethnal Green alongside "toasts to the republicans of St. Domingo and to future insurrections in the West Indies."[125] There was never only one form of abolitionist speech, and making spoken words work politically required careful attention to audiences and contexts, and to both conforming to and challenging formal and informal rules on who could speak, where, when, and how. In addition, spreading the abolitionist message was never just about speech, but worked it together with texts, images, and objects to open debate and change people's minds. For example, Dickson wrote to Phillips from Edinburgh in 1792 to reminded him to send some of Josiah Wedgewood's "Am I Not a Man and a Brother" cameos as soon as possible and told him that "The Slave ships"—the famous print of the *Brookes* that illustrated how the enslaved were packed below decks—"has *lately* been put up in the Banks, Publick offices[,] Coffee-houses &c here, with an excellent effect."[126] Since Dickson had only been in Scotland for four days, and no petitions had been started, those effects—the winning of hearts and the changing of minds—would only have been evident in the conversation of people with whom he talked. In the building of a popular movement, whose collective "voice" would compel Parliament to act, the proliferation of speech—that the slave trade might, as Clarkson put it, "become the general subject of conversation"—was an end in itself.[127] Thus, Thomas Hardy, in a private letter on the allied question of parliamentary reform, argued that the aim of "Humbly petitioning" parliament was to generate a debate that would be widely reported in the newspapers so that "in different parts of the country thousands of people will make it the subject of conversation and inquiry who never thought of it before."[128] That is how talk would drive political change.

However, while setting out this proliferation of talk, equal attention has been paid to the ways in which the spaces and forms of abolitionist speech, like all aspects of abolitionist politics, were shaped by class, gender, and race as this multistranded movement developed. Whiteness and masculinity were as important as class and religion in enabling men like William Dillwyn, Granville Sharp, Thomas Clarkson, and William Dickson to find ways of speaking about slavery and the slave trade. And, while evangelicalism or nonconformity, and the different claims of domesticity and sympathy, or the experience of slavery itself, could provide a basis for white women from Lady Middleton to Lydia Hardy and black men such as Olaudah Equiano to engage in abolitionist debate, there appeared little room for black women to represent themselves and find a voice as activists, at least in Britain in the late eighteenth century.[129] These forms of difference, and the ways they shaped the relationship between speech and antislavery, are also evident when attending to the freedom talk of the enslaved in the Caribbean.

Freedom Talk: Conspiracy and Revolt in the Age of Abolition

Conspiracies and revolts among the enslaved, and slaveholders' interpretations of these actions against slavery, were always, and perhaps increasingly, attuned to questions of imperial geopolitics—of international rivalry; the fortunes of war and trade; and local military and political strength—and of what was happening elsewhere (see chapter 2). Animated by the interisland and circumatlantic circulation of news and rumor carried by "the common wind," the enslaved attempted to evaluate their chances of gaining freedom. Beginning in the late eighteenth century, this evaluation became ever more tightly entwined with the transatlantic debate over abolition. As Julius Scott shows, changes in Spanish imperial policy, the French Revolution's implications for slavery in its colonies, and the British Parliament's debates over the slave trade from 1788 were all discussed among the free and the unfree across the Caribbean. Starting in 1791, news from St. Domingue, which became the independent state of Haiti in 1804, had profound effects among the still-enslaved in Jamaica, Cuba, and farther afield, provoking talk and inspiring insurgent action.[130] As a result, early nineteenth-century revolts in Barbados in 1816, Demerara in 1823—accompanied by a series of conspiracies and smaller-scale revolts in northern Jamaica from later that year—and western Jamaica in 1831–32 need to be understood as part of the politics of transatlantic antislavery. On the one hand, as Michael Craton argues, these uprisings were politicized because slaveholders sought to shift the blame for such evident signs of resistance from the system of slavery itself to the ideas of its metropolitan opponents and the actions of missionaries seen as stirring up trouble among the enslaved on their behalf (as we saw in chapter 4). On the other hand, Craton also notes that "an ever-widening group of slaves became aware not only that they had friends and allies in the imperial metropolis but that their masters felt increasingly embittered and besieged."[131] There was a reciprocal relationship between metropolitan abolitionist activism and Caribbean slave revolts, with each shaping the other. Crucial to this relationship were forms of talk. As Gelien Matthews argues, the interpretation by abolitionists of each subsequent Caribbean revolt shaped "British antislavery discourse" toward "an aggressive justification for demanding slave emancipation."[132] For slaveholders, it was the dangerous talk of abolitionists and missionaries that they identified as fomenting revolt, and it was rumors—particularly of forms of freedom granted by the king or the government but denied to the slaves by planters and colonial assemblies—which they thought impelled the enslaved to action. For the enslaved, it was through the reports of those who traveled the oceans, through what might be overheard, and through the newspapers— from the presses of Britain and the islands—being read aloud, that the name of "Wilberforce" entered how they talked and sang about freedom in the early nineteenth century.[133]

However, closer attention to the forms of talk involved, particularly how the enslaved talked politically and how the authorities responded to what they imagined they heard, can develop this general picture in important ways. First, examination of the modes and spaces of speech among the enslaved indicates how their talk of abolition worked to form various sorts of collectives that might, or might not, move from discussion to action by interpreting news and rumor (as we saw in chapter 2). Second, attention to the forms of speech that imperial and colonial authorities tried to use to manage rumor—particularly proclamations—demonstrates that they were also engaged in attempts to speak across the Atlantic about slavery. They understood the important interpretative role of slave talk and tried to manage it. Doing so both continued to construct the enslaved as active speaking and listening subjects and was used to justify the execution of those who sought to free themselves.

Unsurprisingly, most attention has been paid to the revolts of 1816, 1823, and 1831–32, as moments when talk became action.[134] Yet, examination of the smaller "conspiracies" investigated in Jamaica in 1823–24 allows, with all the necessary caution about the interpretation of evidence from courts of law wielding the death penalty for conspiratorial talk, the differentiation of ways of talking about freedom among the enslaved, or, at the very least, what was said about those ways of talking within a terror-laden judicial process.

Around Christmas of 1823, "rebellious conspiracies" were investigated by magistrates in the parishes of St. Mary, St. James, and St. George on the north side of the island.[135] What happened in St. Mary was, or was represented to be, the grounding of a parish-wide, possibly island-wide, revolt. William, an enslaved boy of fifteen, had revealed to his master "the intention of the negroes to begin to burn and destroy the houses, trash houses and estates, and when such fire took place, to murder all the inhabitants." In court he implicated his father, James Sterling, and said that he "had twice observed the negroes [both men and women] assembled in large bodies near a bridge between Frontier estate and Port Maria, where he heard them speak of an intended rising, and at that time they were flourishing their cutlasses, declaring that they would destroy all the white people." Others corroborated this account, arguing that the alleged ringleaders—Sterling, Charles Brown, Charles Watson, William Montgomery, Richard Cosley, Rodney Wellington, Henry Nibbs, and Morrice Henry—had met to plan the rising and said that "when they had done that they would be free."[136] The magistrates either did not solicit evidence of rumors spread by others, or of talk of abolition or abolitionists, or they were given none because it could be hidden, or because there was none. Watson had, it was testified, told an enslaved woman called Mary, "a native of St. Domingo," and her husband that "they were all to be free," but no one tied this statement to any rumor, or external influence, being more interested in whether the conspirators had obtained firearms.[137] All those tried were sentenced to be hanged, and it was reported that "none would mention any other negroes concerned with them, or shew any symptoms of religion or repentance." One of the condemned even "laughed at the clergyman Mr. Cook, when he attempted to exhort him under the gallows."[138]

In contrast, it was reported by some witnesses that the name of Wilberforce, and rumors of changes in the law, even the granting of freedom, were on many lips in St. James in late 1823. In October the governor was notified that, although "the negroes are every where behaving well," nonetheless "it appears there is a general expectation among the[m] . . . of freedom being given shortly by government at home." The report came with seven instances of things said or overheard to that effect. By December, the magistrates were investigating information from a free black man, Robert Bartibo, that those enslaved at Unity Hall had met and were going to rise at Christmas, along with as many other estates as they could muster, to "kill every white person they met with." Bartibo also said that "they said the King had made them free."[139]

Bartibo's evidence identified Saturday night gatherings—for eating, drinking, and dancing—at the "negro houses" on Unity Hall, half a mile from the overseer's house, as the occasions for such talk between those enslaved on neighboring plantations. Of one dance at John Cunningham's house, he testified that they "said they would rise at Christmas, unless they got Friday and Saturday [for themselves], or freedom"; at another, at Mary Ann Reid's house, "they drank their toast, 'here is your health Mr. Wilberforce,'—this was the main word with them, . . . and every time his name was mentioned they hurra'd."[140] The evidence given, and undoubtedly drawn out by the magistrates according to what they wanted to hear, disputed who was present, what was said, and by whom; what was heard; and, of course, what it all meant. These discussions of freedom involved women as well as men. They involved those who said that freedom would come by decree, or "that Wilbe[r]force had a law for free," and those who said they doubted it. They involved those who said they should take freedom for themselves, those who argued it would come via Mr. Barrett—a local planter—who "had taken his passage to go to England to speak to the King about it," and those who expressed fears that freedom would make matters worse. There was evidence that this was part of a more general discussion, of "news on the road" or "a flying news that all the negroes were to get free." Others stated that "the negroes must have heard about freedom at market, when it was all the talk at Montego Bay." There was also much debate over the reliability of witnesses, particularly Robert Bartibo and his brother Peter. Could what they said be believed over the word of others?[141]

In the end, the majority of the magistrates concluded that "there does not appear to have been any fixed or general place, or any great preparation, if any, for a rebellion." Yet they were also aware that those examined "have agreed not to betray each other," and it is clear the accused knew about the eight hanged in St. Mary and the high price of loose talk.[142] Although there was evidence given of other nighttime meetings and conversations—

including "a large party" meeting "out of hearing of the white people; just past the wharf gate, near the morass"—the magistrates concluded that "the present disposition is entirely owing to the reports of what is going on in England; and had it not been for the arrests, it cannot be doubted there would have been overt acts of rebellion; but, luckily for the country, and the negroes, it was nipped in the bud." The governor agreed that, unlike in St. Mary, "there seems to have been a very active spirit of inquiry, which may be naturally accounted for without attributing to them any criminal intentions." The problem was the rumors of freedom that "led them to make use of inconsiderate and, in some instances, intemperate expressions." Those judged to have engaged in dangerous talk, and in illegal meetings, were sentenced to terms of hard labor, and the governor stressed the "mercy" shown by the potentially death-dealing colonial state. As utterances that were only weakly tied to speaking subjects, rumors could provide the grounds for deniability on all sides.[143]

In St. George, the forms of speech were different again. As in St. Mary, there was evidence given, this time by witnesses saving their own necks, of clandestine gatherings on Balcarres plantation to plan a rising the night after Christmas, and communication with other plantations "to rise at the same time, and proceed to murder all the white people in their reach." As in St. James, there was testimony of discussion among the enslaved "that the King had given them Friday and Saturday, but the buckras would not give them."[144] Yet the contexts and forms of speech were significantly different. Much of the discussion had taken place at Richard Montagnac's house, which was known as the "court house." Men had been mustered there with wooden swords and had escorted Dennis Kerr, the "governor," to the house of "the king," James Thompson, where discussions continued. In his own evidence, Montagnac said this "was all meant in fun" and that his house only had that name because it was where the men met to drink the rum paid for by the fines of five pence levied "when one negro cursed another." It was, he claimed, a place of alternative legality, not resistant sovereignty. Yet the court also heard, from Jean Baptiste Corberand of Mullett Hall plantation, the man who had turned king's evidence, that Corberand had been made "clerk" at Montagnac's, that "he had put on an old blue coat; to which he put epaulets of Osnaburgs" and that "the law was read, that they were to have three days in the week," and that "they then marched from one house to another." Questioned by the magistrates, he affirmed that those at the "court house" and Thompson—the "king" of Mullett Hall— had affirmed "the same law" and that "Henry Oliver said that the King [of England] had given them Friday and Saturday, and they must have it, or they would take it by strong."[145] It sounded as though the political talk of the enslaved, organized through an appropriation of the forms of legal talk and sovereignty talk characteristic of the masculine military modes of empire, and of the slaveholders' assemblies, had reshaped a rumor, or even a piece of news, into a potential call to arms.

In addition, Corberand gave evidence of an external agitator, "Baptiste," who "came from St. Domingo on purpose to stir up the negroes," of guns run up to the rebels from Kingston, and of an obeah man, "Jack," who, invited by Henry Oliver—"who was never at work, [and] had walked every where, arranging with different persons how they were to act"—had come to mobilize the enslaved. Jack, it was said, had practiced divination at Montagnac's house, including with two little figures, one white and one black, which he put on a plate to see whether they would fall. He had administered a binding oath, or "swear," and "prepared a quantity of pounded bush, with which he anointed the bodies of the conspirators, affirming that was to render them invulnerable." This ritual, Corberand alleged, was done in a cow pen at Balcarres, with the men—women, he said, having not been told anything of the plan—standing in a ring. After the men had been "rubbed" and told how they were thus protected, John Smith, a "major," "gave the order, and said they must all have one word, one mouth, one tongue and one desire."[146] For his part, Jack, arrested and charged with both rebellious conspiracy and practicing obeah, and having already been convicted once, testified to the intimate involvement of Corberand in the plot, saying "that fellow fit for hang same as me and Henry," and that if Corberand "had been at his work, buckra could not have taken and done him [Jack] harm." Jack was also unrepentant about his skills in the face of death, saying that "he knew that his doctoring was what buckra called obeah. Buckra had their own fashion; in Guinea, negro could doctor." And he hinted at how this doctoring opened up other worlds beyond death, saying that he did not begrudge Corberand his freedom, "but he hoped to meet him by and by." Jack was condemned to hang, as were Oliver and four others, for what was judged "a most diabolical plot."[147] Others, including Montagnac and Kerr, were transported. Neither the guns nor the revolutionary agitator "Baptiste" were found.

In these three parallel instances there is evidence of distinct ways in which the enslaved talked about freedom in early nineteenth-century Jamaica. One closely follows the long-standing pattern of clandestine collective meetings of men and women to plan a rising. Another demonstrates the power of the ways in which abolitionist news and rumor might be taken up within the collective life of the plantations to express the hopes and fears of the enslaved about their own freedom, and to explore, as the discussion of rumors can, contradictory relationships "between ideas about the personal and the political, the local and the [trans]national."[148] The third suggests a combination of modes of political speech—an appropriation of imperial and colonial proclamation and deliberation, obeah's binding of believers under "one word," and at least the possibility of interisland revolutionary agitation—into a potentially potent, and certainly feared, form of action. How much the presence or absence of these forms of talk—particularly talk of transatlantic abolitionism, and of its key symbol, "Wilberforce"—was a matter of what happened, what those accused could stay silent about, or what the magistrates were interested in, is impossible to determine. Yet the differences do suggest that the transformation of rumor into revolt was not a straightforward process, and that the agency of the enslaved, shaped through particular, and particularly gendered, forms of talk made within the networked social relations of the plantation, was crucial.[149] Indeed, comparing the revolts in Barbados in 1816 and Jamaica in 1831–32 also reveals important differences between forms of organization, as well as similarities, with the former reportedly deploying a distinctive gendered iconography of both respectability and transgression to bind the rebels to the cause, and the latter operating through the intimate connections, and oratorical forms, of the Baptist church.[150]

Examining these revolts through talk also reveals differences, as well as continuities, in the official response. In Barbados in 1816, the governor, James Leith, was in no doubt that the revolt's causes lay in "the discussions which have so generally taken place of late on the question of Slavery, attended by the misconceptions, heat, and exaggeration of many individual opinions, and followed by the mischievous delusions of those who have availed themselves of every circumstance to inflame the minds of the Slaves." In response, he sought to go beyond "abrupt and unexplained acts of Legal power"—the executions of the rebels—"to give them greater effect by some illustration," although without entering into what he called "abstract questions." He set out to personally "see as many of the Slaves as can conveniently be collected in the different parts of the Colony, for the purpose of speaking to them."[151]

Leith's "Address to the Slave Population of the Island of Barbados," printed on April 26, 1816, is remarkable for its professed desire to explain matters to the enslaved, although there were clearly other audiences on both sides of the Atlantic that he was addressing too. He sought to "remove all misconception on a subject of so great importance for the tranquillity of this Colony; and for the well-being of the Slaves themselves." He discussed the historical and geographical ubiquity of slavery, implicated Africans in the slave trade, and argued that amelioration was the only answer since even abolitionists saw immediate emancipation as "morally impracticable." The enslaved had, he said, been misled by "Enemies of the State" into believing that he was returning to the island with their freedom. He told them "to return with cheerfulness to your Duties" and to be assured that everything was being done by "the paternal Government of the Prince Regent"— George Augustus Frederick, ruling in place of his father, King George III, during the king's incapacitation—to protect them, as long as they obeyed the law. If they did not, he promised to "crush the Refractory and punish the Guilty."[152] In September 1816 a further proclamation was delivered which included statements from the Houses of Parliament, and from the Prince Regent himself, that no emancipation orders had been passed or sent. This was also to "be audibly read and distinctly explained to the Slaves."[153]

There were further uses of the sovereign power of proclamation to speak across the Atlantic, and across the sugar colonies, in the 1820s and early 1830s. It was noted in summer 1831 that a proclamation, in the name of King George IV, that freedom had not been declared had been issued in 1824. Indeed, it was then re-sent to the colonies' governors with instructions from King William IV that "it be again published in His name, and as emanating directly from Him." Yet transoceanic speech was no easy matter when it was uncertain how the words proclaimed would be heard. Governors were to decide if the proclamation "might kindle the very excitement which, under the circumstances, it would contribute to allay." Where it was read out, governors were to exercise "the power of modifying the introductory language of the Proclamation, so as to render it conformable with the real facts of the case." They were to put themselves in the position of the enslaved as an active, interpretative audience, animated by both affect and reason, "avoiding the use of exaggerated terms, either in describing the state of mind which exists among the slaves, or in referring to the causes in which their discontents may be supposed to have originated."[154] While these proclamations enabled a sovereign division between those who conformed to the law and those who did not, who might then be put to death, they also constructed their audience as one of active political subjects considering their prospects for freedom.

Conclusion

Along with recording the words of the song the "King of the Eboes" and others sang about William Wilberforce, Matthew Lewis also noted that he had been told that there was another song those enslaved on his own plantation sang, and that had circulated around the neighboring parishes. This one "declar[ed] that I was come over to set them all free." While he denied the song's existence, this recently arrived planter did give credence to his agent's reports that "my negroes really have spread the report that I intend to set *them* free in a few years." In line with his *Journal's* attempt to construct a paternalist view of slavery, and his own implication in it, Lewis noted that this was "in consequence of my over-indulgence to my negroes" and the desire of the enslaved "to give themselves and their master the greater credit upon other estates." He concluded, "As to the truth of an assertion, that is a point which never enters into negro consideration."[155]

It is worth noting, however, that what Lewis reported the enslaved had said about their freedom was not a lie but something that had not yet come true. Saying that Lewis would set them free in a few years was an anticipatory, perhaps even perlocutionary, speech act that, by saying or singing it, might make something happen.[156] As with interpretations of how those Lewis enslaved used what he saw as his paternalism to gain everyday advantages in the world of the plantation, saying that he was different from other masters and that he would set them free, at least opened the possibility of freedom and challenged him to deny that that was what paternalism meant in the age of abolition.[157] Yet this speech act was also different from those everyday weapons of the weak, which might work best to individual advantage. As a song and a rumor, it was a collective speech act, attributed by him to all of those on the plantation, making a collective, even if he called them "my negroes." The rumors of their song about him, paired for the planters with the song of the "King of the Eboes," can be seen as offering a choice for the future: give us our freedom, or we will "Take by force by force!"

An understanding of talk about abolition and emancipation as anticipatory speech acts which attempted to shape the future through making new collectives via what was said, who said it, and where, when, and how it was said can treat mobilization against slavery and the slave trade in Britain and in the Caribbean on the same common ground. As the discussion of the slave's voice in abolitionist writing showed, the abolition movement worked through distinct ways of having the enslaved heard across the Atlantic, in either standard English or some form of creole, to move audiences to "the slave's cause." In addition, popular mobilization, by Quakers and evangelicals, by men and women, and by black and white activists, was conducted through face-to-face talk in debating rooms, at committee meetings, in schoolmasters' studies, around breakfast and tea tables, in parlors, and in coffeehouses, taverns, and banks, which sought to win hearts and minds. Talk, the invisible wiring of the abolition movement, provided the basis for the production and consumption of printed texts, visual images, and material objects that, in turn, aimed to provoke more talk to produce change. Finally, talk about emancipation also took varied forms in the early nineteenth-century Caribbean and was also intimately interwoven with print. At clandestine meetings in out-of-the-way places in St. Mary; within the intermediate space of sociability of dances at the "negro houses" of Unity Hall; through the colonial mimicry of the Mullett Hall "court house"; or in the use of Jack's obeah to conjure "one word, one mouth, one tongue and one desire," the transatlantic discussion on slavery and emancipation was engaged to different effect. Rather than a separation of violent resistance in the Americas, on the one hand, and a polite politics of Parliament, print, and the public sphere in Europe, on the other, when enslaved Africans in Jamaica and Barbados heard newspaper reports of Wilberforce's speeches and "talked free" among themselves, they weren't doing anything qualitatively different from abolitionist activists in Britain.

However, the differences between these forms of talk about abolition and emancipation are important since they each shaped the nature of the collectives that might be formed. Choices between representing the enslaved as speaking standard English or creole, particularly in the context of competing proslavery representations, shaped the imagined relationship between the free and the unfree in terms of both emotion and action. In turn, while talking could engage the twists and turns of debate much more flexibly than using print, it is clear that matters of religion, class, gender, and race intersected in the spaces and forms of talk through which abolition and emancipation were debated to both enable speech and to limit what could be said. Speaking out meant challenging and changing expectations of who could engage in such talk, as well as conforming to them. Thus, while men like Thomas Clarkson and William Dickson, women like Hannah More and Mary Knowles, and black activists such as Olaudah Equiano were making new forms of politics through speech, and being opposed in doing so, these spaces offered little room for black women to speak out rather than be spoken for. Talk of freedom among the enslaved on the sugar islands also took distinct, and gendered, forms. Thus, women might be excluded from modes of speech defined through a martial masculinity, as in St. George in 1823, or they might be part of the spheres of collective debate and mobilization, as in St. James. Indeed, at times, women's words in these contexts could be crucial to moving the enslaved to action. In Barbados in early 1816—just as Matthew Lewis was interpreting

Jamaican songs of freedom—it was, one enslaved witness testified, what Nanny Grig of the Simmons plantation had said that had sparked a revolt. She had told the others that "she had read it in the Newspapers" that there was an order from England to free them on New Year's Day. When that did not happen, the witness said that she spoke of seizing the future by learning from what happened elsewhere: "she was always talking about it to the negroes, and told them that they were all damned fools to work for that she would not, as freedom they were sure to get . . . , and the only way to get it was to fight for it, otherwise they would not get it; and the way they were to do, was to set fire, as that was the way they did in Saint Domingo."[158] Nanny Grig's words were ones that sought to make freedom happen: to talk free.

Last Words

On August 1, 1838, a vast crowd gathered in Spanish Town, Jamaica, to hear some words spoken. Standing on the steps of King's House, facing the building housing the Jamaican assembly, the governor, Sir Lionel Smith, flanked by the bishop of Jamaica, the chief justice, and the Baptist minister James Phillippo, faced the cheering crowd and "read to them the PROCLAMATION of FREEDOM" made by Queen Victoria, which finally abolished slavery in the British Empire. As Phillippo later put it, the end to the punitive "apprenticeship" system, which for five years had succeeded slavery in binding black workers to the plantations, was English history's most momentous event:

> When a century shall have passed away—when statesmen are forgotten—when reason shall regain her influence over prejudice and interest, and other generations are wondering at the false estimate their forefathers formed of human glory—"on the page of history one deed shall stand out in whole relief—one consenting voice pronounce" that the greatest honour England ever attained was when, with her Sovereign at her head, she proclaimed THE SLAVE IS FREE, and established in practice what even AMERICA recognizes in THEORY: *that all men are created equal—that they are endowed by their Creator with certain unalienable rights—that among these are life, liberty, and the pursuit of happiness.*[1]

Many of the formerly enslaved across the island had been up all night raising their voices, Phillippo recalled, "in some instances performing different acts of devotion until the day of liberty dawned, when they saluted it with the most joyous acclaim."[2] At the Baptist chapel in Falmouth, a coffin inscribed with the words "Colonial Slavery, died July 31st, 1838, aged

276 years" and the names of two proslavery Jamaican newspapers, was buried along with a chain, handcuffs, and an iron collar. As the clock struck midnight, William Knibb, the chapel's minister, now back in Jamaica, proclaimed, "THE MONSTER IS DEAD! THE NEGRO IS FREE! THREE CHEERS FOR THE QUEEN!" and the congregation sang.[3] In other towns and villages, Phillippo recorded, the newly freed celebrated by "singing the national anthem and devotional hymns, occasionally rending the air with their acclamations of 'Freedom's come;' 'We're free, we're free; our wives and our children are free.'" When morning came, they attended churches and chapels to hear "sermons . . . preached applicable to the event," and seven thousand adults and two thousand children walked in solemn procession from Spanish Town's Baptist chapel to hear the governor address them "in a speech characterized by much simplicity, affection, and energy." While the governor's speech was followed by many celebratory events across the island, Phillippo was keen to stress that order was strictly maintained: "At none of their repasts was there anything Bacchanalian. Their behaviour was modest, unassuming, and decorous in a high degree. There was no crowding, no vulgar familiarity; all were as courteous, civil, and obliging to each other as members of one harmonious family." Indeed, despite what many planters had predicted, "not a *single instance* of violence or insubordination, of serious disagreement or of intemperance, so far as could be ascertained, occurred in any part of the island," and everyone went quietly and willingly back to work the next day.[4] Freedom had been proclaimed—the previously unutterable had been uttered—and everything was going to be fine.

Things were not, of course, quite so simple. As historians of the post-emancipation Caribbean have shown, the fault lines soon began to show between the disparate ways of imagining the future of these islands that were already evident before the 1830s. Missionaries' dreams of a society based on Christianity, the patriarchal family, and a gendered division of labor; the plantation owners' desire for a labor force compelled by necessity to work the cane and the boiling house; and attempts by the newly freed population to construct a society of independent peasant proprietors who entered the market for labor and produce on their own terms—all foundered on incomplete control over land, labor, and the state. This instability prompted new modes of organized violence to discipline labor and new forms of resistance and rebellion. Crucially, as Catherine Hall and Thomas Holt have each shown, forms of racial difference were increasingly mobilized to justify and enforce continued economic exploitation and political and legal marginalization.[5] As the previous chapters have demonstrated, the "freedom of speech" in the Anglo-Caribbean world—the forms of freedom made through speech across multiple arenas—was, from the development of slave codes in the mid-seventeenth century onward, a matter of attempts to make the distinction between the free and the enslaved synonymous with racialized distinctions between white and black. Thus, "slave evidence" and "Negro evidence" were often used interchangeably and similarly restricted; conspiratorial talk among black men and women was heard as "the timid murmurs of slaves" rather than as calls for freedom; and the capacity to generate and communicate botanical or medical knowledge or to hear and speak as a Christian subject were understood through racialized constructions of civilization and barbarism. One result of all this work making racial difference through speech across the many spaces of everyday life was that when freedom was proclaimed, black freedom was not the same as white freedom.

Yet that is only one side of the central contradiction of racial slavery that this book has set out to explore by considering speech as crucial to the ways in which the humanity of the enslaved was necessary to slavery at the same time as it was denied. The preceding chapters have attended to many types of speech practices to uncover the production, reproduction, and contestation of freedom and bondage in the law, politics, natural knowledge, and religion, as well as in struggles for abolition and emancipation. While there is much that is new within the detailed historical geographies of talk in Barbados and Jamaica that have been set out here—particularly in the attention to the forms of oral communication of the slaveholders as well as those of the enslaved—the aim has not simply been to isolate speech as a particular, and previously neglected, strand of the complex history of Caribbean slavery. Examining speech historically and geographically can do more than that. As a crucial form of everyday practice (understood as contextual, embodied, located, and practiced speech acts that animate broader modes of veridiction) talk is a way into important questions of identity, social relations, representation, violence, and materiality—of power and resistance—that are debated across the humanities and social sciences. Studying speech opens up a world of ongoing practical activity in the making and unmaking of slavery and freedom that is both radically asymmetrical and always and inevitably dialogic. This book does not claim that the history of slavery can be reduced to a history of orality, but considering talk as practice, or as many differentiated practices, can offer a way of rethinking that history—and the histories of colonialism and empire more broadly—that avoids the problems of historical accounts that start, either overtly or not, from the primacy of economic relations, textual or visual representations, or acts of violence. Attending to speech, with its ubiquity and ephemerality, and its absent presence in the archive, is to be forced to reckon with the always ongoing and partial construction of worlds by, for, and between people.

A central element of this attention to talk is to insist on the fundamental importance of speech for European imperial and colonial powers, and for the enslavers, planters, merchants, and others who transformed lives and landscapes elsewhere in the world. Regardless of their investment in literacy, it is important to recognize that empires were oral cultures too. Here, as the book's chapters have shown, Enlightenment definitions of the human, at once universal and structured by racial difference, informed notions of the freedom of speech that were embedded in gendered and nationally defined senses of the liberties of the freeborn Englishman. In turn, these Englishmen's liberties were worked out through definitions of who might speak legally and politically to bind utterances to subjects and form sovereign and representative collectives. Ideas of who was free to speak also shaped speaking subjects in the arenas of natural knowledge and spiritual practice: constructing gentlemanly conversationalists and public speakers who could make matters of fact about nature across Atlantic worlds; and preaching and praying clergymen who could bring God to the sugar islands through rational Christianity. Freedom was formed through speech, and empires were made as words were spoken.

These speech acts required communication across forms of difference. The preceding chapters have been attentive to questions of exchange, interaction, and engagement, even under the radical asymmetries of plantation slavery, evident in the necessary recognition, in different ways, by all people on these islands that others were oath-taking subjects, and in the ways in which those speech practices might be made commensurate—as in treaties with the Maroons—as well as condemned or denied. The book has also demonstrated that conspiratorial speech was heard as the utterances of deliberative political subjects; the healer–patient dialogue as a mode of communicating natural knowledge; and the "broken Accents" of black Christians as evidence of a spirituality shared with nonconformists and evangelicals. Indeed, as the chapter on abolition and emancipation has shown, speech formed the ground on which action in "the slave's cause" might be built, negotiating difference in attempts to bring freedom.

Yet difference remains. The approach to speech taken here has emphasized distinct sorts of speech practices and speech acts by using Bruno Latour's delineation of "modes of existence" to identify and differentiate the ways speech worked in law, politics, science, and religion, and showing how freedom and slavery were produced differently in each.[6] Yet it also pays much closer attention than Latour does to questions of race, gender, and social position to differentiate between speech practices within these modes. In part, this attention has been drawn to demonstrate the different, and unequal, ways in which voices might be raised and heard. For women, both black and white, this often meant being silenced—in the islands' assemblies, for example, or as participants in gentlemanly botanical conversation, or as preachers—or having to negotiate heavily circumscribed ways of speaking that were also shaped by race, freedom, and social position: as participants in conspiratorial talk, through household medicine, at Methodist "love feasts," or as abolitionist activists in public or around the tea table. These inequalities were true of the speech practices that traced their routes through the black Atlantic as well as those that were shaped by an imperial geography. Each chapter has endeavored to show how in speaking in all these domains, enslaved Africans might draw on other speech practices that bound utterances to their speakers, formed collectives, constructed chains of reference, and brought transformative beings close. Indeed, the existence of these modes of speech—often inadequately represented under the single term "obeah"—and their transgression of the boundaries "the Whites" sought to establish between the domains, and modes of veridiction, of law, politics, science, and religion, both made possible forms of exchange (as in the treaties with the Maroons or forms of syncretic religion) and provided the basis for alternative ways of understanding freedom and slavery (which included other ways of comprehending the relationships between humans, other animals, and spirit beings that questioned dominant assumptions about who might speak, and to what purpose).

So, despite multiple forms of silencing—that were ultimately, and sometimes immediately, backed by deadly violence—speech could never simply be contained to produce racialized notions of freedom and bondage. There was a freedom of speech in that its necessity, ubiquity, and ephemerality opened the possibility that the unutterable might be uttered, and that things could change if others spoke. It was there in Francis Williams's insistence that he should be heard, and his father's refusal to be a "black negro"; in conspiratorial imaginings of political futures that drew on different notions of collectives and their common ground; in assertions of the knowledge and power of the enslaved to effect cures; in challenges to wealth and power in this world on the basis of the claims of other spirits; and in voices raised against slavery and the slave trade around the Atlantic.

Taken together, these close historical and geographical examinations of talk and slavery can rework the more familiar, but continually contested, notion of the freedom of speech. While "free speech" can certainly be asserted as a foundational principle, it always, in the end, has to be enacted in practice as contextual, embodied, and located speech acts that draw on particular forms of talk. It has to deploy speech practices that have power-laden histories and geographies—particularly of race, gender, and status—that carry meanings and forms of action into the world. This means that, in practice, the exercise of "free speech" can simultaneously be experienced and understood as an expression of liberty and an illegitimate exertion of power that binds others. *The Freedom of Speech* argues for closer attention to the histories and geographies of the practicalities of speech—in this and other contexts, and at every scale from the face-to-face to the global—to demonstrate how speech practices are embedded in long histories and deep geographies of power and conflict that shape their meanings, and which have to be part of discussions over who speaks, where, and how. Doing so here has meant working across the ground between more familiar, and often polarized, positions on the relationship between speech and slavery: silencing through violence and the abundant proliferation of creolized linguistic forms. One final attempt to understand speech in practice on these islands—some last words—demonstrates what working this way can reveal.

Violence and silence pervade a sketch by the Swiss-born artist and collector Pierre Eugene du Simitière of an execution during the Jamaican slave rebellion of 1760 (figure 19).[7] It depicts a gibbet erected in Kingston's "Savannah," a rare open space within the city's tight grid (figure 20). From the fatal tree hang two metal cages, shaped into human form, into which the condemned prisoners—named as Fortune and Kingston—were welded by a blacksmith. Each man, standing on a small board, had his hands shackled across his chest and was encased in iron bands. They were then suspended side by side to die of thirst in the public square, however long that took.[8] The sketch also shows a carter, whip in hand, looking up at the gallows; a woman—probably a higgler selling ground provisions—seated to the left; a cloaked figure standing to the right; and, on the porch of their watch-house, armed soldiers from the militia company assigned to guard the execution site and the magazine.

The slave rebels had, in rising up against the many forms of violence inflicted on them, brought their own terror to Jamaica. Fortune and Kingston were executed in May 1760 for their part in the revolt's initial stages, particularly the killing of the overseer and other white men on Esher plantation, in St. Mary (*HJ*, 2:448–49). Their execution, while the revolt was still in motion, aimed, in turn, to terrorize those who had taken up arms and those who might have thought to join them.[9] As the planter historian

Figure 19. The killing of Fortune and Kingston, 1760. Sketch by Pierre Eugène du Simitière, Library Company of Philadelphia, du Simitière Papers, Series VI, Papers Relating to the West Indies [968], box 4, folder 19h. Reprinted by permission of the Library Company of Philadelphia.

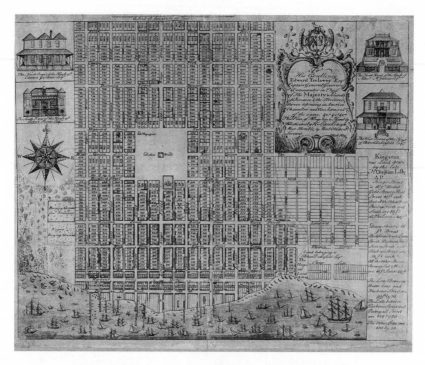

Figure 20. Kingston, Jamaica, 1745. Michael Hay, *Plan of Kingston*, 1745. Reprinted by
permission of the Library of Congress, Geography and Map Division.

Bryan Edwards had it, "In countries where slavery is established, the leading principle on which the government is supported is fear: or a sense of
that absolute coercive necessity which, leaving no choice of action, super-
sedes all questions of right" (*HCC*, 36). This regime of fear and violence
dealt in death. As du Simitière recorded below his sketch, one night a
member of the "mulatto" militia company saw a figure moving across the
square. Repeatedly calling out "Who goes there?" he received no response.
Fearing the worst, he shot and killed the nightwalker, who turned out to be
an unfortunate "new Negro," perhaps still unable to understand English,
who had come into the city with a bundle of wood to sell.[10] It was just
another black life taken among so many as the planters' courts condemned
captured rebels to be burned, hanged, and decapitated, with their heads
displayed on poles. In the square, Fortune and Kingston died slowly as a
warning to others—one lasted seven days, the other nine—and it was said
that they "scarcsely Complain'd."[11] Once dead, their bodies were displayed
at major crossroads to all who passed along the roads to and from Rock-
fort, to the east, and Half-way Tree, to the north.

As I have argued, the counterpart of this violent silencing of enslaved black men and women is the invocation of white speech. This invocation involved both the right to speak and the delineation of appropriate forms of speech so that, in the eyes of the planters themselves, the system of slavery that they built was not simply a world of shouted orders and proclamations of death. These forms of speech—oath taking, evidence giving, political deliberation, polite conversation, speeches, sermons, and prayers—had to be continually made and remade to guarantee the identities, relationships, and modes of freedom that they underpinned. This was a long process. Thomas Walduck, no doubt exaggerating for effect, depicted seventeenth-century Barbados as "a Babel of all Nations and Conditions of men" whose bad language defined who they were. For him, "The English they brought with them drunkenness & Swearing[,] the Scotch Impudence & Falshood, The Welch Covetousness & Revenge[,] The Irish Cruelty & Perjury[,] the Dutch & Deans Craft & Rusticity and the French Dissimulation & Infidelity and here they have Intermarr'd & blended together that of what Quality this present Generation must be of you shall Judge." Yet, despite their differences they were all, Walduck wrote, "unmercifully cruel to their poor slaves," and, in time, they all became white and free in opposition to the enslaved "Negro."[12] As such, they all came to share an investment in forms of speech that defined liberty against slavery, and opposed civility to barbarity, and white to black, even in the face of evidence to the contrary and critique from the metropole.

So, when it came to justifying slavery in the age of abolition, speech was crucial. Edward Long, William Clark, and many others imagined the whip being replaced by the carefully chosen word (see figure 1 in the introduction). In another of Clark's proslavery images of Antigua, published after the end of the slave trade, the sugar boiling house—represented in the abolition debate as a hellhole of violence, cruelty, and death—becomes a space of calm, efficient labor (figure 21).[13] In this print, boiling sugars are stirred, poured, and evaluated by an attentive, and predominantly male, labor force and their masters in a light and well-ordered setting. In the background, a seated white man in a frock coat and top hat, casually, and with an open-handed gesture, explains something—presumably about the quality and worth of the product for market—to an enslaved worker whose scales speak not of justice in the balance but of comparative value and monetary profit (figure 22). White and black are juxtaposed, and rendered unequal, through speech and silence, but the image is one of productive commodity capitalism, of orderly and paternalist social relations, and not of violence and bondage.

Figure 21. *Interior of a Boiling House,* from William Clark, *Ten Views in the Island of Antigua* (London, 1823). Courtesy of the John Carter Brown Library at Brown University.

Figure 22. Boiling house talk. Detail from William Clark, *Interior of a Boiling House,* 1823. Courtesy of the John Carter Brown Library at Brown University.

Alongside these images of labor and commodity production were pro-slavery visions of reproduction that focused on women and imagined the sugar islands as populated by a black peasantry. These visions were as much a part of the "missionary dream," and indeed the evangelical dream, for the Anglo-Caribbean as they were elements of a continued investment in slavery.[14] For example, the artist William Berryman, who lived in Jamaica from 1808 to 1815 and probably worked as a bookkeeper on various Clarendon plantations, presents pastoral landscapes, wooded scenes, images of people traveling, and sketches from life of individual men and women, character types, or something in between. Where work is depicted, it is primarily the work of reproduction, and often of women.[15] One tiny sketch, in pencil and watercolor, has been given, by its cataloguers at the Library of Congress, the title *Four Jamaicans Talking at a Fence* (figure 23). Its figures, three black adults and a girl, her elbows raised and with her back to the viewer, are arrayed along a simple fence. One of the women sits, leaning forward, her arm draped over the top rail. The other two adults seem to attend to her, although the faces are either obscured or impossible to discern, so whether they are talking or not is hard to determine. There is no background. The picture, one among hundreds in Berryman's portfolio, suspends these people and the fence in space and time. The black figures relate only to one another, existing in a moment of everyday interaction

Figure 23. *Four Jamaicans Talking at a Fence.* Pencil and watercolor sketch by William Berryman, 1808–1815. Reprinted by permission of the Library of Congress, Prints and Photographs.

that speaks not of production and plantation slavery but of the reproduction of Jamaican rural life.

While both these images, in their depictions of talk, present an amelio-rationist version of plantation slavery from the early nineteenth century, they also point to something other than the silence that dominates Du Simitière's depiction of the execution of Fortune and Kingston. The ideal-ized figures in Clark's boiling house are a reminder that there was talk to be heard between black and white. The killings for which the men were ex-ecuted were conducted face-to-face. They were an extension of the intimacy of everyday plantation violence.[16] When the enslaved marched on Esher plantation, after exchanging gunfire, the overseer and four other white men barricaded themselves into a room in the house. As one of them, Robert Gordon, who had only been in Jamaica seven months, later recalled, the rebels told them through the door that they would be killed if they re-sisted but would be safe if they came out and laid down their arms. After two of them "ventured out" and were killed, the rebels stormed the room. Judging that all the white men were dead, and after discussing whether to cut the head from Gordon's body, the rebels left him lying in the planta-tion yard, where he was later discovered and nursed back to health. His face was, however, so distinctively scarred that he was thereafter known as "Coromantee Gordon."[17]

Yet these were only some among many deliberations, negotiations, and decisions on life and death during the 1760 revolt. It became a well-known story that "the Esher negroes" were so "strongly attached" to another Gor-don—John Gordon—that "when overtures were first made to them in the conspiracy they refused except on the Express condition that the life of that Gentleman should be spared." When matters took a turn for the worse for the rebels in St. Mary, especially after the killing of their leader Tacky by one of the Scot's Hall Maroons, many sought to find ways to surrender. As Edward Long reported, "the Esher negroes" then "contrived to send an Embassy" to Gordon, "professing they imposed implicit confidence in his honour." He, knowing that they had argued for his life, "was not afraid to trust his person in a congress with their Leaders" and heard them express "their readiness to submit and suffer death by hanging, provided he would assure them of being exempted from Torture."[18] Retold by Long, and later inscribed on a memorial to Gordon in England, this mutual trust in each other's words, and these reciprocal sovereign interventions, became a one-sided act of paternalism. It was said that Gordon was able to secure from the lieutenant governor not only the guarantee that they would escape the bodily mutilations that Esher's black population may have seen as threatening their passage between worlds, but also sentences of transportation from the island rather than death.

There was, however, one exception, and that was the rebel called Kingston. He had been among the group that appealed to John Gordon, but, having been charged with assisting in the murder of the overseer at Esher, he "was reserved for death in pursuance of a Determination that none of them guilty of shedding Blood should be pardoned." He was parted from his companions and taken to the city that shared his name to be gibbeted alongside Fortune. While Long noted in his *History of Jamaica* that "the murders and outrages they had committed, were thought to justify this cruel punishment inflicted upon them *in terrorem* to others" and reasoned "that they appeared to be very little affected by it themselves; behaving all the time with a degree of hardened insolence, and brutal insensibility" (*HJ*, 2:458n), he was later forced to admit that, in the case of Kingston, "it afterwards but too late appeared that he was entirely innocent of the murder."[19]

Speech and silence were both evident in the space of death. One observer reported that Kingston, who lived the longer of the two, "kept his Speech and Senses."[20] Bryan Edwards, who was there too—seventeen years old and recently arrived from England—related, as part of his discussion of the character of the "Koromantyns," the execution of another man captured along with Fortune and Kingston, who "uttered not a groan" as he "saw his legs reduced to ashes" but threw a lighted brand from the fire in the face of his executioner. In contrast, Kingston and Fortune seemed to want to speak to their tormentors. Edwards reported that, on the seventh day of their ordeal, one of them "wished to communicate an important secret to his master," Edwards's uncle and benefactor Zachary Bayly. Bayly being in St. Mary, Edwards was sent for, and "endeavoured, by means of an interpreter, to let him know that I was present; but I could not understand what he said in return. I remember that both he and his fellow sufferer laughed immoderately at something that occurred,—I know not what." If there had been a secret to be told, it was not shared. Instead, the two dying men laughed at something the white man could not understand and communicated nothing, an act Edwards put down to the "savage and sanguinary" manners of such people (*HCC*, 78–80).

But all was not silence and incomprehension. Edwards also reported that while suspended in their iron cages to die, the two men had "never uttered the least complaint, except only of cold in the night." Instead, they had "diverted themselves all day long in discourse with their countrymen, who were permitted, very improperly to surround the gibbet" (*HCC*, 79). They did not die alone or in silence but as part of a collective gathered through speech. It is impossible to say for sure, but might the carter in Du Simitière's sketch be speaking to the condemned rather than observing their deaths? Might the higgler and the cloaked figure be listening to what they had to say? We cannot know, but even this tiny fragment of evidence of the "discourse with their countrymen" turns a depiction of violence and silence into one that must also be of speech. Like the four Jamaicans talking at a fence, it opens the question of what the presence of speech might mean even when it cannot reveal what was said. It cannot be known what words were spoken, how they were spoken, or even who other than the dying men spoke, only that they and others did speak. Such evidence of talk is, as I have argued, poised between silence and the proliferating "miracles of creolization" in Caribbean speech and other sounds.[21] Such speech acts demonstrate how the contradictions of racialized slavery can be heard in a version of the freedom of speech which acknowledges both that freedom and slavery were made by regulating speech and that such constraints could never, finally, be imposed on something at once so fleeting, so multiple, and so powerful.

Fortune and Kingston had an afterlife. Du Simitière recorded that ten years after their executions, there was no trace at all of their bones at the crossroads, only the empty iron cages that had once held them swinging in the Caribbean wind. It was said, he wrote, that "the negroes who passed by there had carried them away as relics."[22] There is, however, no record of what words were said over these sacred objects by those who still lived.

ACKNOWLEDGMENTS

This book has taken a while to write, but then again, books should. It would not have been possible without the support of the Leverhulme Trust, Queen Mary University of London (QMUL), and the many people who have helped me along the way. The Leverhulme Trust's award of one of the inaugural Philip Leverhulme Prizes in 2001 allowed me to start the project with only the sketchiest of ideas. A Leverhulme Major Research Fellowship in 2014 provided time to finish working in the archives and write a first full draft. Sabbatical leave from QMUL also provided much-needed time to research and write.

I have presented the ideas and material here in many places, and I thank all those who came to listen and talk. Particularly intensive discussions were made possible through extended visits to the Rutgers Center for Historical Analysis and Rutgers Center for British Studies in November 2012, organized by James Delbourgo and Toby Jones, and to the Max Planck Institute for the History of Science, Berlin, in March 2017, arranged by Ohad Parnes. For other welcome invitations to cross borders to talk about this work, I would like to thank Romain Bertrand, John Brewer, Luisa Calè, Adam Clulow, Adriana Craciun, Mona Domosh, Karen Morin, Ravi Sundaram, Mary Terrall, and Stephane van Damme. One early seminar at the University of the West Indies in Mona, Jamaica, in March 2003, was particularly important, as it was where I was gently told that working on script and print in the Caribbean would not do and that I had to attend as closely to orality. I thank Susan Mains, Steeve O. Buckridge, and Elizabeth Thomas-Hope for being the catalysts for what followed.

For turning all these words into print, I'd like to thank Christie Henry for her guidance, and Priya Nelson, my editor at the University of Chicago Press, for helping me through the review process. At Chicago, Dylan

Montanari got the final manuscript into good order, and I am very grateful to Lori Meek Schuldt for her careful copyediting. Tamara Ghattas in the manuscript editing department and Kristen Raddatz in marketing have skillfully ushered this book out into the world. Thanks are also due to the publisher's two readers for their appreciation of what I was trying to do, and for critical comments on both the proposal and the manuscript that made me try harder to do it better. Parts of chapter 1 have been previously published as "The Power of Speech: Orality, Oaths and Evidence in the British Atlantic World, 1650–1800," *Transactions of the Institute of British Geographers* 36, no. 1 (2011): 109–25, and are reprinted by permission of John Wiley and Sons. Much of chapter 3 previously appeared as "Talking Plants: Botany and Speech in Eighteenth-Century Jamaica," *History of Science* 51, no. 3 (2013): 251–82, copyright © 2013 by Science History Publications Ltd., and is reprinted here by permission of SAGE Publications, Ltd.

I am lucky to have colleagues (past and present) in the School of Geography at QMUL who make working there a pleasure. My three years as Head of School (2011–14) may have interrupted research on this book, but staff and students made sure it did not stop me thinking! In particular, I have been helped in various ways by Alison Blunt, Tim Brown, Kavita Datta, Anna Dulic-Sills, Shereen Fernandez, Kerry Holden, Al James, Azeezat Johnson, Regan Koch, Roger Lee, Simon Lewis, Caron Lipman, Andrew Loveland, Jon May, Cathy McIlwaine, Will Monteith, Jenny Murray, Alastair Owens, David Pinder, Simon Reid-Henry, Olivia Sheringham, Adrian Smith, Marta Timoncini, and Kathryn Yusoff. Special thanks go to Edward Oliver for drawing the maps (figures 5 and 6 in the introduction), and for calm technical help with images. I have also been fortunate to work with a great group of PhD students in recent years. Thank you to Lamees Casling, Elin Jones, Anna Kretschmer, Shaun Owen, Victoria Pickering, Tim Riding, Hannah Stockton, and Gabriel Wick.

The research would not have been possible without the expertise and assistance of staff at the Barbados Archives in Black Rock, the Barbados Public Library, the Barbados Museum and Historical Society, the Bodleian Library, the Boston Public Library, the British Library, the Historical Society of Pennsylvania, the Jamaica Archives (Spanish Town), the John Carter Brown Library (particularly the director Neil Safier), Lambeth Palace Library, the Library Company of Philadelphia (particularly Krystal Appiah), the Library of Congress (particularly Martha Kennedy), the Library of the Society of Friends (particularly Melissa Atkinson), the National Archives (Kew), the National Library of Jamaica (particularly Yvonne Fraser-Clarke), the Royal Society Archives (particularly Joanna Corden and Robert Baker), and the Library of the School of Oriental and African Studies.

My interpretations of this material are the result of many conversations. For discussions on Jamaican plants and much more, I thank the *Reconnecting Sloane* team and fellow travelers: Charlie Jarvis, Victoria Pickering, Alice Marples, Felicity Roberts, Anne Goldgar, Kim Sloan, Arnold Hunt, Lizzie Eger, Martha Fleming, Julie Harvey, Lisa Smith, Mark Spencer, and Alison Walker. For discussing other matters botanical, and beyond, I thank Daniela Bleichmar, Sarah Easterby-Smith, Kate Murphy, Emily Senior, April Shelford, Emma Spary, and, at the Max Planck Institute, Elaine Leong, Marius Buning, Clare Griffin, and Hansun Hsiang. On matters Caribbean I have benefited from talking with Jean Besson, Ian Foster, Joy Gregory, Gad Heuman, Peter Hulme, Dian Kriz, Patricia Noxolo, Christer Petley, Sarah Thomas, and Nuala Zahedieh. James Robertson, Susan Mains, and Andy Taitt provided intellectual stimulation and hospitality during visits to Jamaica and Barbados. Other wise and useful words have come from Bob Batchelor, Tita Chico, Zirwat Chowdhury, Zoe Laidlaw, Chris Reid, and Simon Schaffer.

My research and writing continue to be enriched by enduring collective arenas for discussion. Convivial conversation has been provided by the London Group of Historical Geographers seminar series at the Institute of Historical Research, and my fellow convenors, Ruth Craggs, Felix Driver, and Innes Keighren, and by the Queen Mary Centre for Eighteenth-Century Studies seminar series, convened by Alice Dolan, Markman Ellis, Colin Jones, Jessica Patterson, Barbara Taylor, and Amanda Vickery. Over the last few years, in what has come to be called the Edward Long Reading Group, Markman Ellis, Catherine Hall, Silvia Sebastiani, and Diana Paton have provided a rare opportunity for sustained and open-ended discussion of race and slavery, and, along with Vincent Brown, James Delbourgo, and Julie Kim, they have read and commented on much of this book, to my great benefit. Indeed, in all these places and others, academic discussion and friendly talk begin to blur. I'd like to thank Catherine, James, Markman, Silvia, Julie, Vince, Di, Innes, Ruth, Felix, Amanda, Barbara, Colin, Alice, and Jessica, along with Michael Bravo, Stephen Daniels, Jon Dennis, Mona Domosh, Margot Finn, Christian Frost, John Horan, David Lambert, Steve Legg, John Maher, Scott McCracken, Lynn Nead, Hester Parr, Chris Philo, Graham Rinaldi, Nicholas Robson, Charlie Smith, Phil Stern, and Charlie Withers for talking about many things other than talk.

I am grateful for the support of those I am fortunate enough to call family: Jon and Claudette; Kate, Nick, and Cora; Harriet, Andrew, Matthew, and Louis; Graínne and David; Bridgit, and memories of stories and laughter with David too; Claire, Sue, and Ada; and Dot, Belle, and Annie. This book is dedicated to my mother, Jane, who first taught me words and showed what they could do; to my daughter, Eve, who learned to speak in a fraction of the time this book took to write, and now weaves poetic magic; and, most of all, to Catherine Nash, who has a way with words I can never match. Thank you, my darling, for all you say and all you do.

NOTES

INTRODUCTION

1. Michel-Rolph Trouillot, *Silencing the Past: Power and the Production of History* (Boston: Beacon Press, 1995); and Saidiya V. Hartman, "Venus in Two Acts," *Small Axe* 26 (2008): 1–14.

2. The National Archives, Kew (hereafter cited as TNA), CO1/53, America and West Indies, Colonial Papers, Extract of a Letter from Barbados, December 18, 1683, f.264v.

3. Trevor Burnard, *Mastery, Tyranny, and Desire: Thomas Thistlewood and His Slaves in the Anglo-Jamaican World* (Chapel Hill: University of North Carolina Press, 2004).

4. Beinecke Library, Yale University (hereafter cited as BLYU), OSB MSS 176/3, Thistlewood Diary, 1752, p. 250 (December 27, 1752).

5. Thistlewood Diary, 1752, pp. 247–49.

6. Thistlewood Diary, 1752, p. 251; and Thistlewood Diary, 1753, p. 4 (January 5, 1753), OSB MSS 176/4, BLYU. Douglas Hall, *In Miserable Slavery: Thomas Thistlewood in Jamaica, 1750–86* (Kingston: University of the West Indies Press, 1989), p. 56, argues that Sam was acquitted. However, Thistlewood only records "Sam" being returned to Egypt on January 10, 1753 (Thistlewood Diary, 1753, p. 7).

7. John Lean and Trevor Burnard, "Hearing Slave Voices: The Fiscal's Reports of Berbice and Demerara-Essequibo," *Archives* 27 (2002): 37–50; and Marisa J. Fuentes, *Dispossessed Lives: Enslaved Women, Violence, and the Archive* (Philadelphia: University of Pennsylvania Press, 2016).

8. Natalie Zacek, "Voices and Silences: The Problem of Slave Testimony in the English West Indian Law Court," *Slavery and Abolition* 24, no. 3 (2003): 24–39, at p. 26. See also Trevor Burnard, *Hearing Slaves Speak* (Caribbean Press, 2010); Dorothy Couchman, "'Mungo Everywhere': How Anglophones Heard Chattel Slavery," *Slavery and Abolition* 36, no. 4 (2015): 704–20; and Shane White and Graham White, *The Sounds of Slavery: Discovering African American History through Songs, Sermons, and Speech* (Boston: Beacon Press, 2005).

9. *Acts of Assembly, Passed in the Island of Jamaica; From 1681 to 1737* (London, 1738), p. 78.

10. Morgan Godwyn, *The Negro's and Indian's Advocate* (London, 1680), pp. 40, 29–30.

11. *The Importance of Jamaica to Great-Britain, Consider'd . . .* (London, [late 1730s]), p. 17. On the forms of "Black geographies," see Katherine McKittrick, *Demonic*

Grounds: Black Women and the Cartographies of Struggle (Minneapolis: University of Minnesota Press, 2006).

12. Robin Blackburn, *The Making of New World Slavery: From the Baroque to the Modern, 1492–1800* (London: Verso, 1998), pp. 12–13.

13. Walter Johnson, "To Remake the World: Slavery, Racial Capitalism, and Justice," *Boston Review*, October 26, 2016; and Saidiya V. Hartman, *Scenes of Subjection: Terror, Slavery, and Self-Making in Nineteenth-Century America* (New York: Oxford University Press, 1997), p. 6.

14. Walter Johnson, *River of Dark Dreams: Slavery and Empire in the Cotton Kingdom* (Cambridge, MA: Belknap Press of Harvard University Press, 2013); and Sowande' M. Mustakeem, *Slavery at Sea: Terror, Sex, and Sickness in the Middle Passage* (Urbana: University of Illinois Press, 2016).

15. [Edward Long], *The History of Jamaica*, 3 vols. (London, 1774), 2:351, 353, 356 (hereafter cited as *HJ*).

16. Silvia Sebastiani, *The Scottish Enlightenment: Race, Gender, and the Limits of Progress* (Basingstoke, UK: Palgrave Macmillan, 2013), pp. 66–67.

17. British Library, Additional Manuscripts (hereafter cited as BL Add. MS), 12405, f.294r.

18. BL Add. MS 12405, ff.291r and 293r; and Charles Burney, *A General History of Music, from the First Ages to the Present Period*, 4 vols. (London, 1776), 1:304.

19. BL Add. MS 12405, f.281r (diagrams at f.285r), quoting Jean-Jacques Rousseau, *A Discourse on the Origin and Foundations of Inequality among Men* (London, 1761), pp. 229–30, n. 10.

20. For fuller discussions of Long and Monboddo, see Miles Ogborn, "Discriminating Evidence: Closeness and Distance in Natural and Civil Histories of the Caribbean," *Modern Intellectual History*, 11, no. 3 (2014): 629–51; and Silvia Sebastiani, "Challenging Boundaries: Apes and Savages in Enlightenment," in *Simianization: Apes, Gender, Class and Race*, ed. Wulf D. Hund, Charles W. Mills, and Silvia Sebastiani (Zurich: Lit, 2015), pp. 105–38.

21. [James Burnet, Lord Monboddo], *Of the Origin and Progress of Language*, 6 vols. (Edinburgh, 1773–92), vol. 1, pp. 1–2, 174–75, 197, 238–39, 272. See also Alan Barnard, "*Orang Outang* and the Definition of Man: The Legacy of Lord Monboddo," in *Fieldwork and Footnotes: Studies in the History of European Anthropology*, ed. Hans F. Vermeulen and Arturo Alvarez Roldan (London: Routledge, 1995), pp. 95–112.

22. *HJ*, 2:370–71 n. f. *Mutum pecus* is from Horace.

23. Roxann Wheeler, *The Complexion of Race: Categories of Difference in Eighteenth-Century British Culture* (Philadelphia: University of Pennsylvania Press, 2000), pp. 209–20; and Suman Seth, "Materialism, Slavery, and *The History of Jamaica*," *Isis*, 105 (2014): 764–72.

24. Sylvia Wynter, "Unsettling the Coloniality of Being/Power/Truth/Freedom: Toward the Human, after Man, Its Overrepresentation—An Argument," *New Centennial Review*, 3, no. 3 (2003): 257–337, at p. 266; and Christopher L. Brown, *Moral Capital: Foundations of British Abolitionism* (Chapel Hill: University of North Carolina Press, 2006), pp. 33–101. For Wynter, the definition of the human through "languaging" remains a crucial political issue as she aims to define humans "as a hybrid-auto-instituting-languaging-storytelling species" (*homo narrans*) with African origins against other "origin myths"; see Sylvia Wynter and Katherine McKittrick, "Unparalleled Catastrophe for Our Species? Or, to Give Humanness a Different Future: Conversations," in *Sylvia Wynter: On Being Human as Praxis*, ed. Katherine McKittrick (Durham, NC: Duke University Press, 2015), pp. 9–89, at p. 25.

25. Godwyn, *Negro's and Indian's Advocate*, p. 12.

26. *A Letter from A Merchant at Jamaica to a Member of Parliament in London Touching the African Trade* (London, 1709), p. 29. See Jack P. Greene, "'A Plain and Natural Right to Life and Liberty': An Early Natural Rights Attack on the Excesses of the Slave System in Colonial British America," *William and Mary Quarterly* 62, no. 4 (2000): 793–808.

27. For example, Philotheos Physilogus [Thomas Tryon], *Friendly Advice to the Gentleman-Planters of the East and West Indies* (London, 1684) includes "A DISCOURSE In way of *Dialogue*, Between an *Ethiopian* or *Negro-Slave* And a CHRISTIAN, That was his *Master* in *America*," pp. 146–222; see also Anonymous, "The Speech of Moses Bon Sàam (1735)," in *Caribbeana: An Anthology of English Literature of the West Indies, 1657–1777*, ed. Thomas W. Krise (Chicago: University of Chicago Press,1999), pp. 101–7. For a proslavery version, see Thomas W. Krise, "True Novel, False History: Robert Robertson's Ventriloquized Ex-Slave in *The Speech of Mr John Talbot Campo-bell* (1736)," *Early American Literature* 30, no. 2 (1995): 153–64.

28. Hume's essay "Of National Characters" (1748) is discussed in Aaron Garrett and Silvia Sebastiani, "David Hume on Race," in *The Oxford Handbook of Philosophy and Race*, ed. Naomi Zack (Oxford: Oxford University Press, 2017), pp. 31–43.

29. BL Add. MS 12405, f.281r.

30. Suman Seth, *Difference and Disease: Medicine, Race and the Eighteenth-Century British Empire* (Cambridge: Cambridge University Press, 2018).

31. Ruth Paley, "After *Somerset*: Mansfield, Slavery and the Law in England, 1772–1830," in *Law, Crime and English Society*, ed. Norma Landau (Cambridge: Cambridge University Press, 2002), pp. 165–84. Lord Monboddo was in the minority objecting against a similar judgment in a later case in Scotland, that of James Knight; see Sebastiani, *Scottish Enlightenment*.

32. C. Brown, *Moral Capital*, pp. 155–206.

33. A Planter [Edward Long], *Candid Reflections upon the Judgement Lately Awarded by the Court of King's Bench, in Westminster-Hall, on What Is Commonly Called the Negroe-Cause* (London, 1772), p. 41; and A West Indian [Samuel Estwick], *Considerations on the Negroe Cause, Commonly So Called, Addressed to the Right Honourable Lord Mansfield* (London, 1772).

34. Samuel Estwick, *Considerations on the Negroe Cause, Commonly So Called, Addressed to the Right Honourable Lord Mansfield*, 2nd ed. (London, 1773), pp. 71–72, 74, 80.

35. [Estwick], *Considerations on the Negroe Cause*, 1st ed., p. 7. This was certainly how Sharp read these pamphlets and Long's *History*; see York Minster Archives COLL 1896/1 Granville Sharp Letterbook, 1768–1773, 1793, Sharp to Colonel James of Jamaica, London, April 8, 1773, pp. 181–84; and Prince Hoare, *Memoirs of Granville Sharp, Esq* (London, 1820), appendix IV.

36. BL Add. MS 12405, f.288v.

37. Noël Antoine Pluche, *Spectacle de la nature; or, Nature Displayed* (London, 1733).

38. On creole "loquaciousness," see also Bryan Edwards, *The History Civil and Commercial of the British Colonies of the West Indies*, 3rd ed., vol. 2 (London, 1801), p. 100 (hereafter cited as *HCC*).

39. On these anxieties, see Brooke N. Newman, "Gender, Sexuality, and the Formation of Racial Identities in the Eighteenth-Century Anglo-Caribbean World," *Gender and History* 22, no. 3 (2010): 585–602.

40. Long considered removing this description, if not the argument, from a later edition, BL Add. MS 12405, f.223.

41. On Humean conversation in England, see Michèle Cohen, *Fashioning Masculinity: National Identity and Language in the Eighteenth Century* (London: Routledge, 1996); and Jon Mee, *Conversable Worlds: Literature, Contention, and Community 1762 to 1830* (Oxford: Oxford University Press, 2011).

42. *HJ*, 2:270–71; and BL Add. MS 12405, f.217. Granville Sharp, *A Representation of the Injustice and Dangerous Tendency of Tolerating Slavery in England* (London, 1769), p. 13 argued that asserting property rights over the enslaved treated "these their *fellow men*, as if they thought them, mere *things*, horses, dogs, &c."

43. Vincent Brown, *The Reaper's Garden: Death and Power in the World of Atlantic Slavery* (Cambridge, MA: Harvard University Press, 2008); and Vincent Brown, "Social Death and Political Life in the Study of Slavery," *American Historical Review* 114, no. 5 (2009): 1231–49.

44. Library Company of Philadelphia (hereafter cited as LCP), du Simitière Papers, Series VI, Papers Relating to the West Indies [968], box 4, folder 10e f.1r. I am grateful to Neil Safier and Ernesto Mercado-Montero for their translations.

45. Toni Morrison, "Unspeakable Things Unspoken: The Afro-American Presence in American Literature," *Michigan Quarterly Review* 28, no. 1 (1989): 1–34; Zora Neale Hurston quoted in Sandra M. Gilbert and Susan Gubar, *No Man's Land: The Place of the Woman Writer in the Twentieth Century*, vol. 3, *Letters from the Front* (New Haven, CT: Yale University Press, 1994), p. 123. See also Patricia A. Broussard, "Black Women's Post-Slavery Silence: A Twenty-First Century Remnant of Slavery, Jim Crow and Systemic Racism—Who Will Tell Her Stories?" *Journal of Gender, Race and Justice* 16 (2013): 373–421.

46. As quoted in Jacques Rancière, *Dis/agreement: Politics and Philosophy*, trans. Julie Rose (Minneapolis: University of Minnesota Press, 1999), p. 1.

47. Jacques Rancière, *The Politics of Aesthetics: The Distribution of the Sensible*, trans. Gabriel Rockhill (London: Continuum, 2004), p. 12.

48. Edouard Glissant, *Caribbean Discourse: Selected Essays*, trans. J. Michael Dash (Charlottesville: University Press of Virginia, 1989), pp. 123–24.

49. Hartman, "Venus in Two Acts"; and Fuentes, *Dispossessed Lives*, pp. 142–43.

50. Richard Cullen Rath, *How Early America Sounded* (Ithaca, NY: Cornell University Press, 2003), pp. 77–80.

51. Erving Goffman, *Forms of Talk* (Oxford: Basil Blackwell, 1981).

52. [Monboddo], *Of the Origin and Progress of Language*, vol. 6, p. 228.

53. [Edward Long], *The Prater: By Nicholas Babble, Esq.* 2nd ed. (London, 1757).

54. Important examples are Laura Gowing, *Domestic Dangers: Women, Words and Sex in Early Modern London* (Oxford: Oxford University Press, 1999); Sandra M. Gustafson, *Eloquence Is Power: Oratory and Performance in Early America* (Chapel Hill: University of North Carolina, 2000); Jane Kamensky, *Governing the Tongue: The Politics of Speech in Early New England* (New York: Oxford University Press, 1997); and Terri L. Snyder, *Brabbling Women: Disorderly Speech and the Law in Early Virginia* (Ithaca, NY: Cornell University Press, 2003).

55. Diana Paton, "Gender, Language, Violence and Slavery: Insult in Jamaica, 1800–1836," *Gender and History*, 18, no. 2 (2006): 246–65.

56. [Joshua Steele], *Letters of Philo-Xylon* (Bridgetown, Barbados, 1789), p. 9.

57. Miles Ogborn, *Global Lives: Britain and the World, 1550–1800* (Cambridge: Cambridge University Press, 2008); Christer Petley, *Slaveholders in Jamaica: Colonial Society and Culture during the Era of Abolition* (London: Pickering and Chatto, 2009);

and Richard S. Dunn, *A Tale of Two Plantations: Slave Life and Labor in Jamaica and Virginia* (Cambridge, MA: Harvard University Press, 2014).

58. Stephen Hornsby, *British Atlantic, American Frontier: Spaces of Power in Early Modern British America* (Hanover, NH: University of New England Press, 2005); Jack P. Greene, *Settler Jamaica in the 1750s: A Social Portrait* (Charlottesville: University of Virginia Press, 2016); and Pedro Welch, *Slave Society in the City: Bridgetown, Barbados, 1680–1834* (Kingston: Ian Randle, 2003).

59. Julius Sherrard Scott III, "The Common Wind: Currents of Afro-American Communication in the Era of the Haitian Revolution" (PhD diss., Duke University, 1986); and Ada Ferrer, *Freedom's Mirror: Cuba and Haiti in the Age of Revolution* (Cambridge: Cambridge University Press, 2014).

60. Kamau Brathwaite, *The Development of Creole Society in Jamaica, 1770–1820* (Oxford: Clarendon Press, 1971). Two specific examples, from a vast literature, are Richard Cullen Rath, "African Music in Seventeenth-Century Jamaica: Cultural Transit and Transition," *William and Mary Quarterly* 50, no. 4 (1993): 700–726; and Steeve O. Buckridge, *The Language of Dress: Resistance and Accommodation in Jamaica, 1760–1890* (Kingston: University of the West Indies Press, 2004).

61. For examples, see Rebecca Shumway, *The Fante and the Transatlantic Slave Trade* (Rochester, NY: University of Rochester Press, 2011), p. 133, on Fante as a new language—combining Akan, Guan, and Etsi dialects, with some words and phrases from Portuguese, English, and Dutch—formed via processes of trade and state formation in eighteenth-century West Africa; Lauren Derby, "Beyond Fugitive Speech: Rumor and Affect in Caribbean History," *Small Axe* 44 (2014): 123–40; Joan M. Fayer, "African Interpreters in the Atlantic Slave Trade," *Anthropological Linguistics* 45, no. 3 (2003): 281–95; and James H. Sweet, "Mistaken Identities? Olaudah Equiano, Domingos Álvares, and the Methodological Challenges of Studying the African Diaspora," *American Historical Review* 114, no. 2 (2009): 279–306.

62. Frederic G. Cassidy, *Jamaica Talk: Three Hundred Years of the English Language in Jamaica* (London: Macmillan, 1961); Mervyn C. Alleyne, "Acculturation and the Cultural Matrix of Creolization," in *Pidginization and Creolization of Languages*, ed. Dell Hymes (Cambridge: Cambridge University Press, 1971), pp. 169–86; and Barbara Lalla and Jean D'Costa, *Language in Exile: Three Hundred Years of Jamaican Creole* (Tuscaloosa: University of Alabama Press, 1990). On naming, see Richard Price and Sally Price, "Saramaka Onomastics: An Afro-American Naming System," *Ethnology* 11, no. 4 (1972): 341–67; and Margaret Williamson, "Africa or Old Rome? Jamaican Slave Naming Revisited," *Slavery and Abolition* 38, no. 1 (2017): 117–34.

63. Mary Caton Lingold, "Peculiar Animations: Listening to Afro-Atlantic Music in Caribbean Travel Narratives," *Early American Literature* 52, no. 3 (2017): 623–50, at p. 638. For an excellent extended example, see Laurent Dubois, *The Banjo: America's African Instrument* (Cambridge, MA: Belknap Press of Harvard University Press, 2016).

64. Miles Ogborn, "Spaces of Speech, Worlds of Meaning," in *The Sage Handbook of Historical Geography*, ed. Mona Domosh, Michael Heffernan, and Charles W. J. Withers (London: Sage, forthcoming). For example, see Jacques Derrida, *Of Grammatology*, trans. Gayatri Chakravorty Spivak (Baltimore: Johns Hopkins University Press, 1976); Adam Fox, *Oral and Literate Culture in England, 1500–1700* (Oxford: Clarendon Press, 2000), Paula McDowell, *The Invention of the Oral: Print Commerce and Fugitive Voices in Eighteenth-Century Britain* (Chicago: University of Chicago Press,

2016); and Keith Thor Carlson, Kristina Fagan, and Natalia Khanenko-Friesen, *Orality and Literacy: Reflections Across Disciplines* (Toronto: University of Toronto Press, 2011).

65. Kamensky, *Governing the Tongue*.

66. Miles Ogborn, "A War of Words: Speech, Script and Print in the Maroon War of 1795-6," *Journal of Historical Geography* 37 (2011): 203–15.

67. For example, discussions of print and slave revolts in Laurent Dubois, "An Enslaved Enlightenment: Rethinking the Intellectual History of the French Atlantic," *Social History* 31, no. 1 (2006): 1–14; Michael P. Johnson, "Demark Vesey and his Co-Conspiritors," *William and Mary Quarterly* 58, no. 4 (2001): 915–76; and Vincent Brown, *Tacky's Revolt: The Story of an Atlantic Slave War* (Cambridge MA: Harvard University Press, 2019).

68. For Caribbean slavery, see Henrice Altink, *Representations of Slave Women in Discourses on Slavery and Abolition, 1780–1838* (London: Routledge, 2007); and Keith A. Sandiford, *The Cultural Politics of Sugar: Caribbean Slavery and Narratives of Colonialism* (Cambridge: Cambridge University Press, 2000).

69. For contrasting examples, see Jon Wilson, *India Conquered: Britain's Raj and the Chaos of Empire* (London: Simon and Schuster, 2017); and Chris Philo, "Discursive Life," in *A Companion to Social Geography*, ed. Vincent J. Del Casino et al., (Oxford: Wiley Blackwell, 2011), pp. 362–84.

70. For an argument for "culture-as-dialogue" rather than "culture-as-text" in the study of slavery, see John Smolenski, "Hearing Voices: Microhistory, Dialogicality and the Recovery of Popular Culture on an Eighteenth-Century Virginia Plantation," *Slavery and Abolition* 24, no. 1 (2003): 1–23.

71. Miles Ogborn, "Francis Williams's Bad Language: Historical Geography in a World of Practice," *Historical Geography* 37 (2009): 5–25.

72. Michel de Certeau, *The Practice of Everyday Life*, trans. Stephen Rendell (Berkeley: University of California Press, 1984), xii. See the idea of "Anansi tactics" in Emily Zobel Marshall, *Anansi's Journey: A Story of Jamaican Cultural Resistance* (Kingston: University of the West Indies Press, 2012).

73. Goffman, *Forms of Talk*, p. 67; and J. L. Austin, *How to Do Things with Words*, 2nd ed. (Cambridge, MA: Harvard University Press, 1975).

74. Erving Goffman, "Felicity's Condition," *American Journal of Sociology* 89, no. 1 (1983): 1–53, at p. 25; Austin, *How to Do Things with Words*, pp. 14–24; Judith Butler, *Excitable Speech: A Politics of the Performative* (London: Routledge, 1997).

75. Bruno Latour, *An Inquiry into Modes of Existence: An Anthropology of the Moderns*, trans. Catherine Porter (Cambridge, MA: Harvard University Press, 2013), p. 18.

76. Bruno Latour, *The Making of Law: An Ethnography of the Conseil D'État* (Cambridge: Polity Press, 2010).

77. Latour, *Inquiry into Modes of Existence*, p. 29.

78. See Judith Butler, "Performative Acts and Gender Constitution: An Essay in Phenomenology and Feminist Theory," *Theatre Journal* 40, no. 4 (1988): 519–31; and Sarah Salih, "On Judith Butler and Performativity," in *Sexualities and Communication in Everyday Life: A Reader*, ed. Karen E. Lovaas and Mercilee M. Jenkins (Thousand Oaks, CA: Sage, 2006), pp. 55–68. For a similar argument on "race," see Louis F. Mirón and Jonathan Xavier Inda, "Race as a Kind of Speech Act," *Cultural Studies: A Research Annual*, 5 (2000): 85–107.

79. Bruno Latour, *We Have Never Been Modern*, trans. Catherine Porter (London: Harvester Wheatsheaf, 1993).

80. Latour, *Inquiry into Modes of Existence*, p. 165; and Hilary McD Beckles, "Capitalism, Slavery, and Caribbean Modernity," *Callaloo*, 20, no. 4 (1997): 777–89.

81. Diana Paton, *The Cultural Politics of Obeah: Religion, Colonialism and Modernity in the Caribbean World* (Cambridge: Cambridge University Press, 2015); and Diana Paton and Maarit Forde, eds., *Obeah and Other Powers: The Politics of Caribbean Religion and Healing* (Durham, NC: Duke University Press, 2012).

82. Diana Paton, "Witchcraft, Poison, Law and Atlantic Slavery," *William and Mary Quarterly* 69, no. 2 (2012): 235–64.

83. Eduardo Viveiros de Castro, *The Relative Native: Essays on Indigenous Conceptual Worlds* (Chicago: Hau Books, 2015).

84. Jerome S. Handler and Kenneth M. Bilby, "On the Early Use and Origin of the Term 'Obeah' in Barbados and the Anglophone Caribbean," *Slavery and Abolition* 22, no. 2 (2001): 87–100; and Walter Rucker, *Gold Coast Diasporas: Identity, Culture, and Power* (Bloomington: Indiana University Press, 2015), 190–92. On obeah and the violence of slavery, see Randy M. Browne, "The 'Bad Business' of Obeah: Power, Authority, and the Politics of Slave Culture in the British Caribbean," *William and Mary Quarterly* 68, no. 3 (2011): 451–80.

85. Kenneth M. Bilby, "How the 'Older Heads' Talk: A Jamaican Maroon Spirit Possession Language and Its Relationship to the Creoles of Suriname and Sierra Leone," *New West India Guide* 57, nos. 1/2 (1983): 32–88; Kenneth M. Bilby, "The Kromanti Dance of the Windward Maroons of Jamaica," *New West India Guide* 55, nos. 1/2 (1981): 52–101; and Kenneth M. Bilby, *True-Born Maroons* (Gainesville: University Press of Florida, 2005).

86. Zobel Marshall, *Anansi's Journey*, p. 75. Thistlewood recorded being told a story of a talking crab in 1751; see Philip D. Morgan, "Slaves and Livestock in Eighteenth-Century Jamaica: Vineyard Pen, 1750–1751," *William and Mary Quarterly* 52, no. 1 (1995): 47–76, p. 70.

87. Ana María Ochoa Guatier, *Aurality: Listening and Knowledge in Nineteenth-Century Columbia* (Durham, NC: Duke University Press, 2014), pp. 31–75; and Marcy Norton, "Going to the Birds: Animals as Things and Beings in Early Modernity," in *Early Modern Things: Objects and Their Histories, 1500–1800*, ed. Paula Findlen (London: Routledge, 2013), pp. 53–83.

88. I am grateful to Silvia Sebastiani for directing me toward earlier examples of willfully silent and work-shy apes in Richard Jobson, *The Golden Trade, or a Discovery of the River Gambra* (London, 1623); and Jacobus Bontius, *Historiae naturalis et medicae Indiae Orientalis* (Amsterdam, 1658).

CHAPTER ONE

1. [Joshua Steele], *Letters of Philo-Xylon* (Bridgetown, Barbados, 1789), p. 13.

2. Bruno Latour, *An Inquiry into Modes of Existence: An Anthropology of the Moderns*, trans. Catherine Porter (Cambridge, MA: Harvard University Press, 2013), pp. 366, 361, 364, and 370–71.

3. John Spurr, "A Profane History of Early Modern Oaths," *Transactions of the Royal Historical Society* 11(2001): 37–63.

4. Harold Love, "Oral and Scribal Texts in Early Modern England," in *The Cambridge History of the Book in Britain*, vol. 4, *1557–1695*, ed. John Barnard and Donald F. McKenzie (Cambridge: Cambridge University Press, 2002), pp. 97–121.

5. Barbara J. Shapiro, *A Culture of Fact: England, 1550–1720* (Ithaca, NY: Cornell University Press, 2000).

6. Jack P. Greene, "Introduction: Empire and Liberty," in *Exclusionary Empire: English Liberty Overseas, 1600–1900*, ed. Jack P. Greene (Cambridge: Cambridge University Press, 2010), pp. 1–24, at p. 3.

7. Michoko Suzuki, "Daughters of Coke: Women's Legal Discourse in Early Modern England, 1642–1689," in *Challenging Orthodoxies: The Social and Cultural Worlds of Early Modern Women*, ed. Sigrun Haude and Melinda S. Zook (London: Routledge, 2014), pp. 165–92.

8. Eliga H. Gould, "Zones of Law, Zones of Violence: The Legal Geography of the British Atlantic, circa 1772," *William and Mary Quarterly* 60, no. 3 (2003): 471–510.

9. Lauren Benton, *Law and Colonial Cultures: Legal Regimes in World History, 1400–1900* (Cambridge: Cambridge University Press, 2002).

10. Lauren Benton, *A Search for Sovereignty: Law and Geography in European Empires, 1400–1900* (Cambridge: Cambridge University Press, 2010).

11. Jack P. Greene, "Liberty and Slavery: The Transfer of British Liberty to the West Indies, 1627-1865," in Greene, *Exclusionary Empire*, pp. 50–76; and Susan D. Amussen, *Caribbean Exchanges: Slavery and the Transformation of English Society, 1640–1700* (Chapel Hill: University of North Carolina Press, 2007).

12. Mary S. Bilder, *The Transatlantic Constitution: Colonial Legal Culture and the Empire* (Cambridge, MA: Harvard University Press, 2004).

13. BL Add. MS 33845, "Notes Relating to Barbados,"f.27r.

14. *Journals of the Assembly of Jamaica* (hereafter cited as *JAJ*), Vol. 2, *From 1st March 1710 to 19th February 1731* (St. Jago de la Vega, Jamaica, 1815), p. 30 (May 18, 1711).

15. BL Add. MS 33845, "Notes Relating to Barbados," ff.87v–88r.

16. BL Add. MS 12416, [Charles Knight], "A History of Jamaica to 1742," f.237.

17. *The Laws of Jamaica* (St. Jago de la Vega, Jamaica, 1792), xlv.

18. Jack P. Greene, "Empire and Identity from the Glorious Revolution to the American Revolution," in *The Oxford History of the British Empire*, vol. 2, *The Eighteenth Century*, ed. Peter J. Marshall (Oxford: Oxford University Press, 1998), pp. 208–30.

19. Michael Guasco, *Slaves and Englishmen: Human Bondage in the Early Modern Atlantic World* (Philadelphia: University of Pennsylvania Press, 2014).

20. Edward B. Rugemer, "The Development of Mastery and Race in the Comprehensive Slave Codes of the Greater Caribbean during the Seventeenth Century," *William and Mary Quarterly* 70, no. 3 (2013): 429–58.

21. William Rawlin, *The Laws of Barbados* (London, 1699), p. 156.

22. Jamaica's 1664 slave act quoted in D. Barry Gaspar, "'Rigid and Inclement': Origins of the Jamaica Slave Laws of the Seventeenth Century," in *The Many Legalities of Early America*, ed. Christopher L. Tomlins and Bruce H. Mann (Chapel Hill: University of North Carolina Press, 2001), pp. 78–96, at p. 90.

23. George P. Hagan, "The Rule of Law in Asante, a Traditional Akan state," *Présence Africaine* 113 (1980): 193–208, at p. 193; and T. C. McCaskie, *State and Society in Pre-Colonial Asante* (Cambridge: Cambridge University Press, 1995).

24. Robert S. Rattray, *Ashanti Law and Constitution* (Oxford: Clarendon Press, 1929), p. 385.

25. Hagan, "Rule of Law in Asante," p. 203.

26. Robert S. Rattray, *Religion and Art in Ashanti* (Oxford: Clarendon Press, 1927), p. 277; and Kofi E. Agorsah, "Women in African Traditional Politics," *Journal of Legal Pluralism* 30–31 (1990–1991): 77–86.

27. Hagan, "Rule of Law in Asante," pp. 199 and 201.

28. Rattray, *Ashanti Law and Constitution*, p. 387; and Marta García Novo, "Islamic Law and Slavery in Premodern West Africa," *Entremons* 2 (2011): 1–20.

29. Rawlin, *Laws of Barbados*, pp. 137 and 206.

30. *The Laws of Jamaica* (London, 1684), p. ix; and *The Acts of Assembly and Laws of Jamaica* (London, 1704), p. 121.

31. *Acts of Assembly, Passed in the Island of Jamaica; from 1681 to 1737* (London, 1738), act no. 328; and Rawlin, *Laws of Barbados*, p. 185.

32. Michael J. Braddick, "Civility and Authority," in *The British Atlantic World, 1500–1800*, ed. David Armitage and Michael J. Braddick (London: Palgrave Macmillan, 2002), pp. 93–113; Steven Shapin, *A Social History of Truth: Civility and Science in Seventeenth-Century England* (Chicago: University of Chicago Press, 1994); and Shapiro, *Culture of Fact*.

33. Rawlin, *Laws of Barbados*, p. 185; and *Acts of Assembly Passed in the Island of Jamaica; from the Year 1681 to the Year 1769* (St. Jago de la Vega, Jamaica, 1771), p. 46.

34. Spurr, "Profane History."

35. *Caribbeana*, vol. 1 (London, 1741), pp. 122 and 149.

36. *Caribbeana*, pp. 128, 142–43, and 234–35.

37. *The Laws of Jamaica* (London, 1683), p. 104; *Acts of Assembly and Laws of Jamaica* (1704), p. 101; and *Acts of Assembly Passed in Jamaica* (1771), p. 90.

38. *Caribbeana*, p. 145.

39. *Acts and Statues of the Island of Barbados* (London, 1654), p. 53.

40. Francis Hanson, preface to *Laws of Jamaica* (1683), sig c.4[3]$^{\text{r-v}}$.

41. *Acts and Statues of Barbados* (1654), p. 136.

42. *Acts of Assembly and Laws of Jamaica* (1704), pp. 141 and 144; and *The Laws of Barbados* (1699), act no. 173.

43. *Acts and Statues of Barbados* (1654), p. 33.

44. *Acts of Assembly, Passed in the Island of Jamaica: From 1681 to 1737* (London, 1743), p. 164.

45. *Laws of Barbados* (1699), pp. 121 and 201. See also the manuscript material on the evidence of Jews inserted into Richard Hall, *Acts Passed in the Island of Barbados: From 1643 to 1762* (London, 1764), pp. 94–95, in Library of Congress, Pinfold Papers 10.

46. *JAJ*, vol. 2, *From 20th January 1664 to 20th April 1709* (St, Jago de la Vega, Jamaica, 1811), p. 439.

47. *Acts of Assembly, Passed in Jamaica* (1743), p. 95. See Daniel Livesay, *Children of Uncertain Fortune: Mixed-Race Jamaicans in Britain and the Atlantic Family* (Chapel Hill: University of North Carolina Press, 2018).

48. *Acts of Assembly, Passed in the Island of Barbadoes; from 1717–18 to 1738, Inclusive* (London, 1739). This slightly reworded an act of 1709; see *Acts of Assembly, Passed in the Island of Barbadoes, from 1648 to 1718* (London, 1732), p. 238.

49. Jacques Derrida, *Of Grammatology*, trans. Gayatri Chakravorty Spivak (Baltimore: Johns Hopkins University Press, 1976).

50. Rattray, *Religion and Art in Ashanti*, pp. 205–15; and Jessica A. Krug, "Social Dismemberment, Social (Re)membering: Obeah Idioms, Kromanti Identities and the Trans-Atlantic Politics of Memory, c. 1675–Present," *Slavery and Abolition* 35, no. 4 (2014): 537–58.

51. Walter Rucker, *Gold Coast Diasporas: Identity, Culture, and Power* (Bloomington: Indiana University Press, 2015), pp. 91–92.

52. E. Kofi Agorsah, "Spiritual Vibrations of Historic Kormantse and the Search for African Diaspora Identity and Freedom," in *Materialities of Ritual in the Black Atlantic,* ed. Akinwumi Ogundiran and Paula Saunders (Bloomington: Indiana University Press, 2014), pp. 87–107.

53. On "Koromantin" as a forbidden word, see Rattray, *Religion and Art in Ashanti,* p. 213.

54. Diana Paton, "Witchcraft, Poison, Law, and Atlantic Slavery," *William and Mary Quarterly* 69, no. 2 (2012): 235–64.

55. Vincent Brown, "Spiritual Terror and Sacred Authority in Jamaican Slave Society," *Slavery and Abolition* 24, no. 1 (2003): 24–53.

56. Helen McKee, "From Violence to Alliance: Maroons and White Settlers in Jamaica, 1739–1795," *Slavery and Abolition* 39, no. 1 (2018): 27–52.

57. Mavis Campbell, *The Maroons of Jamaica, 1655–1796: A History of Resistance, Collaboration and Betrayal* (Bergin and Garvey, Granby MS, 1988); and Kathleen Wilson, "The Performance of Freedom: Maroons and the Colonial Order in Eighteenth-Century Jamaica and the Atlantic Sound," *William and Mary Quarterly* 66, no. 1 (2009): 45–86.

58. Barbara Klamon Kopytoff, "Colonial Treaty as Sacred Charter of the Jamaican Maroons," *Ethnohistory* 26, no. 1 (1979): 45–64; and Werner Zips, "Laws in Competition: Traditional Maroon Authorities within Legal Pluralism in Jamaica," *Journal of Legal Pluralism* 37–38 (1996): 279–305.

59. Kenneth Bilby, "Swearing by the Past, Swearing to the Future: Sacred Oaths, Alliances, and Treaties among the Guianese and Jamaican Maroons," *Ethnohistory* 44, no. 4 (1997): 655–89.

60. Robert C. Dallas, *The History of the Maroons, from Their Origin to the Establishment of Their Chief Tribe at Sierra Leone,* vol. 1 (London: Longman and Rees, 1803), p. 55.

61. BL Add. MS 12431, Colonel Guthrie to John Gregory, February 21, 1739, f.91r, and [Knight], "A History of Jamaica to 1742," f.273r.

62. BL Add. MS 12431, Copy of letter from J. Lewis to unknown recipient, Westmoreland, Jamaica, December 20, 1743, f.99r–v.

63. Ian Baucom, *Spectres of the Atlantic: Finance Capital, Slavery, and the Philosophy of History* (Durham, NC: Duke University Press, 2005); Achille Mbembe, "Necropolitics," *Public Culture* 15 (2003): 11–40; Marisa J. Fuentes, *Dispossessed Lives: Enslaved Women, Violence, and the Archive* (Philadelphia: University of Pennsylvania Press, 2016); Bernard Marshall, *Slavery, Law and Society in the British Windward Islands, 1673–1823: A Comparative Study* (Kingston: Arawak Publications, 2007); Mindie Lazarus-Black, "Slaves, Masters, and Magistrates: Law and the Politics of Resistance in the British Caribbean, 1736–1834," in *Contested States: Law, Hegemony, and Resistance,* ed. Mindie Lazarus-Black and Susan F. Hirsch (New York: Routledge, 1994), pp. 252–81; Nicole N. Aljoe "'Going to law': Legal Discourse and Testimony in Early West Indian Slave Narratives," *Early American Literature* 46, no. 2 (2011): 351–381; and Natalie Zacek, "Voices and Silences: The Problem of Slave Testimony in the English West Indian Law Court," *Slavery and Abolition* 24, no. 3 (2003): 24–39, which demonstrates that where slave evidence was taken against white people, it was part of attempts to preserve white solidarity against hated and disruptive figures.

64. Elsa V. Goveia, *The West Indian Slave Laws of the Eighteenth Century* (Barbados: Caribbean Universities Press, 1970), p. 25. On the implications of the law's failure to acknowledge gender difference, see Fuentes, *Dispossessed Lives.*

65. Diana Paton, "Punishment, Crime, and the Bodies of Slaves in Eighteenth-Century Jamaica," *Journal of Social History* 34, no. 4 (2001): 923–54.

66. [Knight], "A History of Jamaica to 1742," ff.184r and 183r.

67. *Acts of Assembly, Passed in Jamaica* (1738), p. 80; and Goveia, *West Indian Slave Laws*, p. 24.

68. *Acts of Assembly, Passed in the Island of Jamaica; from 1681 to 1754* (St. Jago de la Vega, Jamaica, 1756), act no. 183.

69. Paton, "Punishment, Crime, and the Bodies of Slaves"; Fuentes, *Dispossessed Lives*; and B. Marshall, *Slavery, Law and Society*.

70. *Laws of Jamaica* (1684), p. 147; and the "General View of the Principles . . . [of] this System of Laws," produced for the House of Commons committee on the slave trade, *House of Commons Accounts and Papers*, vol. 26 (1789), no. 646a, part III.

71. Giorgio Agamben, *Homo Sacer: Sovereign Power and Bare Life* (Stanford, CA: Stanford University Press, 1998); and V. Brown, "Social Death and Political Life." For the counterargument, see Fuentes, *Dispossessed Lives*, pp. 108–9.

72. Fuentes, *Dispossessed Lives*, p. 111.

73. *Laws of Barbados* (1699), p. 160.

74. *Laws of Jamaica* (1684), p. 144; *Acts of Assembly, Passed in Jamaica* (1738), p. 78; and R. Hall, *Acts Passed in Barbados*, p. 117n.

75. *Acts of Assembly, Passed in the Island of Jamaica, from the Year 1784 to the Year 1788* (Kingston, Jamaica, 1789), p. 202; and Lazarus-Black, "Slaves, Masters, and Magistrates."

76. *Acts of Assembly Passed in Jamaica* (1771), p. 61; and *Acts of Assembly, Passed in Jamaica* (1789), p. 204.

77. See "An Act for the Incouragement of all Negroes and Slaves, that shall discover any Conspiracy," in *Laws of Barbados* (1699), pp. 186–87.

78. *Acts of Assembly, Passed in Jamaica* (1756), p. 323.

79. *Laws of Barbados* (1699), p. 160; and Rugemer, "Development of Mastery and Race."

80. Jill Lepore, *New York Burning: Liberty, Slavery, and Conspiracy in Eighteenth-Century Manhattan* (New York: Alfred A. Knopf, 2005); and Thomas J. Davis, "Conspiracy and Credibility: Look Who's Talking, about What—Law Talk and Loose Talk," *William and Mary Quarterly* 59, no. 1 (2002): 167–74.

81. R. Hall, *Acts Passed in Barbados*, p. 117n.

82. R. Hall, *Acts Passed in Barbados*, p. 325.

83. Described in David Lambert, *White Creole Culture, Politics and Identity during the Age of Abolition* (Cambridge: Cambridge University Press, 2005).

84. [Steele], *Letters of Philo-Xylon*, pp. 6–7 and 10.

85. [Steele], *Letters of Philo-Xylon*, pp. 13, 16–17, and 27; and David N. Livingstone, *Adam's Ancestors: Race, Religion and the Politics of Human Origins* (Baltimore: Johns Hopkins University Press, 2008).

86. [Steele], *Letters of Philo-Xylon*, pp. 9, 15, 17, and 28.

87. [Steele], *Letters of Philo-Xylon*, pp. 13, 15, and 22.

88. Christer Petley, "'Legitimacy' and Social Boundaries: Free People of Colour and the Social Order in Jamaican Slave Society," *Social History* 30 (2005): 481–98; and Livesay, *Children of Uncertain Fortune*, p. 24.

89. *Acts of Assembly, Passed in Jamaica* (1738), p. 162; and *Acts of Assembly Passed in Jamaica* (1771), pp. 56 and 60.

90. TNA CO139/12 Jamaica, Acts, 1728–1730, f.79r; and *Acts of Assembly, Passed on the Island of Jamaica; from 1770 to 1783* (Kingston, Jamaica, 1786), p. 99.

91. *Acts of Assembly, Passed in Jamaica* (1738), p. 79.

92. *Acts of Assembly, Passed in Jamaica* (1756), p. 290.

93. Brooke N. Newman, *A Dark Inheritance: Blood, Race, and Sex in Colonial Jamaica* (New Haven, CT: Yale University Press, 2018).

94. *Acts of Assembly, Passed in Jamaica* (1738), pp. 260–62. On the context for this law, see Livesay, *Children of Uncertain Fortune.*

95. *Acts of Assembly, Passed in Jamaica* (1756), p. 290.

96. Petley, "'Legitimacy' and Social Boundaries."

97. *Acts of Assembly. Passed on Jamaica* (1786), p. 125.

98. For example, *Acts of Assembly, Passed in Jamaica* (1738), p. 258. See Linda L. Sturtz, "Mary Rose: 'White' African Jamaican Woman? Race and Gender in Eighteenth-Century Jamaica," in *Gendering the African Diaspora; Women, Culture, and Historical Change in the Caribbean and Nigerian Hinterland,* ed Judith A. Byfield et al. (Bloomington: Indiana University Press, 2010), pp. 59–87.

99. *Acts of Assembly Passed in Jamaica* (1771), pp. 10 and 18. See Samuel J. Hurwitz and Edith F. Hurwitz, "A Token of Freedom: Private Bill Legislation for Free Negroes in Eighteenth-Century Jamaica," *William and Mary Quarterly* 24, no. 3 (1967): 423–31; and Livesay, *Children of Uncertain Fortune.*

100. Brooke N. Newman, "Contesting 'Black' Liberty and Subjecthood in the Anglophone Caribbean," *Slavery and Abolition* 32, no. 2 (2011): 168–83.

101. *JAJ,* vol. 1, pp. 437, 439, and 446 (January 30 and February 12 and 19, 1708).

102. *House of Lords Journal,* vol. 19, May 7, 9, 11, 16, 23, and 25, 1711.

103. *JAJ,* vol. 2, pp. 152–217 (November 25 and 30 and December 3 and 7, 1715; and September 28 and October 11, 13, 16, and 20, 1716).

104. William J. Gardner, *The History of Jamaica* (New York: Appleton, 1909), p. 207, although I could find no mention of this in the *JAJ* for 1708 to 1716. See Miles Ogborn, "Francis Williams's Bad Language: Historical Geography in a World of Practice," *Historical Geography* 37 (2009): 5–25.

105. See also Suman Seth, "Materialism, Slavery, and *The History of Jamaica,*" *Isis,* 105 (2014): 764–72.

106. Vincent Carretta, "Who Was Francis Williams?" *Early American Literature* 38, no. 2 (2003): 213–37, gives his christening date as December 26, 1697. See also John T. Gilmore, "Francis Williams," *Oxford Dictionary of National Biography* (hereafter cited as *ODNB*).

107. Royal Society Archive, Journal Books, JBO/12/86, pp. 128–29 (October 25, 1716). See also "Supplement to the Gentleman's Magazine: For the Year 1771," *Gentleman's Magazine* 41 (1771): 595–96. I am grateful to Simon Schaffer for identifying Robert Smith, Plumian professor of astronomy from 1716.

108. Royal Society Archive, Council Minutes, CMO/2/268, p. 236 (November 8, 1716).

109. *The Family Memoirs of the Rev. William Stukeley, M.D.* (Durham, UK: Surtees Society, 1882), p. 100; and David Boyd Haycock, "Martin Folkes," *ODNB.*

110. This well-known portrait, which may have been commissioned by Williams and was owned by the Long family, is insistent on matters of racialized bodies, dress, politeness, and learning (the open book is titled *Newton's Philosophy*). What it meant to viewers dependended on who the viewers were. Although not a group portrait, it might be understood as a particular sort of "conversation piece." See Marcia Pointon, *Hanging the Head: Portraiture and Social Formation in Eighteenth-Century England*

(New Haven, CT: Yale University Press, 1997) on such pictures as quasi-legal genea-
logical statements.

111. See also "Supplement to the Gentleman's Magazine"; and *An Inquiry into the Origin,
Progress, and Present State of Slavery: with a Plan for the Gradual, Reasonable, & Secure
Emancipation of Slaves* (London, 1789), pp. 26–27.

112. *JAJ*, vol. 2, p. 512 (November 20, 1724).

113. *JAJ*, vol. 2, p. 113 (December 18, 1713) and p. 233 (November 10, 1716). On
Brodrick's "immoral vicious Irreligious Atheistical & hellish Principles and practices" as a descendant, it was supposed, of Oliver Cromwell's trumpeter, see National Library of Jamaica (hereafter cited as NLJ), MS 2050, Correspondence of
Hugh Totterdell: Hugh Totterdell to Captain H. Gawne, Spanish Town, June 18,
1712, f.9; and *Groans of Jamaica* (London, 1714).

114. *JAJ*, vol. 2, p. 512 (November 20 and 21, 1724).

115. TNA CO139/12 no. 22; and Carretta, "Who Was Francis Williams?"

116. *Acts of the Privy Council of England*, Colonial Series, vol. 3, *1720–1745*, p. 345.

117. See Caretta, "Who Was Francis Williams?" and John T. Gilmore, "Parrots, Poets and
Philosophers: Language and Empire in the Eighteenth Century," *EnterText* 2, no. 2
(2003): 84–102.

118. John T. Gilmore, "The British Empire and the Neo-Latin Tradition: The Case of
Francis Williams," in *Classics and Colonialism* ed. Barbara Goff (London: Duckworth,
2005), pp. 92–106; and E. J. Chinook's translation of Williams's poem in Gardner,
History of Jamaica, p. 510, although Judith Hallett translates it as, "All eloquence is
lacking in our hearths," in Caretta, "Who Was Francis Williams?" p. 234.

119. Quoted in James Robertson, "A 1748 'Petition of Negro Slaves' and the Local Politics of Slavery in Jamaica," *William and Mary Quarterly* 67, no. 2 (2010): 319–46, at
pp. 323 and 338.

120. *JAJ*, volume 4, *18th March 1746 to 22ⁿᵈ December 1756*, p. 122 (May 18, 1748) and
p. 127 (May 26, 1748). The manuscript original is reproduced in Robertson, "A
1748 'Petition of Negro Slaves.'"

121. *JAJ*, vol. 4, p. 123 (May 18, 1748).

122. K. Wilson, "Performance of Freedom."

123. *JAJ*, vol. 4, p. 122 (May 18, 1748). The manuscript original has "faction" for "patron," but the effect is the same; see Robertson, "A 1748 'Petition of Negro Slaves,'"
p. 343.

124. [James Smith], *A Letter from a Friend at J[amaica]* (London, 1747), pp. 4 and 12.

125. *JAJ*, vol. 4, p. 127 (May 25 and 26, 1748).

126. Robertson, "A 1748 'Petition of Negro Slaves,'" p. 341.

127. *JAJ*, vol. 4, p. 122 (May 18, 1748).

128. Miles Ogborn, "A War of Words: Speech, Script and Print in the Maroon War of
1795–6," *Journal of Historical Geography* 37 (2011): 203–15.

129. TNA CO137/96 Jamaica, Original Correspondence, Governor Balcarres to the Duke
of Portland, Jamaica, May 9, 1796, f.260; and CO139/48 Jamaica, Acts, no. 971.

CHAPTER TWO

1. [Nicholas Bourke], *A Letter Concerning the Privileges of the Assembly of Jamaica* (Kingston, 1765), p. 20.

2. Jack P. Greene, "The Jamaica Privilege Controversy, 1764–66: An Episode in the Process of Constitutional Definition in the Early Modern British Empire," *Journal of Imperial and Commonwealth History*, 22, no. 1 (1994): 16–53; Jack P. Greene, "Liberty,

Slavery and the Transformation of British Identity in the Eighteenth-Century West Indies," *Slavery and Abolition* 21, no. 1 (2000): 1–31; and Andrew J. O'Shaughnessy, *An Empire Divided: The American Revolution and the British Caribbean* (Philadelphia: University of Pennsylvania Press, 2000).

3. Jürgen Habermas, *The Structural Transformation of the Public Sphere: An Inquiry into a Category of Bourgeois Society* (London: Polity Press, 1992), p. 36.

4. Craig Calhoun, ed., *Habermas and the Public Sphere* (Cambridge MA: MIT Press, 1992); and Harold Mah, "Phantasies of the Public Sphere: Rethinking the Habermas of Historians," *Journal of Modern History* 72, no. 1 (2000): 153–82.

5. See Michael Warner, *The Letters of the Republic: Publication and the Public Sphere in Eighteenth-Century America* (Cambridge, MA: Harvard University Press, 1990); Elizabeth Maddock Dillon, *New World Drama: The Performative Commons in the Atlantic World, 1649-1849* (Durham, NC: Duke University Press, 2014); and Robert Fanuzzi, *Abolition's Public Sphere* (Minneapolis: University of Minnesota Press, 2003).

6. Jürgen Habermas, *The Theory of Communicative Action*, 2 vols. (London: Polity Press, 1986).

7. For attempts to historicize speech and politics in the coffeehouse, see Markman Ellis, *The Coffee House: A Cultural History* (London: Wiedenfeld and Nicholson, 2004); Brian Cowan, *The Social Life of Coffee: The Emergence of the British Coffeehouse* (New Haven, CT: Yale University Press, 2005); and Eric Laurier and Chris Philo, "'A Parcel of Muddling Muckworms': Revisiting Habermas and the English Coffee-Houses," *Social and Cultural Geography* 8, no. 2 (2007): 259–81.

8. Bruno Latour, *An Inquiry into Modes of Existence: An Anthropology of the Moderns*, trans. Catherine Porter (Cambridge, MA: Harvard University Press, 2013) p. 340.

9. Bruno Latour, "What if We Talked Politics a Little?" *Contemporary Political Theory* 2 (2003): 143–63, at pp. 145–46, 148–49, and 155.

10. Barbados Public Library, Lucas Transcripts of the Barbados Council Minutes (hereafter cited as LT), vol. 14, p. 212 (December 8, 1719).

11. LT, vol. 14, p. 210 (December 8, 1719).

12. *The Barbadoes Packet* (London, 1720); and "Robert Lowther (1681–1745)," History of Parliament, Institute of Historical Research, accessed December 26, 2018, http://www.historyofparliamentonline.org/volume/1690-1715/member/lowther-robert-1681-1745.

13. [William Gordon], *A Representation of the Miserable State of Barbadoes, under the Arbitrary and Corrupt Administration of His Excellency, Robert Lowther Esq.* (London, 1719), and LT, vol. 13, p. 395 (October 8, 1719); *Barbadoes Packet*, p. 25, and LT, vol. 13, pp. 434–35 (October 28, 1719), vol. 13, p. 518 (February 16, 1720), and vol. 14, p. 69 (December 1719).

14. *JAJ*, vol. 4, p. 123 (May 18, 1748); and [James Smith], *A Letter from a Friend at J[amaica]* (London, 1747). For "politics-out-of-doors" in Britain, see John Brewer, *Party Ideology and Popular Politics at the Accession of George III* (Cambridge: Cambridge University Press, 1976); and Kathleen Wilson, *The Sense of the People: Politics, Culture and Imperialism in England, 1715-1785* (Cambridge: Cambridge University Press, 1998).

15. Jacques Rancière, *Dis/agreement: Politics and Philosophy*, trans. Julie Rose (Minneapolis: University of Minnesota Press, 1999); and Jacques Rancière, *The Politics of Aesthetics: The Distribution of the Sensible*, trans. Gabriel Rockhill (London: Continuum, 2004).

16. Edouard Glissant, *Caribbean Discourse: Selected Essays*, trans. J. Michael Dash (Charlottesville: University Press of Virginia, 1989); and Marisa J. Fuentes, *Dispossessed*

Lives: Enslaved Women, Violence, and the Archive (Philadelphia: University of Pennsylvania Press, 2016).

17. TNA CO138/3, Lords of Trade Entry Books: Jamaica, April 3, 1674, to July 15, 1681, Earl of Carlisle's Instructions (March 30, 1678), p. 218.

18. Lisa Ford, *Settler Sovereignty: Jurisdiction and Indigenous People in America and Australia, 1788–1836* (Cambridge, MA: Harvard University Press, 2011).

19. "A journal kept by Sir William Beeston, from his first coming to Jamaica," in *Interesting Tracts, Relating to the Island of Jamaica* (St. Jago de la Vega, Jamaica, 1800), pp. 271–300, quotations from pp. 274 (August 15, 1660) and 286 (February 27, 1666); see also pp. 287 (June 14 and July 2, 1670), 293 (April 5, 1678), 296 (July 12, 1679), and 298 (September 26, 1679).

20. *JAJ*, vol. 1, pp. 7–8 (March 17, 1675), and appendix pp. 10 and 12–13, "Instructions for Colonel Modyford" (February 18, 1664) and "Sir Thomas Lynch's commission" (January 13, 1671).

21. For example, *JAJ*, vol. 1, appendix pp. 3, 10 and 46, "Instructions to Edward D'Oyley," "Instructions for Colonel Modyford," and "The State of Jamaica under Sir Thomas Lynch"; and NLJ MS 159, "History and State of Jamaica under Lord Vaughn," p. 41.

22. Fuentes, *Dispossessed Lives*, pp. 37–45.

23. Barbados Museum and Historical Society (hereafter cited as BMHS), Minutes of the Meeting of the Governor and Council, vol. 2, October 22, 1667 to October 31, 1682, pp. 103 (December 23 and 30, 1669), 188 (January 25, 1672), and 216 (September 17, 1672). For later examples, see Barbados Archives (hereafter cited as BA), Act Book, 1723–1733, f.79v; Act Book, November 20, 1733, to December 27, 1744, ff.99v–100r, and Act Book, December 23, 1771, to May 16, 1775. For a Jamaican example, see TNA CO137/32, Jamaica, Original Correspondence, Minutes of the Council Meeting, St. Jago de la Vega, April 10, 1760, f.4r.

24. LT, vol. 5, p. 155 (December 11, 1693).

25. LT, vol. 1, p. 190 (February 26, 1656), vol. 6, pp. 104–5 (December 7, 1697), and vol. 6, p. 113 (December 30, 1697).

26. BMHS, Minutes of the Council, vol. 2, pp. 12 and 17 (November 12 and 13, 1667). For another unofficial "Proclamation," delivered from a white horse on Spanish Town parade to challenge the authority of Jamaica's governor and council, see NLJ MS 52D, Copies of Letters of John Mouat, Orkney Hall, Clarendon, 1763–1786, f.4r (November 20, 1786).

27. BMHS, Minutes of the Council, vol. 2, p. 269 (April 9, 1673).

28. Abigail L. Swingen, *Competing Visions of Empire: Labor, Slavery, and the Origins of the British Atlantic Empire* (New Haven, CT: Yale University Press, 2015).

29. LT, vol. 2, pp. 294 and 299, and vol. 6, p. 189 (September 6, 1698).

30. This was in the earliest governors' commissions; see *JAJ*, vol. 1, appendix p. 2, "Colonel D'Oyley's Commission to be Governor of Jamaica."

31. LT, vol. 5, pp. 421–28 (August 7, 1696). For oaths in the Jamaican assembly, see *JAJ*, vol. 1, pp. 4 (February 1, 1672), 7 (January 15, 1674), 11 (April 9, 1677), and 30 (September 23, 1678). A later assembly oath book survives as NLJ MS 286, "Oaths &c. Taken and Subscribed by Members of the Assembly, 1803–c.1833," 2 vols.

32. "Instructions for Colonel Modyford," p. 10; BMHS, Minutes of the Council, vol. 2, pp. 220–21 (October 28, 1672); and TNA CO138/3, Earl of Carlisle's Instructions (March 30, 1678), p. 221.

33. Edwards, *HCC*, 419–20; and Greene, "Liberty, Slavery and the Transformation of British Identity."

34. *JAJ*, vol. 1, appendix p. 22, "Thomas Modyford's Answers to the Inquiries of His Majesty's Commissioners" [1670]; and NLJ MS 159, "History and State of Jamaica under Lord Vaughn," p. 30.

35. This argument is taken from [Nicholas Bourke], *The Privileges of the Island of Jamaica Vindicated* (London, 1766), p. 33, which Long may have had a hand in writing.

36. Quotations are from TNA CO31/2, Barbados Assembly, 1670–1683, pp. 127–30 (December 2, 1674). These were not the earliest rules, which must, in some form, have begun with the first assemblies. For other "Orders for regulating debates," see "Some Records of the House of Assembly of Barbados," *Journal of the Barbados Museum and Historical Society* 10, no. 4 (August 1943): 173–87.

37. *JAJ*, vol. 1, p. 24 (September 2, 1678).

38. For new rules, see *JAJ*, vol. 1, pp. 60 (September 22, 1682), 82 (June 2, 1686), 98 (September 17, 1686), and 117 (April 11, 1688); and LT, vol. 5, p. 490 (November 19, 1696). The rules seem to have stabilized by the early eighteenth century— see *JAJ*, vol. 1, pp. 314–15 (October 8, 1703) and 325–26 (April 12, 1704)—but were displayed earlier: *JAJ*, vol. 1, p. 120 (July 21, 1688).

39. "Some Records of the House of Assembly of Barbados," p. 175; and "Journal kept by Sir William Beeston," p. 284 (August 1664).

40. See, for example, the dispute over whether Thomas Martyn's words had been him "reflecting very unworthily or scandalously on divers members" of the assembly and its proceedings, or were justified by his understanding "that a member had liberty to move and speak freely, without exception or prejudice"; *JAJ*, vol. 1, p. 17 (May 25 and June 7, 1677).

41. LT, vol. 5, pp. 228–32 (October 9 and 10, 1694) and vol. 1, p. 12 (April 19, 1677) for bills refused by Jamaica's governor in 1675 because reading them only twice did not conform to the practice of "English parliaments or assemblies." On which, see Christopher Reid, *Imprison'd Wranglers: The Rhetorical Culture of the House of Commons, 1760–1800* (Oxford: Oxford University Press, 2012).

42. Ellis, *Coffee House*.

43. *JAJ*, vol. 1, pp. 23 (September 2, 1678) and 41 (August 19, 1679), and "Some Records of the House of Assembly of Barbados," pp. 183 and 185.

44. James Robertson, *Gone Is the Ancient Glory: Spanish Town Jamaica, 1534–2000* (Kingston: Ian Randle, 2005). The Barbados assembly was still meeting at a private house in 1759; see LT, vol. 25, p. 315 (July 11, 1759). The key contention in Jamaica was whether the assembly should be in Kingston or Spanish Town. See NLJ MS 1644 No. 3, Petition of the Planters and Residents of St. Jago de la Vega to the Lords Commissioners for Trade and Plantations, November 21, 1754.

45. For example, *JAJ*, vol. 1, p. 11 (April 9, 1677).

46. [John Gay Alleyne], *Remarks Upon a Book, intitled, A Short History of Barbados* (Barbados, 1768).

47. *JAJ*, vol. 1, p. 351 (October 6, 1704).

48. Swingen, *Competing Visions of Empire*; and Nuala Zahedieh, "Regulation, Rent-Seeking, and the Glorious Revolution in the English Atlantic Economy," *Economic History Review* 63, no. 4 (2010): pp. 865–90.

49. See, for example, an earlier delineation of these positions in BL Add. MS 11411, Thomas Povey, Register of Letters Relating to the West Indies, 1655–1661, ff.3–5.

50. "A Report of Right Honourable the Lords of the Committee for Trade and the Plantations to the King," May 28, 1679, and "An Address of the Assembly of Jamaica to

the Earl of Carlisle," [November 1679], in BL Add. MS 12429, A Collection of Tracts Relating to Jamaica, ff.91v and 98r.

51. BL Egerton MS 2395, Letter from Lord Vaughn to the Lords of Trade, January 25, 1675, ff.523–24.

52. TNA CO138/3, Report of the Lords of Trade on Jamaica, p. 162. For the effect of Poynings' Law (1495) in the seventeenth and eighteenth centuries, see James Kelly, *Poynings' Law and the Making of Law in Ireland, 1660–1800* (Dublin: Four Courts Press, 2007).

53. NLJ MS 159, "History and State of Jamaica under Lord Vaughn," p. 68. The Lords of Trade thought the Jamaican assembly had assumed unwarranted powers; see TNA CO138/3, Report of the Lords of Trade on Jamaica, and Earl of Carlisle's Instructions, pp. 162 and 222. The assembly was aware of the danger of voting perpetual revenue bills from the 4.5 percent act passed by Barbados in 1663; see TNA CO138/3, Earl of Carlisle to Lords of Trade, St. Jago, June 10, 1679, pp. 318–19.

54. TNA CO138/3, "Order of Council Touching the Earl of Carlisle's Commission and the Laws of Jamaica," February 15, 1678, p. 182, and Earl of Carlisle's Instructions, p. 241.

55. BL Add. MS 12429, The Earl of Carlisle's Answer to Samuel Long's Petition, f.107v, and "Journal kept by Sir William Beeston," p. 299.

56. *JAJ*, vol. 1, p. 46 (August 26, 1679).

57. TNA CO138/3, Earl of Carlisle to Secretary Coventry, Port Royal, July 31, 1678, pp. 251–53. Carlisle was seen as profiting from pirates and privateers against the interests of the planters; see BL Add. MS 12429, Deposition of Samuel Long, ff.113v and 116r; and Zahedieh, "Regulation, Rent-Seeking, and the Glorious Revolution."

58. TNA CO138/3, Earl of Carlisle to the Lords of Trade, November 23, 1679, p. 364; *JAJ*, vol. 1, pp. 24 and 32 (September 2 and 26, 1678), 47 (October 30, 1679), and 55 (November 20, 1679); and BL Add. MS 12429, Deposition of Peter Beckford, f.120v.

59. TNA CO138/3, Lords of Trade to the Earl of Carlisle, January 16, 1680, p. 350, and Earl of Carlisle to the Lords of Trade, St. Jago, September 15, 1679, p. 328.

60. *JAJ*, vol. 1, pp. 52–53 (November 14, 1679). The proclamation was reprinted in *Laws of Jamaica* (1792), pp. i–ii.

61. BL Egerton MS 2395, "The State of Jamaica with a Proposal Concerning the Method of Making Laws," f.604r; and *JAJ*, vol. 1, p. 53 (November 14, 1679) and pp. 37–38 (October 4 and 9, 1678). For Barbadian conflict with governors, see TNA CO31/2, Letter to William Willoughby, Barbados, November 17, 1670, pp. 12–13.

62. TNA CO138/3, Earl of Carlisle to Secretary Coventry, St. Jago de la Vega, October 24, 1678, p. 280.

63. TNA CO138/3, Earl of Carlisle to the Lords of Trade, November 23, 1679, and February 23, 1680, p. 390, and Earl of Carlisle's Charge against Colonel Long, read September 16, 1680, pp. 365 and 422; and TNA CO391/3, Minutes of the Lords of Trade, p. 152 (March 5, 1680).

64. TNA CO138/3, "King in Privy Council Order to the Judges about the Question of Jamaica," June 23, 1680, p. 382.

65. TNA CO391/3, Minutes of the Lords of Trade, 1679–1682, pp. 111 (December 22, 1679), 115–16 (January 13, 1680), 159–62 (March 11, 1680), and 164–69 (April 27, 1680), quotation at p. 167. The former governors' accounts of Jamaican

government are in TNA CO1/43 ff.323–26 and 349 (read December 22, 1679). See Swingen, *Competing Visions of Empire*, p. 118.

66. TNA CO391/3, Minutes of the Lords of Trade, pp. 214–16 (October 14, 1680) and 219–20 (October 21, 1680).

67. TNA CO138/3, Petition of Jamaican Planters, read October 28, 1680, Report touching some Proposals Concerning Jamaica made by the Merchants and Planters, December 18, 1680, and Lords of Trade to Earl of Carlisle, with the Copy of the Planters Paper, pp. 442–43, 455–60, and 476–79; and TNA CO1/46, Planters of Jamaica to Lords of Trade, read November 4, 1680, no. 32. See Swingen, *Competing Visions of Empire*, pp. 118–21.

68. TNA CO391/3, Minutes of the Lords of Trade, pp. 219–25 (October 21, 27, 28, and 30, and November 1, 1680), quotation at p. 224. See also TNA CO138/3, King's Private Instructions to the Earl of Carlisle, November 3, 1680, pp. 453–54; and, for the extension to twenty-one years, *JAJ*, vol. 1, pp. 78–79 (October 19, 1683).

69. *JAJ*, vol. 1, pp. 65 (September 5, 1683) and 59 (September 21, 1682).

70. NLJ MS 472, Governor Edward Trelawny to Pelham, Jamaica, May 1747, ff.1v–2r. At the same time, another observer described "Great squabbling amongst the Members": NLJ MS 1579, Jonathan Tyler, Kingston, to Captain David Hamilton, Norfolk, April 31, 1747, f.1r.

71. LT, vol. 27, p. 53 (December 30, 1767).

72. [Bourke], *Privileges of the Island of Jamaica Vindicated*, appendix.

73. For Long's role, see Greene, "Jamaica Privilege Controversy." Bourke's text was later reprinted as part of an 1810 political conflict over the rights of the assembly to call military officers before it; see *The Privileges of Jamaica Vindicated* (Jamaica, 1810).

74. See *Slave Law of Jamaica: With Proceedings and Documents Relative Thereunto* (London, 1828), p. 252, which argued for the right to "meet as the unfettered representatives of Englishmen legally assembled to deliberate on the state of their country." For Barbados, see *Memoirs of the First Settlement of the Island of Barbadoes* (London, 1742); [Samuel Frere], *A Short History of Barbados, from its First Discovery and Settlement to the Year 1767* (London, 1768); and [Alleyne], *Remarks Upon a Book*.

75. Roderick Cave, *Printing and the Book Trade in the West Indies* (London: Pindar Press, 1987).

76. LT, vol. 14, pp. 7–11 (February 1720). See William Gordon, *A Sermon Preach'd before the Governor, Council and General Assembly of the Island of Barbados . . . on Friday the 18th of August, 1716* (London, 1718); [Gordon], *A Representation of the Miserable State of Barbadoes*; and [William Gordon], *The Barbadoes Packet* and *The Self-Flatterer* (London, 1720), which included discussions of material printed in the London newspapers.

77. *A Letter from an Apothecary in the West-Indies* (London, 1720), pp. 11–12 and 15.

78. *The Groans of Jamaica* (London, 1714), ii; and NLJ MS 2050, Correspondence of Hugh Totterdell: R. Mackenzie[?] to P. Beckford and H. Totterdell, London, August 21, 1714, f.1r–v.

79. NLJ MS 2050, Hugh Totterdell to P. Castelman, Spanish Town, May 27, 1715, f.1v.

80. *A View of the Proceedings of the Assemblies of Jamaica* (London, 1716), v.

81. [Gordon], *Barbadoes Packet*, pp. 41–42 and 46. *Letter from an Apothecary*, p. 24, alleged that, in court, Lowther "swore and storm'd . . . worse than any Boatswain of a Privateer."

82. *Letter from a Gentleman at Barbados to his Friend now in London* (London, 1740), p. 39.

83. *A Letter from a Citizen at Port-Royal in Jamaica to a Citizen in New York* (London, 1756), pp. 8 and 14.
84. *Groans of Jamaica*, p. 29.
85. [Smith], *Letter from a Friend at J[amaica]*, pp. 16 and 33.
86. [Alleyne], *Remarks Upon a Book*, p. 67. For example, the problems caused in 1688 when a Jamaican assemblyman, John Towers, keen to get away to a horse race, quipped that *salus populi suprema lex* (the good of the people is the supreme law) and was accused of uttering an "expression fitter for a commonwealth than a monarchy," *JAJ*, vol. 1, pp. 116–17 (April 6, 7, 10, and 11, 1688).
87. *JAJ*, vol. 1, pp. 334 (May 12, 1704), 351–52 (October 6, 1704), and 363 (July 10, 1705).
88. For political insults for which women were the audience, and which were acted against by the Barbados council, see LT, vol. 5 (December 18 and 19, 1695), and vol. 14, p. 181 (February 17, 1720). For political discussion between men and women, see NLJ MS 610, Donald Campbell to Lady Campbell, Spanish Town, May 4, 1796. For the political implications of "private" life, see *Groans of Jamaica*, p. 17, where Governor Hamilton's carriage was said to be "notoriously common to . . . a certain Cook-Maid, who pretended to command the same at Pleasure, and also to her chief Gossip, a certain Tavern-keeper, who (tho' never married) has had several Bastard Children to as many different Fathers." In addition, a "Satire sur quelques personnes des plus eminentes & des plus Connues de l'isle de la Jamaique" (satire on some of the most eminent and well-known persons of the island of Jamaica) mixes men and women, and public and private life; see LCP, du Simitière Papers, Series VI, Papers Relating to the West Indies [968], box 4, folder 15.
89. This quotation related to the Jamaican governor's position on the management of the Spanish trade, as overheard in a private conversation and then spread. See Jamaica Archives, Spanish Town, 1b/5/2/1, p. 216 (August 4, 1686). For political contentions between men in and around the Bridgetown coffeehouse, see LT, vol. 3, pp. 232–34 (November 1686).
90. *Barbadoes Packet*, pp. 50–51.
91. *Self-Flatterer*, sig. A2[1]v.
92. [Bourke], *Letter Concerning the Privileges of the Assembly of Jamaica*, pp. 2–3.
93. [Frere], *Short History of Barbados*, p. 84; and [Bourke], *Privileges of the Island of Jamaica Vindicated*, xix.
94. On the complex relationship between "political slavery" and "chattel slavery," particularly in the work of John Locke, who wanted to argue for opposing "tyrants" while owning slaves, see Mary Nyquist, *Arbitrary Rule: Slavery, Tyranny, and the Power of Life and Death* (Chicago: University of Chicago Press, 2013). See also Susan Buck-Morss, *Hegel, Haiti, and Universal History* (Pittsburgh, PA: University of Pittsburgh Press, 2009).
95. [Bourke], *Letter Concerning the Privileges of the Assembly of Jamaica*, pp. 24, 27, and 28.
96. [John Dickinson], *An Address to the Committee of Correspondence in Barbados* (Philadelphia, 1766), pp. 8 and 10. On the context for this, see O'Shaughessy, *Empire Divided*, pp. 101–4.
97. [Frere], *Short History of Barbados*, p. 117.
98. *An Essay Concerning Slavery, and the Danger Jamaica is Expos'd to from the Too Great Number of Slaves* (London, 1746), pp. 12 and 34.

99. Fuentes, *Dispossessed Lives*, p. 143.

100. Maddock Dillon, *New World Drama*, pp. 9 and 182.

101. See, for example, Jerome Handler, "Slave Revolts and Conspiracies in Seventeenth-Century Barbados," *New West Indian Guide* 56, nos. 1/2 (1982): 5–42; Renee K. Harrison, *Enslaved Women and the Art of Resistance in Antebellum America* (New York: Palgrave Macmillan, 2009); Richard Cullen Rath, "Drums and Power: Ways of Creolizing Music in South Carolina and Georgia, 1730–1790," in *Creolization in the Americas: Cultural Adaptations to the New World*, ed. Stephen Reinhardt and David Buisseret (Arlington: Texas A&M Press, 2000), pp. 90–130; Jason Sharples, "Weapons of the Weak," in *The Princeton Companion to Atlantic History*, ed. Joseph C. Miller (Princeton, NJ: Princeton University Press, 2015), pp. 491–95; and John Thornton, *Africa and Africans in the Making of the Atlantic World, 1400–1800* (Cambridge: Cambridge University Press, 1992), pp. 221–29.

102. K. Wilson, "Performance of freedom"; and Ogborn, "War of Words."

103. BL Add. MS 12431, f.95; and Giorgio Agamben, *Homo Sacer: Sovereign Power and Bare Life* (Stanford, CA: Stanford University Press, 1998). Jessica Krug interprets these as differences between Maroon leaders over how far to oppose the plantation system; see Jessica A. Krug, "Social Dismemberment, Social (Re)membering: Obeah Idioms, Kromanti Identities and the Trans-Atlantic Politics of Memory, c. 1675–Present," *Slavery and Abolition* 35, no. 4 (2014): 537–58.

104. *Memoirs and Anecdotes of Philip Thicknesse* (Dublin, 1790), p. 74.

105. Lynda Day, *Gender and Power in Sierra Leone: Women Chiefs of the Last Two Centuries* (New York: Palgrave Macmillan, 2012); Walter Rucker, *Gold Coast Diasporas: Identity, Culture, and Power* (Bloomington: Indiana University Press, 2015); Flora E. S. Kaplan, ed., *Queens, Queen Mothers, Priestesses, and Power: Case Studies in African Gender* (New York: New York Academy of Sciences, 1997); Edna G. Bay, *Wives of the Leopard: Gender, Politics and Culture in the Kingdom of Dahomey* (Charlottesville: University of Virginia Press, 1998); and Emmanuel K. Akyeampong and Pashington Obeng, "Spirituality, Gender and Power in Asante History," *International Journal of African Historical Studies* 28, no. 3 (1995): 481–508.

106. Monica Schuler, "Ethnic Slave Rebellions in the Caribbean and Guianas," *Journal of Social History* 3, no. 4 (1970): 374–85; and Thornton, *Africa and Africans*.

107. *Great Newes from the Barbadoes; Or, A True and Faithful Account of the Grand Conspiracy of the Negroes against the English* (London, 1676), p. 9.

108. Thomas Thistlewood's diary (June 7, 1760), quoted in Maria Alessandra Bollettino, "Slavery, War and Britain's Atlantic Empire: Black Soldiers, Sailors, and Rebels in the Seven Years' War" (PhD thesis, University of Texas at Austin, 2009), p. 204.

109. *JAJ*, vol. 5, pp. 233–34 (December 5 and 6, 1760); and *HJ*, 2:455.

110. Thomas Thistlewood's diary (September 2, 1760), quoted in Bollettino, "Slavery, War and Britain's Atlantic Empire," p. 237n580.

111. See Vincent Brown, "Spiritual Terror and Sacred Authority in Jamaican Slave Society," *Slavery and Abolition* 24, no. 1 (2003): 24–53.

112. *Report of the Lords of the Committee of Council Appointed for the Consideration of all Matters Relating to Trade and Foreign Plantations . . . Concerning the Present State of the Trade to Africa, and Particularly the Trade in Slaves* (London, 1789, hereafter cited as *LTST*), part 3, Jamaica A, no. 22-6 (repeated in *HCC*, p. 118).

113. For an intellectual history of this sort of "political system," see James H. Sweet, *Domingos Álvares, African Healing, and the Intellectual History of the Atlantic World* (Chapel Hill: University of North Carolina Press, 2011), p. 231.

114. In 1776, the Jamaican governor Basil Keith noted of "the Creole Negroes" that "there was every reason to fear that they had connections throughout the whole Island" and recommended that the Lords of Trade read Long's *History of Jamaica* to understand the Maroons. See TNA CO137/71 Jamaica, Original Correspondence, Basil Keith to the Lords of Trade, August 6, 1776, ff.228r and 230r.

115. Quoted in Richard B. Sheridan, "The Jamaican Slave Insurrection Scare of 1776 and the American Revolution," *Journal of Negro History* 61, no. 3 (1976): 290–308, at pp. 300–301.

116. Julius Sherrard Scott III, "The Common Wind: Currents of Afro-American Communication in the Era of the Haitian Revolution" (PhD diss., Duke University, 1986); and Ada Ferrer, *Freedom's Mirror: Cuba and Haiti in the Age of Revolution* (Cambridge: Cambridge University Press, 2014).

117. Luise White, *Speaking with Vampires: Rumor and History in Colonial Africa* (Berkeley: University of California Press, 2000), p. 58; and Ada Ferrer, "Speaking of Haiti: Slavery, Revolution, and Freedom in Cuban Slave Testimony," in *The World of the Haitian Revolution*, ed. David Patrick Geggus and Norman Fiering (Bloomington: Indiana University Press, 2009), pp. 223–47.

118. Quoted in Scott, "Common Wind," p. 20.

119. Sarah E. Johnson, *The Fear of French Negroes: Transcolonial Collaboration in the Revolutionary Americas* (Berkeley: University of California Press, 2012).

120. Houghton Library, Harvard University, Vassall Correspondence III, HOU b MS Am 1250, Folder 84, Leonard Stedman to William Vassall, St. Elizabeth, Jamaica, June 7, 1760, f.1r.

121. NLJ MS 1861, Robert Rollo Gillespie to Earl Balcarres, Spanish Town, September 24, 1796.

122. Ogborn, "War of Words."

123. Sheridan, "Jamaican Slave Insurrection Scare of 1776"; and Michael Craton, *Testing the Chains: Resistance to Slavery in the British West Indies* (Ithaca, NY: Cornell University Press, 1982).

124. Craton, *Testing the Chains*; David Barry Gaspar, *Bondmen and Rebels: A Study of Master-Slave Relations in Antigua* (Baltimore: Johns Hopkins University Press, 1985); Jason T. Sharples, "Hearing Whispers, Casting Shadows: Jailhouse Conversation and the Production of Knowledge during the Antigua Slave Conspiracy Investigation of 1736," in *Buried Lives: Incarcerated in Early America*, ed. Michele Lise Tarter and Richard Bell (Athens: University of Georgia Press, 2012), pp. 35–59; and Michael P. Johnson, "Denmark Vesey and his Co-Conspirators," *William and Mary Quarterly* 58, no. 4 (2001): 915–76.

125. TNA CO28/1, Barbados, Original Correspondence, Governor James Kendall to the Lords of Trade, Barbados, November 3, 1692, f.205r; and Jason T. Sharples, "Discovering Slave Conspiracies: New Fears of Rebellion and Old Paradigms of Plotting in Seventeenth-Century Barbados," *American Historical Review* 120, no. 3 (2015): 811–43.

126. TNA CO137/71, Examination of Sam, July 19, 1776, f.252.

127. [Nathaniel Saltonstall], *A Continuation of the State of New-England . . . . Together with an Account of the Intended Rebellion of the Negroes in Barbadoes* (London, 1675), p. 19 ("Councel"); TNA CO31/1, Journal of the Proceedings of the Governor and Council of Barbados, May 29, 1660, to November 30, 1686, and February 16, 1686, f.385v ("Consult & Contrive"); final quote from copy of all "Judicial Proceedings relative to the Trial and Punishment of Rebels, or alleged Rebels, in the Island of Jamaica,

since the 1st of January 1823," in *Papers Relating to the Manumission, Government and Population of the Slaves in the West-Indies, 1822–1824*, House of Commons, British Parliamentary Papers, 1825 (66) xxv: 37–132 (hereafter cited as JPTPR), at p. 64.

128. TNA CO137/71, Hanover Magistrates to Sir Basil Keith, July 29 and August 2, 1776, and General Palmer to Sir Basil Keith, July 30, 1776, ff.247, 280, and 296. Edward Long was convinced that the Maroons were involved; see BL Add. MS 12405, f.376v. For St. James's in 1823, see JPTPR, p. 82.

129. Those giving evidence under threat of death certainly tried to use their testimony to keep themselves alive and, in some ways, to limit the effects of their words on others. See TNA CO137/71, Hanover Magistrates to Sir Basil Keith, Lucea, August 4, 7, 16, 21, and 29, 1776, ff.340v, 345v, 350, 352, and 357r. For constrasting approaches, see Jason T. Sharples, "The Flames of Insurrection: Fearing Slave Conspiracy in Early America, 1670–1780" (PhD thesis, Princeton University, 2010); and Aisha K. Finch, *Rethinking Slave Rebellion in Cuba: La Escalera and the Insurgencies of 1841–1844* (Chapel Hill: University of North Carolina Press, 2010).

130. Walter Johnson, "On Agency," *Journal of Social History* 37, no. 1 (2003): 113–24, quotation at p. 118.

131. TNA CO137/71, Examination of the Negro Adam, July 17, 1776, f.234v.

132. Finch, *Rethinking Slave Rebellion in Cuba*.

133. On the revolts on the Danish island of St. John and on Antigua, see Rucker, *Gold Coast Diasporas*; Gaspar, *Bondmen and Rebels*; and Saidiya Hartman, *Lose Your Mother: A Journey Along the Atlantic Slave Route* (New York: Farrer, Strauss and Giroux, 2007), pp. 91–95. Vincent Brown, *Tacky's Revolt: The Story of an Atlantic Slave War* (Cambridge MA: Harvard University Press, 2019) reconstructs the complex Atlantic history of Apongo/Wager, whom Thistlewood called "King of the rebels," and said had been "a Prince in Guinea, tributary to the King of Dome [Dahomey]"; Douglas Hall, *In Miserable Slavery: Thomas Thistlewood in Jamaica, 1750–86* (Kingston: University of the West Indies Press, 1989), pp. 105–6

134. In 1776, evidence was given that "Ebos," "Coromantees," and "Creoles" were each led by those they "Elected King"; TNA CO137/71, Governor Basil Keith to the Lords of Trade, Jamaica, August 6, 1776, and Examination of Sam, ff.228 and 253. For the translation of nonstatist political repertoires between Africa and the Americas, see Jessica A. Krug, *Fugitive Modernities: Kisama and the Politics of Freedom* (Durham NC: Duke University Press, 2018).

135. John Thornton, "'I Am the Subject of the King of Congo': African Political Ideology and the Haitian Revolution," *Journal of World History* 4, no. 2 (1993): 181–214.

136. BL Add. MS 12431, Copy of letter from J. Lewis to unknown recipient, Westmoreland, December 20, 1743, f.99.

137. Helyer Letters, Somerset Record Office, DD/WHh 1809 part 3, William Helyer to Jonathan Ryan, June 8, 1678. On the question of the woman's role in these events, see Finch, *Rethinking Slave Rebellion in Cuba*.

138. Evidence given by Pontack and Charles, enclosed in TNA CO137/71, Hanover Magistrates to Sir Basil Keith, 29 and 31, 1776, ff.276v and 288r.

139. Kwesi Yankah, *Speaking for the Chief: Okyeame and the Politics of Akan Royal Oratory* (Bloomington: Indiana University Press, 1995).

140. Rucker, *Gold Coast Diasporas*, p. 8.

141. Thornton, "'I Am the Subject of the King of Congo,'" p. 210, and Kate Ramsey, *The Spirits and the Law: Vodou and Power in Haiti* (Chicago: University of Chicago Press, 2011), p. 17, both note the enduring power of secret societies (*sosyete sekrè*) of this

sort in Haiti. Buck-Morss, *Hegel, Haiti, and Universal History*, stresses an Atlantic history of Freemasonry.

142. These practices are understood in terms of political philosophy in Kwame Gyekye, *Tradition and Modernity: Philosophical Reflections on the African Experience* (New York: Oxford University Press, 1997), chap. 4; and Richard H. Bell, *Understanding African Philosophy: A Cross-Cultural Approach to Classical and Contemporary Issues* (London: Routledge, 2002), pp. 109–14.

143. Far fewer people were taken from the Upper Guinea Coast to Jamaica and Barbados than from the Gold Coast—37,963 enslaved Africans arrived from there in the period 1626–1800, compared with 282,839 from the Gold Coast; see Transatlantic Slave Trade Database (website), Emory University, copyright 2013, http://www.slavevoyages.org—but they were still a significant element of the population, and there were also comparable forms of social organization in other areas of West Africa. For "power associations," see George E. Brooks, *Landlords and Strangers: Ecology, Society and Trade in Western Africa, 1000–1630* (Boulder, CO: Westview Press, 1994).

144. John Newton, *Thoughts upon the African Slave Trade* (London, 1788), p. 26.

145. Kenneth Little, *The Mende of Sierra Leone* (London: Routledge, 1967), p. 184.

146. Day, *Gender and Power in Sierra Leone*.

147. Marcus Rediker, *The "Amistad" Rebellion: An Atlantic Odyssey of Slavery and Freedom* (New York: Penguin Books, 2013).

148. Jean Besson, *Transformations of Freedom in the Land of the Maroons: Creolization in the Cockpits, Jamaica* (Kingston: Ian Randle, 2016).

149. Walter Johnson, "Time and Revolution in African America: Temporality and the History of Atlantic Slavery," in *A New Imperial History: Culture, Identity and Modernity in Britain and the Empire, 1660–1840*, ed. Kathleen Wilson (Cambridge: Cambridge University Press, 2004), pp. 197–215; and Finch, *Rethinking Slave Rebellion in Cuba*.

150. V. Brown, *Tacky's Revolt*. For the 1690s, see TNA CO28/1, Governor James Kendall to the Lords of Trade, Barbados, November 3, 1692, f.205r. For the 1730s, see NLJ MS 1020, M. Bladen to Sir Robert Walpole, London, October 31, 1734, f.5v, in which Bladen argues that the Maroons were in correspondence with the French and the Spanish and would assist them "on the first Declaration of War," hence the need to make the treaties. For the 1760s and 1770s, see Bollettino, "Slavery, War and Britain's Atlantic Empire"; and Gerald Horne, *The Counter-Revolution of 1776: Slave Resistance and the Origins of the United States of America* (New York: New York University Press, 2014). For the 1790s, see Ogborn, "War of Words"; and Laurent Dubois, *Avengers of the New World: The Story of The Haitian Revolution* (Cambridge, MA: Harvard University Press, 2005).

151. Governor Woodley to Secretary of State, April 20, 1770, quoted in Gaspar, *Bondmen and Rebels*, p. 212.

152. Laurent Dubois, "An Enslaved Enlightenment: Rethinking the Intellectual History of the French Atlantic," *Social History* 31, no. 1 (2006): 1–14; and Greene, "Liberty, Slavery and the Transformation of British Identity."

153. TNA CO137/71, Governor Keith to the Lords of Trade, Jamaica, August 6, 1776, ff.228v–9r.

154. Rucker, *Gold Coast Diasporas*, p. 16, argues that these were not "liberalized or liberal subjects moulded by Europe's Age of Reason" because they "developed ideas of liberty and autonomy that favored the corporate or collective over the individual." On the gendering of such speech, see Finch, *Rethinking Slave Rebellion in Cuba*.

155. Maddock Dillon, *New World Drama*, p. 19.

156. TNA CO29/1, Barbados Entry Books, 1627–1674, Lord William Willoughby to the Lords of His Majesty's Council, July 9, 1668, f.58v.

157. Rucker, *Gold Coast Diasporas*; and Thornton, *Africa and Africans*.

158. Rebecca Shumway, *The Fante and the Transatlantic Slave Trade* (Rochester, NY: University of Rochester Press, 2011), p. 114.

159. *Great Newes from the Barbadoes*, p. 12.

160. TNA CO29/1, Willoughby to His Majesty's Council, f.58v.

CHAPTER THREE

1. Edinburgh Royal Botanic Gardens archive, William Wright, *Hortus Jamaicensis* (3 vols.), vol. 1, specimen no. 39, and vol. 2, no. 226. See also William Wright, "An Account of the Medicinal Plants Growing in Jamaica," *London Medical Journal* 8 (1787): 217–95.

2. Steven Shapin, *A Social History of Truth: Civility and Science in Seventeenth-Century England* (Chicago: University of Chicago Press, 1994); and Margaret C. Jacob and Larry Stewart, *Practical Matter: Newton's Science in the Service of Industry and Empire, 1687–1851* (Cambridge, MA: Harvard University Press, 1994).

3. Adrian Johns, *The Nature of the Book: Print and Knowledge in the Making* (Chicago: University of Chicago Press, 1998); James A. Secord, *Victorian Sensation: The Extraordinary Publication, Reception, and Secret Authorship of Vestiges of the Natural History of Creation* (Chicago: University of Chicago Press, 2000); and James A. Secord, "How Scientific Conversation Became Shop Talk," *Transactions of the Royal Historical Society* 17 (2007): 129–56.

4. James Delbourgo, "Science," in *The British Atlantic World, 1500–1800*, 2nd edition, ed. David Armitage and Michael J. Braddick (London: Palgrave Macmillan, 2002) pp. 92–110, at p. 93.

5. Bruno Latour, *Science in Action: How to Follow Scientists and Engineers through Society* (Cambridge, MA: Harvard University Press, 1987).

6. Bruno Latour, *An Inquiry into Modes of Existence: An Anthropology of the Moderns*, trans. Catherine Porter (Cambridge, MA: Harvard University Press, 2013), pp. 109 and 123–49.

7. This chapter uses more material from Jamaica, due to the more extensive transformation of the landscape of Barbados and its lack of a botanical garden in the eighteenth century.

8. Richard Drayton, *Nature's Government: Science, Imperial Britain, and the "Improvement" of the World* (New Haven, CT: Yale University Press, 2000), p. 93; and Londa Schiebinger and Claudia Swan, eds., *Colonial Botany: Science, Commerce, and Politics in the Early Modern World* (Philadelphia: University of Pennsylvania Press, 2007), p. 13.

9. Londa Schiebinger, *Plants and Empire: Colonial Bioprospecting in the Atlantic World* (Cambridge, MA: Harvard University Press, 2004); and Harold J. Cook, "Global Economies and Local Knowledge in the East Indies: Jacobus Bontius Learns the Facts of Nature," in Schiebinger and Swan, *Colonial Botany*, pp. 100–118.

10. Susan Scott Parrish, *American Curiosity: Cultures of Natural History in the Colonial British Atlantic World* (Chapel Hill: University of North Carolina Press, 2006); and Susan Scott Parrish, "Diasporic African Sources of Enlightenment Knowledge," in *Science and Empire in the Atlantic World*, ed. James Delbourgo and Nicholas Dew (New York: Routledge, 2008), pp. 281–310.

11. Pratik Chakrabarti, *Materials and Medicine: Trade, Conquest and Therapeutics in the Eighteenth Century* (Manchester, UK: Manchester University Press, 2010), p. 157.

12. Henry Barham, *Hortus Americanus* (Kingston, 1794), pp. 148–49.

13. Barham, *Hortus Americanus*, p. 149.

14. Michel Foucault, *The Birth of the Clinic: An Archaeology of Medical Perception* (London: Tavistock, 1976), xviii.

15. Nicholas Jewson, "The Disappearance of the Sick Man from Medical Cosmology, 1770–1870," *Sociology* 10 (1976): 225–44.

16. Mary E. Fissell, "The Disappearance of the Patient's Narrative and the Invention of Hospital Medicine," in *British Medicine in an Age of Reform*, ed. Roger French and Andrew Wear (London: Routledge, 1991), pp. 92–109; Roy Porter, "The Rise of Physical Examination," in *Medicine and the Five Senses*, ed. W. F. Bynum and Roy Porter (Cambridge: Cambridge University Press, 1993), pp. 179–97; and Stanley Joel Reiser, *Medicine and the Reign of Technology* (Cambridge: Cambridge University Press, 1978), chap. 1.

17. Vincent Brown, *The Reaper's Garden: Death and Power in the World of Atlantic Slavery* (Cambridge, MA: Harvard University Press, 2008); Suman Seth, *Difference and Disease: Medicine, Race and the Eighteenth-Century British Empire* (Cambridge: Cambridge University Press, 2018); and Emily Senior, *The Caribbean and the Medical Imagination, 1764–1834: Slavery, Disease and Colonial Modernity* (Cambridge: Cambridge University Press, 2018).

18. Richard B. Sheridan, *Doctors and Slaves: A Medical and Demographic History of Slavery in the British West Indies, 1680–1834* (Cambridge: Cambridge University Press, 1985).

19. *Memoir of the late William Wright M.D.* (Edinburgh, 1828), pp. 45–46.

20. Thomas Dancer, *The Medical Assistant: Or Jamaica Practice of Physic; Designed Chiefly for the Use of Families and Plantations* (Kingston, 1801), p. 253.

21. BL Sloane MS 4078, Henry Fuller to Hans Sloane, Jamaica, August 21, 1696, f.114.

22. Hans Sloane, *A Voyage to the Islands Madera, Barbadoes, Nieves, St Christophers, and Jamaica*, 2 vols. (London, 1707 and 1725), 2:xiv (hereafter cited as *Natural History of Jamaica*).

23. Sloane, *Natural History of Jamaica*, 1:cxxviii, xcix, cxl, and cxlviii; and Wendy D. Churchill, "Bodily Differences? Gender, Race, and Class in Hans Sloane's Jamaican Medical Practice, 1687–1688," *Journal of the History of Medicine and Allied Sciences* 60 (2005): 391–444.

24. [William King], *The Present State of Physick in the Island of Cajamai* (London? 1710?), p. 4.

25. Andrew Wear, *Knowledge and Practice in English Medicine, 1550–1680* (Cambridge: Cambridge University Press, 2000).

26. BL Sloane MS 4043, Henry Barham to Hans Sloane, St. Jago de la Vega, May 10, 1712, f.45r.

27. Bodleian Library, Oxford University (hereafter cited as BLOU), Rawlinson MS C. 833, "A catalogue of herbs, shrubs, and trees which grows in the island of Jamaica by H. B., 1715," f.15r.

28. James Robertson, "Knowledgeable Readers: Jamaican Critiques of Sloane's Botany," in *From Books to Bezoars: Sir Hans Sloane and His Collections*, ed. Alison Walker, Arthur MacGregor, and Michael Hunter (London: British Library, 2012), pp. 80–89.

29. Roy Porter "The Patient in England, c. 1660–c. 1800," in *Medicine in Society: His-*

torical Essays, ed. Andrew Wear (Cambridge: Cambridge University Press, 1992), pp. 91–118.

30. Elaine Leong, "Making Medicines in the Early Modern Household," *Bulletin of the History of Medicine* 82, no. 1 (2008): 145–68.

31. Sasha Turner, *Contested Bodies: Pregnancy, Childrearing, and Slavery in Jamaica* (Philadelphia: Pennsylvania University Press, 2017).

32. Barham to Sloane, May 10, 1712.

33. BL Sloane MS 4045, Henry Barham to Hans Sloane, London, April 17, 1718, f.108r. See BLOU Rawlinson MS C. 833.

34. Sloane, *Natural History of Jamaica*, 2:386, quoting Barham's manuscript; Barham, *Hortus Americanus*, p. 90.

35. BL Sloane MS 4045, Henry Barham to Hans Sloane, London, December 11, 1717, f.77v; Sloane, *Natural History of Jamaica*, 1:liv.

36. Barham to Sloane, December 11, 1717, f.77v.

37. Barham to Sloane, December 11, 1717, 78r.

38. Barham, *Hortus Americanus*, pp. 18, 86, 113, 121, 146, and 210. Peanuts and sesame were used as medicines too. See also BL Sloane MS 2302, Thomas Walduck to James Petiver, Barbados, September 17, 1712, f.28, informing him that he was able to "know the Qualities of Herbs Roots, barks &c as I have negros and other Opportunities to imploy my time in those things."

39. NLJ, Natural History Department, Anthony Robinson Notebooks, vol. 5, Transcribed Letter to Robinson, no date, f.277. Sloane, *Natural History of Jamaica*, 1:xxii, discusses "the Bichy or Buzzee tree in *Guinea*." See also Kathleen S. Murphy, "Collecting Slave Traders: James Petiver, Natural History, and the British Slave Trade," *William and Mary Quarterly* 70, no. 4 (2013): 637–70.

40. Londa Schiebinger, *Secret Cures of Slaves: People, Plants, and Medicine in the Eighteenth-Century Atlantic World* (Stanford, CA: Stanford University Press, 2017).

41. Hans Sloane, *Catalogus plantarum quæ in insula Jamaica* (London, 1696); Schiebinger, *Plants and Empire*.

42. Barham, *Hortus Americanus*, pp. 52, 88, 94, 115–16, 171–72, and 122. See also BLOU Rawlinson MS C. 833; and Boston Public Library MS Eng. 179, Joseph Smallwood, Memorandum on the Flora and Fauna of Jamaica, *c.* 1765.

43. Wright, *Hortus Jamaicensis*, vol. 1, no. 119; and Barham, *Hortus Americanus*, p. 9.

44. For example, Barham, *Hortus Americanus*, pp. 19, 77, and 196; Patrick Browne, *The Civil and Natural History of Jamaica* (London, 1756), p. 344. On the communication of such knowledge, see Kathleen S. Murphy, "Translating the Vernacular: Indigenous and African knowledge in the Eighteenth-Century British Atlantic," *Atlantic Studies* 8 (2011): 29–48.

45. Barham, *Hortus Americanus*, p. 168.

46. Manuscript notes from Barham on Vervain inserted after p. 170 of the Natural History Museum (London) copy of Sloane, *Natural History of Jamaica*, vol. 1. See also p. 171. For thumbnails, see Barham, *Hortus Americanus*, p. 34; *HJ*, 3:779; and Matthew Gregory Lewis, *Journal of a West India Proprietor, Kept during a Residence in the Island of Jamaica* (London, 1834), p. 330.

47. Robert Voeks, "African Magic and Medicine in the Americas," *Geographical Review* 83 (1993): 66–78; James H. Sweet, *Domingos Álvares, African Healing, and the Intellectual History of the Atlantic World* (Chapel Hill: University of North Carolina Press, 2011); and Pablo F. Gómez, *The Experiential Caribbean: Creating Knowledge and Heal-*

ing in the Early Modern Atlantic (Chapel Hill: University of North Carolina Press, 2017).

48. Robert Voeks, "Spiritual Flora of Brazil's African Diaspora: Ethnobotanical Conversations in the Black Atlantic," *Journal for the Study of Religion, Nature and Culture* 6, no. 4 (2012): 501–22; and Pablo F. Gómez, "The Circulation of Bodily Knowledge in the Seventeenth-Century Black Spanish Caribbean," *Social History of Medicine* 26, no. 3 (2013): 383–402.

49. Philip J. Havik, "Hybridising Medicine: Illness, Healing and the Dynamics of Reciprocal Exchange on the Upper Guinea Coast (West Africa)," *Medical History* 60, no. 2 (2016): 181–205; and Tinde van Andel, "The Reinvention of Household Medicine by Enslaved Africans in Suriname," *Social History of Medicine* 29, no. 4 (2016): 676–94.

50. Jerome S. Handler, "Slave Medicine and Obeah in Barbados, circa 1650 to 1834," *New West Indian Guide* 74, nos. 1/2 (2000): 57–90; Michael Laguerre, *Afro-Caribbean Folk Medicine* (South Hadley, MA: Bergin and Garvey, 1987); Sheridan, *Doctors and Slaves*, chap. 3. Gómez understands words as "a sensual sonic phenomenon that project a force and a physical, phenomenological power independent of [their] meaning" (*Experiential Caribbean*, p. 102).

51. Scott Parrish, *American Curiosity*; and Philip D. Morgan, "Slaves and Livestock in Eighteenth-Century Jamaica: Vineyard Pen, 1750–1751," *William and Mary Quarterly* 52, no. 1 (1995): 47–76.

52. Sloane, *Natural History of Jamaica*, 1:cxlii.

53. Sheridan, *Doctors and Slaves*, p. 89; P. Browne, *Civil and Natural History of Jamaica*, p. 300; Sloane, *Natural History of Jamaica*, 2:383.

54. Barham, *Hortus Americanus*, p. 101. See also Sloane, *Natural History of Jamaica*, 1:lv; BL Sloane MS 4045, Henry Barham to Hans Sloane, London, November 21, 1717, f.69r.

55. *Memoir of the late William Wright*, pp. 27 and 92.

56. Lewis, *Journal of a West India Proprietor*, p. 321. On these questions of "ethics," see Schiebinger, *Secret Cures of Slaves*.

57. For an earlier metropolitan version, see Richard Coulton, "Curiosity, Commerce, and Conversation in the Writing of London Horticulturalists during the Early-Eighteenth Century" (PhD thesis, Queen Mary University of London, 2005).

58. April G. Shelford, "'Birds of a Feather': Natural History and Male Sociability in Eighteenth-Century Jamaica" (paper presented to the seventh symposium of the Social History Project, University of the West Indies, Jamaica, March 2006).

59. State Library of New South Wales. Papers of Sir Joseph Banks, Section 5: Gardeners and collectors. Series 14 (hereafter cited as SLNSW Banks Papers), William Wright to Joseph Banks, Edinburgh, April 28, 1788, and March 6, 1789; and BL Add. MS 32439, William Wright to Robert Brown, Edinburgh, February 15, 1801, f.27r.

60. SLNSW Banks Papers, William Wright to Joseph Banks, Edinburgh, November 6, 1778, Port Royal, October 23, 1783, and Edinburgh, November 23, 1790.

61. Scott Parrish, *American Curiosity*, chap. 4. For a British instance, see Victoria R. M. Pickering, "Putting Nature in a Box: Hans Sloane's 'Vegetable Substances' Collection" (PhD thesis, Queen Mary University of London, 2016), chap. 5.

62. SLNSW Banks Papers, William Wright to Joseph Banks, Trelawny, Jamaica, August 20, 1784 (on visiting Dr. Drummond of Westmoreland) and March 29, 1785.

63. Scott Parrish, *American Curiosity*.

64. Margaret Meredith, "Friendship and Knowledge: Correspondence and Communication in Northern Trans-Atlantic Natural History, 1780–1815," in *The Brokered World: Go-Betweens and Global Intelligence, 1770–1820*, ed. Simon Schaffer et al. (Sagamore Beach, MA: Science History Publications, 2009), pp. 151–91, at p. 158.

65. Barham to Sloane, May 10, 1712. For Sloane's "network," see James Delbourgo, *Collecting the World: The Life and Curiosity of Hans Sloane* (London: Allen Lane, 2017).

66. BL Sloane MS 4045, Barham to Sloane, London, October 21, November 6 and 21, and December 11, 1717, and January 29 and April 17 and 29, 1718, ff.55, 58r–60r, 68r–71r, 77r–79v, 89r–91v, 108r–v, 110r–v; and BL Sloane MS 4047, Henry Barham to Hans Sloane, April 17, 1725, f.337r.

67. Schiebinger, *Plants and Empire*, chap. 5.

68. Wright to Banks, October 23, 1783, and November 18, 1790.

69. William Wright to Joseph Banks, Kingston, Jamaica, July 12, 1783, and Edinburgh, January 21, 1789, and May 5, 1790, SLNSW Banks Papers; and *Memoir of the Late William Wright*, p. 43.

70. BL Add. MS 18275A, Long Papers, Thomas Thistlewood to Edward Long, ff.111v, 113v, and 115v (including a poem of Robinson's entitled "Man of Consequence"); Trevor Burnard, *Mastery, Tyranny, and Desire: Thomas Thistlewood and His Slaves in the Anglo-Jamaican World* (Chapel Hill: University of North Carolina Press, 2004), p. 102; and Douglas Hall, *In Miserable Slavery: Thomas Thistlewood in Jamaica, 1750–86* (Kingston: University of the West Indies Press, 1989), p. 165.

71. NLJ MS 178 II, Anthony Robinson's *Jamaica Birds*, p. 52, depicts a "Crabcatcher" bought from "a Negro-Boy."

72. NLJ, Natural History Department, Anthony Robinson Notebooks, vol. 5, George Spence to Anthony Robinson, Lucea, May 17 and June 22, 1765, ff.280, 281, and 283.

73. Andrew Cunningham, "The Culture of Gardens," in *Cultures of Natural History*, ed. Nicolas Jardine, James A. Secord, and Emma C. Spary (Cambridge: Cambridge University Press, 1996), pp. 3–13; and Thomas Hallock, "Male Pleasure and the Genders of Eighteenth-Century Botanic Exchange: A Garden Tour," *William and Mary Quarterly* 62, no. 4 (1996): 697–718.

74. D. Hall, *In Miserable Slavery*, pp. 164, 236, 264, and 309.

75. Burnard, *Tyranny, Mastery, and Desire*, p. 126.

76. D. Hall, *In Miserable Slavery*, p. 229.

77. Burnard, *Tyranny, Mastery, and Desire*, p. 85; and Barry W. Higman, *Plantation Jamaica: Capital and Control in a Colonial Economy, 1750–1850* (Kingston: University of West Indies Press, 2005).

78. BL Add. MS 18275A, "Plants growing in Thos. Thistlewoods garden at Bread Nutt Island Pen, June 1776," ff.117v–121r, and Thomas Thistlewood to Edward Long, Westmoreland, Jamaica, June 17, 1776, f.128; and D. Hall, *In Miserable Slavery*, pp. 125, 164, 228, 238, and 244. See also Douglas Hall, "Planters, Farmers and Gardeners in Eighteenth-Century Jamaica," in *Slavery, Freedom and Gender: The Dynamics of Caribbean Society*, ed. Brian L. Moore et al. (Kingston: University of West Indies Press, 2001), pp. 97–114; and Douglas Hall, "Botanical and Horticultural Enterprise in Eighteenth-Century Jamaica," in *West Indies Accounts*, ed. Roderick A. McDonald (Kingston: University of West Indies Press, 1996), pp. 101–25.

79. Miles Ogborn, "Vegetable Empire," in *Worlds of Natural History*, ed. Helen Anne Curry et al. (Cambridge: Cambridge University Press, 2018), pp. 271–86.

80. "Plants growing in Thos. Thistlewoods garden," ff.120v and 121r.

81. D. Hall, *In Miserable Slavery*, p. 229.
82. Elizabeth DeLoughrey, "Globalizing the Routes of Breadfruit and Other Bounties," *Journal of Colonialism and Colonial History* 8, no. 3 (2007), no pagination.
83. D. Hall, *In Miserable Slavery*, p. 238; Jill H. Casid, *Sowing Empire: Landscape and Colonization* (Minneapolis: University of Minnesota Press, 2005); and Judith A. Carney and Richard Nicholas Rosomoff, *In the Shadow of Slavery: Africa's Botanical Legacy in the Atlantic World* (Berkeley: University of California Press, 2009).
84. D. Hall, *In Miserable Slavery*, p. 312n8.
85. D. Hall, *In Miserable Slavery*, pp. 269, 300, and 309.
86. Roderick A. McDonald, *The Economy and Material Culture of Slaves: Goods and Chattels on the Sugar Plantations of Jamaica and Louisiana* (Baton Rouge: Louisiana State University Press, 1993).
87. D. Hall, *In Miserable Slavery*, p. 231; Casid, *Sowing Empire*, p. 209; Burnard, *Tyranny, Mastery, and Desire*, pp. 228–40.
88. D. Hall, *In Miserable Slavery*, pp. 160 and 263; and P. Morgan, "Slaves and Livestock."
89. D. Hall, *In Miserable Slavery*, p. 307. See BLYU OSB MSS 176/37, Thistlewood Diary, 1786, pp. 47 and 48 (March 20 and 23, 1786).
90. Jennifer Reed calculates that Thistlewood committed at least 147 rapes in the garden. Jennifer Reed, "'Sites of Terror' and Affective Geographies on Thomas Thistlewood's Breadnut Island Pen," *Caribbeana: The Journal of the Early Caribbean Society* 1, no. 1 (2016): 34–62.
91. Burnard, *Tyranny, Mastery, and Desire*.
92. Drayton, *Nature's Government*; and Richard H. Grove, *Green Imperialism: Colonial Expansion, Tropical Island Edens and the Origins of Environmentalism, 1600–1860* (Cambridge: Cambridge University Press, 1995).
93. Drayton, *Nature's Government*, p. 121.
94. Alan Eyre, *The Botanic Gardens of Jamaica* (London: Andre Deutsch, 1966). In the 1760s, Anthony Robinson had lamented the "neglect" of "the Culture of Plants" in Jamaica and thought that nothing would be done "Unless leading Men of Publick Spirit wou'd encourage it by their own Examples"; see NLJ, Natural History Department, Anthony Robinson Notebooks, vol. 4, f.121.
95. William Wright to Joseph Banks, Edinburgh, June 1, 1788, SLNSW Banks Papers.
96. BL Add. MS 22678, Long Papers, Thomas Dancer to Edward Long, [Jamaica], July 24, 1789, f.10v.
97. Wright noted in 1790 that "Dr Rutherford gives lectures in a truly learned and Philosophical Stile" at the Edinburgh garden; William Wright to Joseph Banks, Edinburgh, November 18, 1790, SLNSW Banks Papers.
98. A manuscript "Abstract of an Oration deliver'd at A Board of Directors under the new Act (1789) respecting the Bath & the Botanic Garden in Jamaica by T. Dancer M.D. Island Botanist. 1st March 1790" and a newspaper report of the "Substance of an Address, Delivered at a Meeting of the New Bath Directors, By Dr. Dancer, Island Botanist" are both in BL Add. MS 22678, ff.35r–43r and 45r.
99. All quotations are from the printed version; BL Add. MS 22678, f.45r.
100. Kate Davies, "A Moral Purchase: Femininity, Commerce and Abolition, 1788–1792," in *Women, Writing and the Public Sphere, 1700–1830*, ed. Elizabeth Eger et al. (Cambridge: Cambridge University Press, 2001), pp. 133–59.
101. Drayton, *Nature's Government*, p. 115.
102. The manuscript version called them "groveling Kinds"; compare BL Add. MS 22678, ff.37r and 45r.

103. BL Add. MS 22678, ff.41r and 45r.

104. NLJ MS 566, Hinton East to Joseph Banks, Kingston, Jamaica, July 19, 1784. See DeLoughrey, "Globalizing the Routes of Breadfruit."

105. BL Add. MS 22678, f.45r, and Thomas Dancer to Edward Long, Jamaica, July 20, 1791, f.60v.

106. Dancer to Long, Jamaica, July 24, 1789, ff.9r, 9v, and 10r.

107. I am grateful to Julie Kim for highlighting this element of Long's botanical work.

108. BL Add. MS 22678, Edward Long to Thomas Dancer, London, July 6, 1789, f.7v.

109. Dancer to Long, July 24, 1789, ff.10v–11r, 14v.

110. BL Add. MS 22678, Thomas Dancer to Edward Long, Jamaica, [1790?], ff.49v–50r, and July 20, 1791, ff.60v–61r.

111. Thistlewood to Long, June 17, 1776, f.128v. He had also heard about the garden from John Cope in January 1775; see Burnard, *Mastery, Tyranny, and Desire*, p. 119.

112. BL Add. MS 22678, Thomas Dancer to Edward Long, April 8, 1787, and April 6, 1788, ff.1–2 and 3–4 (quotation at f.4r), July 24, 1789, [1790?], December 20, 1790, and April 13, 1791.

113. BL Add. MS 22678, Dancer to Long, Jamaica, [1790?], f.48v; Edward Long to Mr Martin, London, [no date], f.6.

114. Dancer to Long, December 20, 1790, f.57r.

115. SLNSW Banks Papers, Bryan Edwards to Joseph Banks, London, June 9, 1793.

116. Dancer to Long, Jamaica, [1790?], f.49r, and December 20, 1790, f.56v.

117. Thomas Dancer, *Some Observations Respecting the Botanical Gardens* (Jamaica, 1804), pp. 3, 5, and 8. Banks' copy is BL B.93(3).

118. Julie Kim, "St. Vincent Botanical Garden in the Age of Revolution" (paper presented to the Queen Mary Centre for Eighteenth Century Studies seminar series, London, December 7, 2016). When Hinton East's garden came into the ownership of the assembly, it was noted that there were "thirty-nine Negroes belonging to it, many of whom are valuable gardeners"; see D. Hall, "Botanical and Horticultural Enterprise in Eighteenth-Century Jamaica," p. 119.

119. Carney and Rosomoff, *In the Shadow of Slavery*, p. 123. See also Casid, *Sowing Empire*, chap. 5.

120. [Thomas Dancer], *Catalogue of Plants, Exotic and Indigenous, in the Botanical Garden, Jamaica* (St. Jago de la Vega, Jamaica, 1792), p. 4, also included the Akee tree as "introduced by Negroes in some of Mr. Hibbet's ships."

121. Sloane, *Natural History of Jamaica*, vol. 1, sig. B[2]r; 2:61.

122. Sloane, *Natural History of Jamaica*, 2:viii.

123. BL Sloane MS 2032, Thomas Walduck to James Petiver, Barbados, September 17, 1712, f.25r.

124. John Southall, *A Treatise of Buggs* (London, 1730), pp. 12–14.

125. BL Sloane MS 4050, John Southall to Hans Sloane, December 24, 1729, f.250; and Southall, *Treatise of Buggs*, v–vi.

126. Neil Safier, *Measuring the New World: Enlightenment Science and South America* (Chicago: University of Chicago Press, 2008). Compare the presentation of the "agency" and "voice" of the "other" as Sloane abstracted material from Barham's manuscript in *Natural History of Jamaica*, 2:367 (Pickering's herb), 2:370 (*Papaver spinosum*), and 32:87 (hog plum).

127. Although see later examples of "botanical prescription" in Schiebinger, *Secret Cures of Slaves*.

128. BL Sloane MS 4045, Henry Barham to Hans Sloane, London, November 21, 1717, f.68v; and Gómez, *Experiential Caribbean*.

129. Wright, *Hortus Jamaicensis*, vol. 2, no. 341.

130. Drayton, *Nature's Government*, p. 93; Chakrabarti, *Materials and Medicine*.

CHAPTER FOUR

1. [Cynric Williams], *Hamel, the Obeah Man*, 2 vols. (London, 1827), 2:315–17.

2. [Williams], *Hamel, the Obeah Man*, 2:321.

3. Bruno Latour, *Rejoicing: Or the Torments of Religious Speech* (Cambridge: Polity Press, 2011), pp. 2, 27, 36, and 55. The speaker is an imagined figure, not exactly Latour himself.

4. Bruno Latour, *An Inquiry into the Modes of Existence: An Anthropology of the Moderns*, trans. Catherine Porter (Cambridge, MA: Harvard University Press, 2013), pp. 303 and 308–10.

5. There are clear parallels with the calls for ontological turns in anthropology and history; see Eduardo Viveiros De Castro, *The Relative Native: Essays on Indigenous Conceptual Worlds* (Chicago: Hau Books, 2015); and Greg Anderson, "Retrieving Lost Worlds of the Past: The Case for an Ontological Turn," *American Historical Review* 120, no. 3 (2015): 787–810. For a critique, see Zoe Todd, "An Indigenous Feminist's Take on the Ontological Turn: 'Ontology' Is Just Another Word for Colonialism," *Journal of Historical Sociology* 29, no. 1 (2016): 4–22.

6. Latour, *Inquiry into Modes of Existence*, pp. 305, 183, and 185. See also Barbara Herrnstein Smith, "Anthropotheology: Latour Speaking Religiously," *New Literary History* 47 (2016): 331–51.

7. Latour, *Inquiry into Modes of Existence*, pp. 192, 197, and 202. See also Roger Luckhurst, *Zombies: A Cultural History* (London: Reaktion, 2015).

8. BL Sloane MS 2302, Thomas Walduck to James Petiver, November 12, 1710, f.15.

9. Nicholas M. Beasley, *Christian Ritual and the Creation of British Slave Societies, 1650–1780* (Athens: University of Georgia Press, 2009), pp. 11 and 22.

10. Carla Gardina Pestana, *Protestant Empire: Religion and the Making of the British Atlantic World* (Philadelphia: University of Pennsylvania Press, 2009).

11. Joseph Glanvill, *A Seasonable Defence of Preaching: And the Plain Way of It* (London, 1703), pp. 10, 28, 33, and 51; and Adrian Johns, "Coleman Street," *Huntington Library Quarterly* 71, no. 1 (2008): pp. 33–54.

12. Joseph Glanvill, *An Essay Concerning Preaching: Written for the Direction of a Young Divine*, 2nd ed. (London, 1703), pp. 68–69 and 23.

13. John Jennings, *Two Discourses: The First of Preaching Christ; The Second of Particular and Experimental Preaching* (London, 1724), p. 19.

14. Glanvill, *Essay Concerning Preaching*, pp. 54–57, 70, 79, and 81.

15. John Jennings, *Two Discourses*, p. 29.

16. Thomas Bray, *The Whole Course of Catechetical Institution* (London, 1704), [sig]c^v and p. 32.

17. William Brion Davis, *The Problem of Slavery in Western Culture* (Oxford: Oxford University Press, 1988).

18. Isaac Watts, *A Discourse in the Way of Instruction by Catechisms* (London, 1753), pp. 57, vii, and 15.

19. Watts, *Discourse in the Way of Instruction by Catechisms*, p. 44; and Bray, *Whole Course of Catechetical Institution*, [sig]A2[2]^v-[3]^r and [sig]C^r.

20. Henry Crossman, *An Introduction to the Knowledge of the Christian Religion* (London, 1783), p. 14.

21. Katharine Gerbner, *Christian Slavery: Conversion and Race in the Protestant Atlantic World* (Philadelphia: University of Pennsylvania Press, 2018).

22. Morgan Godwyn, *The Negro's and Indian's Advocate* (London, 1680), p. 10.

23. Travis Glasson, *Mastering Christianity: Missionary Anglicanism and Slavery in the Atlantic World* (Oxford: Oxford University Press, 2011).

24. "Letter 1: The Bishop of London's Letter to the Masters and Mistresses of Families in the English Plantations abroad; Exhorting them to encourage and promote the Instruction of their Negroes in the Christian Faith, 19th May 1727," in *Two Letters of the Bishop of London* (London, 1729), p. 25.

25. *Instructions for the Clergy Employ'd by the Society for The Propagation of the Gospel in Foreign Parts* [London, 1710], p. 3.

26. Glanvill, *Seasonable Defence of Preaching*, p. 10. See also *A Sermon Preach'd at St Mary-le Bow, Feb. 16. 1704/5 Before the [SPG] . . . By the Lord Bishop of Litchfield and Coventry* (London, 1705); and *A Sermon Preach'd before the . . . [SPG] At the Parish Church of St Mary le Bow, February 21st, 1706/7 By the . . . Lord Bishop of St. Asaph* (London, 1707). On such sermons, see Bob Tennant, *Corporate Holiness: Pulpit Preaching and the Church of England Missionary Societies, 1760–1870* (Oxford: Oxford University Press, 2013).

27. Glasson, *Mastering Christianity*, p. 133.

28. Thomas Wilson, *Knowledge and Practice of Christianity Made Easy to the Meanest Capacities: Or, an Essay towards an Instruction for the Indians*, 3rd ed. (London, 1742), pp. ii and 214.

29. Glasson, *Mastering Christianity*; and Lambeth Palace Library, London (hereafter cited as LPL), Society for the Propagation of the Gospel (hereafter cited as SPG), Minutes, vol. 3, April 15, 1737, ff.127–30.

30. Frank J. Klingberg, ed. *Codrington Chronicle: An Experiment in Altruism on a Barbados Plantation, 1710–1834* (Berkeley: University of California Press, 1949); and Glasson, *Mastering Christianity*.

31. LPL, SPG, Minutes, vol. 4, November 21, 1740, f.41r, December 19, 1740, f.46r, and March 16, 1743, f.189v. *The Whole Duty of Man* was a seventeenth-century devotional text by Henry Hammond.

32. Glasson, *Mastering Christianity*; and LPL, SPG, Minutes, vol. 4, November 16, 1744, f.295r.

33. LPL, SPG, Minutes, vol. 4, April 17, 1741, f.76v, and vol. 5, November 8, 1745, f.46v.

34. LPL, MS 1124/1, SPG, Minutes, 1758–1761, September 15, 1759, f.10v, and November 21, 1760, ff.138–39.

35. The SPG's own summary of its record at Codrington up to 1782, is BLOU, United Society for the Propagation of the Gospel (hereafter cited as USPG), C/WIN/BAR 5 Item 76, "Papers concerning Instruction of Negroes."

36. LPL, SPG, Minutes, vol. 17, South Carolina: Richard Tabor and Thomas Lloyd, Spanish Town, Jamaica, to John Chamberlayne, December 5, 1707, f.282r–v, and LPL, Fulham Papers XVII, James White, Kingston, to the Bishop of London, June 3, 1726, ff.246v–247r.

37. LPL, MS1124/1, November 21, 1760, f.139v.

38. Vincent Caretta and Ty M. Reese, eds., *The Life and Letters of Philip Quaque, the First African Anglican Missionary* (Athens: University of Georgia Press, 2010); and Travis Glasson, "Missionaries, Methodists, and a Ghost: Philip Quaque in London and

Cape Coast, 1756–1816," *Journal of British Studies* 48, no. 1 (2010): 29–50. For their examination, see LPL, MS1124/1, December 15, 1758, ff.18v–19r. For the tensions between Quaque and the SPG, see Edward E. Andrews, *Native Apostles: Black and Indian Missionaries in the British Atlantic World* (Cambridge, MA: Harvard University Press, 2013), chap. 4; and Glasson, *Mastering Christianity*, chap. 6.

39. BLOU, USPG X15, Barbados Journal, 1738–1764, February 20, 1761, p. 290.

40. LPL, MS1124/1, April 17, 1761, f.161; BLOU, USPG B6, Letters Sent and Received, box 1 (Barbados), Item 78: Rev. James Butcher to SPG, August 30, 1768, f.181r, and Item 86: James Butcher to SPG, May 4, 1769, f.194.

41. BLOU, USPG B6, Item 80: Thomas Wharton to SPG, August 30, 1768, f.184, and reiterated in Item 122: Codrington Attornies to SPG, September 14, 1775, f.247r.

42. LPL, SPG, Minutes, vol. 17, Tabor and Lloyd to Chamberlayne, December 5, 1707, f.282v.

43. LPL, Fulham Papers XV, Arthur Holt, Barbados, to Bishop of London, December 21, 1727, f.266r.

44. Tabor and Lloyd to Chamberlayne, December 5, 1707, f.285v.

45. Martha Warren Beckwith, *Black Roadways: A Study of Jamaican Folk Life* (Chapel Hill: University of North Carolina Press, 1929); and Robert J. Stewart, *Religion and Society in Post-Emancipation Jamaica* (Knoxville: University of Tennessee Press, 1929).

46. Kenneth M. Bilby and Jerome S. Handler, "Obeah: Healing and Protection in West Indian Slave Life," *Journal of Caribbean History* 38, no.2 (2004): 153–83, at p. 154. On *bayi*, see Walter Rucker, *Gold Coast Diasporas: Identity, Culture, and Power* (Bloomington: Indiana University Press, 2015). Indeed, *dibia* is identified as the practitioner, not the knowledge; see Jerome S. Handler and Kenneth M. Bilby, "On the Early Use and Origin of the Term 'Obeah,' in Barbados and the Anglophone Caribbean," *Slavery and Abolition* 22, no. 2 (2001): 87–100.

47. See Toni Wall Jaudon, "Obeah's Sensations: Rethinking Religion at the Transnational Turn," *American Literature* 84, no. 4 (2012): 715–41; and Pablo F. Gómez, *The Experiential Caribbean: Creating Knowledge and Healing in the Early Modern Atlantic* (Chapel Hill: University of North Carolina Press, 2017), who emphasizes ritual practitioners creation of a wider sensorium including vocalized sounds.

48. Diana Paton, "Obeah Acts: Producing and Policing the Boundaries of Religion in the Caribbean," *Small Axe* 13, no. 1 (2009): 1–18, at p. 16; and Diana Paton, *The Cultural Politics of Obeah: Religion, Colonialism and Modernity in the Caribbean World* (Cambridge: Cambridge University Press, 2015).

49. TNA BT6/10, "Copies of Evidence Submitted to the Committee of Council for Trade and Plantations in the Course of their Enquiry into the State of the African Slave Trade, 1788," ff.188–89.

50. *LTST* (1789), Part III: Jamaica, Answers no. 22-26. See Jill H. Casid, "'His Master's Obi': Machine Magic, Colonial Violence, and Transculturation," in *The Visual Culture Reader*, 2nd ed., ed. Nicholaz Mirzoeff (London: Routledge, 2002), pp. 533–45.

51. *LTST* (1789), Part III: Barbados, Answer no. 23, and Jamaica, Answers nos. 22–26.

52. School of Oriental and African Studies, London, Wesleyan Methodist Missionary Society Archive (hereafter cited as WMMS), Biographical, West Indies, Fiche Box Number (hereafter cited as FBN) 2, John Shipman, "Thoughts on the Present State of Religion among the Negroes in Jamaica" [1820], pp. 5–6.

53. George Pickard, *Notes on the West Indies*, vol. 1 (London, 1806), p. 273. See Peter A. Roberts, *From Oral to Literate Culture: Colonial Experience in the English West Indies* (Kingston: University of the West Indies Press, 1997).

54. [John Stewart], *An Account of Jamaica and Its Inhabitants* (London, 1808), p. 251.
55. Vincent Brown, *The Reaper's Garden: Death and Power in the World of Atlantic Slavery* (Cambridge, MA: Harvard University Press, 2008), p. 66.
56. See also Griffith Hughes, *The Natural History of Barbados* (London, 1750), p. 15; and, for West Africa, John Matthews, R.N., *A Voyage to the River Sierra-Leone, on the Coast of Africa* (London, 1788), pp. 122–24.
57. James H. Sweet, *Domingos Álvares, African Healing, and the Intellectual History of the Atlantic World* (Chapel Hill: University of North Carolina Press, 2011); Gómez, *Experiential Caribbean*; Randy M. Browne, "The 'Bad Business' of Obeah: Power, Authority, and the Politics of Slave Culture in the British Caribbean," *William and Mary Quarterly* 68, no. 3 (2011): 451–80; and Juanita de Barros, "'Setting Things Right': Medicine and Magic in British Guiana, 1803–38," *Slavery and Abolition* 25, no. 1 (2004): pp. 28–50.
58. Luise White, *Speaking with Vampires: Rumor and History in Colonial Africa* (Berkeley: University of California Press, 2000), p. 59.
59. *LTST* (1789), Part III: Jamaica, Answers no. 22–26.
60. *LTST* (1789), Part III: Jamaica, Answers no. 22–26. William Shepherd's 1804 poem "The Negro Incantation" used these sources to present "the professors of a species of incantation, known among the blacks by the name of Obi," see William Shepherd, *Poems Original and Translated* (London, 1829), p. 9.
61. TNA CO139/21, "An Act to Remedy the Evils Arising from the Irregular Assemblies of Slaves." See Diana Paton, "Witchcraft, Poison, Law, and Atlantic Slavery," *William and Mary Quarterly* 69, no. 2 (2012): 235–64.
62. *LTST* (1789), Part III: Jamaica, Answers no. 22–26.
63. James Grainger, *The Sugar Cane: A Poem* (London, 1764), book 4, line 392; and Benjamin Moseley, *A Treatise on Sugar*, 2nd ed. (London, 1800), pp. 191–92.
64. Moseley, *Treatise on Sugar*, p. 190; and Shipman, "Thoughts on the Present State of Religion," p. 14.
65. R. Browne, "'Bad Business' of Obeah."
66. Douglas Hall, *In Miserable Slavery: Thomas Thistlewood in Jamaica, 1750–86* (Kingston: University of the West Indies Press, 1989), p. 61 (January 6, 1754).
67. *LTST* (1789), Part III: Jamaica, Answers no. 22–26.
68. Diana Paton and Maarit Forde, eds., *Obeah and Other Powers: The Politics of Caribbean Religion and Healing* (Durham, NC: Duke University Press, 2012); and Jerome S. Handler and Kenneth M. Bilby, *Enacting Power: The Criminalization of Obeah in the Anglophone Caribbean, 1760-2011* (Kingston: University of The West Indies Press, 2012).
69. *LTST* (1789), Part III: Jamaica, Answers no. 22–26.
70. BL Sloane MS 2302, Thomas Walduck to James Petiver, November 24, 1710, f.20r, and September 17, 1712, f.28v.
71. Katharine Gerbner, "'They Call Me Obea': German Moravian Missionaries and Afro-Caribbean Religion in Jamaica, 1754–1760," *Atlantic Studies* 12, no. 2 (2015): 160–78.
72. *LTST* (1789), Part III: Jamaica, Answers no. 22–26.
73. Gerbner, "'They Call Me Obea,'" p. 172.
74. Moseley, *Treatise on Sugar*, p. 203.
75. WMMS, West Indies Correspondence General, 1823–24, FBN5, John Crofts to WMMS, Spanish Town, October 16, 1824, f.2.

76. Gerbner, *Christian Slavery*; and Maya Jasanoff, *Liberty's Exiles: The Loss of America and the Remaking of the British Empire* (London: Harper Press, 2011).

77. WMMS FBN2, Letters Respecting the Wesleyan Methodist Missionaries in the Colonies, 1818, no. 26, "Letter from Stephen Drew, St. Ann's, Jamaica, 12th March 1818," p. 118; Mary Turner, *Slaves and Missionaries: The Disintegration of Jamaican Slave Society, 1787–1834* (Kingston: University of West Indies Press, 1998); Catherine Hall, *Civilizing Subjects: Metropole and Colony in the English Imagination, 1830–1867* (Cambridge: Polity Press, 2002); and Sylvia R. Frey and Betty Wood, *Come Shouting to Zion: African-American Protestantism in the American South and British Caribbean to 1830* (Chapel Hill: University of North Carolina Press, 1998).

78. WMMS FBN2, William Ratcliffe to WMMS, Kingston, October 20, 1817, f.3.

79. WMMS, West Indies Correspondence General, 1803–1817, FBN1, William Gilgrass to WMMS, Bridgetown, May 25, 1810, f.3.

80. Shipman, "Thoughts on the Present State of Religion," pp. xviii and 28.

81. Shipman, "Thoughts on the Present State of Religion," pp. v and 61.

82. WMMS FBN1, John Wiggins to WMMS, Kingston, March 10, 1814, and George Johnston to WMMS, Kingston, May 1, 1809, f.2; William White to WMMS, Spanish Town, January 29, 1816, ff.1 and 3; and Isaac Bradnock, William Gilgrass, and James Knowlan, Kingston, to WMMS, August 4, 1807, f.7.

83. Johnston to WMMS, May 1, 1809, f.3; and WMMS FBN1, John Wiggins to WMMS, Kingston, January 27, 1814, f.1. The requested works were probably Thomas Hannam, *The Pulpit Assistant* (London, 1810), and Charles Simeon, *Helps to Composition, or Five Hundred Skeletons of Sermons* (London, 1801).

84. Jennifer Farooq, *Preaching in Eighteenth-Century London* (Woodbridge, UK: Boydell Press, 2013).

85. WMMS FBN1, Isaac Bradnock, Morant Bay, to WMMS, June 7, 1806, f.1; and WMMS FBN2, Biographical, West Indies, Series of Tracts for Slaves in the West Indies, part 3, "The Marriage of Quaco and Quasheba," p. 9.

86. WMMS FBN1, Isaac Bradnock to WMMS, Jamaica, 1807, f.2.

87. Shipman, "Thoughts on the Present State of Religion," pp. 64–65.

88. Crofts to WMMS, October 16, 1824, f.2.

89. Ratcliffe to WMMS, October 20, 1817, f.2.

90. Shipman, "Thoughts on the Present State of Religion," pp. 56–57; and WMMS FBN2, Biographical, West Indies, John Shipman, "Plan of Instruction" [1820], p. 24.

91. WMMS FBN4, West Indies Correspondence General, 1821–1823, William Horne, Kingston, to WMMS, February 12, 1822, f.4; FBN6, West Indies Correspondence General, 1824–1827, Peter Duncan to WMMS, September 3, 1824, ff.1–3; and FBN1, Gilgrass to WMMS, May 25, 1810, f.2.

92. Ratcliffe to WMMS, October 20, 1817, f.1.

93. WMMS FBN5, Robert Young, Stony Hill, to WMMS, August 5, 1825, f.3; and M. Turner, *Slaves and Missionaries*, p. 93.

94. Frey and Wood, *Come Shouting to Zion*. In 1804, the Kingston Methodist Society contained 340 women and 180 men; WMMS FBN1, William Fish, Kingston, to WMMS, April 26, 1804.

95. "Letter from Stephen Drew, 12th March 1818," p. 134; WMMS FBN4, Robert Young, Kingston, to WMMS, March 6, 1823, f.2; Crofts to WMMS, October 16, 1824, f.1; Duncan to WMMS, September 3, 1824.

96. Shipman, "Plan of Instruction," p. 3.

97. Shipman, "Plan of Instruction," p. 9; and John Wesley, *Instructions for Children* (Dublin, 1749). For the catechism in use, see Duncan to WMMS, September 3, 1824, f.2.

98. Bradnock, Gilgrass, and Knowlan to WMMS, August 4, 1807; and WMMS FBN7, West Indies Correspondence General, 1827–1828, Isaac Whitehouse, Bath, Jamaica, to WMMS, July 16, 1827, f.2.

99. WMMS FBN1, James Wiggins, Kingston, to WMMS, April 29, 1813, f.1, and FBN2, James Wiggins, Morant Bay, to WMMS, August 28, 1817, ff.1–2.

100. "Letter from Stephen Drew, 12th March 1818," p. 121; and Evidence of Reverend William Knibb to Select Committee of the House of Lords on the State of the West India Colonies, *British Parliamentary Papers*, 1831–32 (127) xxxvi, vols. 11 and 12, p. 766. Methodists also referred to "Local Preachers"; Wiggins to WMMS, April 29, 1813, f.1.

101. M. Turner, *Slaves and Missionaries*.

102. Shipman, "Plan of Instruction," p. 16; Shipman, "Thoughts on the Present State of Religion," p. 60; and WMMS Biographical, West Indies, FBN2, Series of Tracts for Slaves in the West Indies, part 4, "An Address to Quaco and Quasheba after their Marriage," p. 23. See also Whitehouse to WMMS, July 16, 1827, f.2; and WMMS FBN8, West Indies Correspondence General, 1828–30, Peter Duncan, Montego Bay, to WMMS, June 1, 1829, f.1.

103. WMMS FBN1, Charges against the Methodist Missionaries in Jamaica by the Common Council and Assembly . . . , from George Johnston, Morant Bay, to WMMS, February 10, 1811, f.1; Stephen Cook, Kingston, to William Johnstone, August 15, 1809, f.1; and C. Hall, *Civilizing Subjects*.

104. Shipman, "Thoughts on the Present State of Religion," pp. 25 and 27.

105. WMMS FBN2, Biographical, West Indies, Series of Tracts for Slaves in the West Indies, part 1, "A Conversation on Marriage," p. 23.

106. WMMS FBN1, Isaac Bradnock, Jamaica, to WMMS, July 12, 1806, f.1; and Shipman, "Plan of Instruction," pp. 52–53. On Methodist tears, see Thomas Dixon, "Enthusiasm Delineated: Varieties of Weeping in Eighteenth-Century Britain," *Litteraria Pragensia* 22 (2012): 59–81.

107. Crofts to WMMS, October 16, 1824, f.3.

108. WMMS FBN2, Biographical, West Indies, Series of Tracts for Slaves in the West Indies, part 2, "A Conversation on Marriage," p. 12; and Crofts to WMMS, October 16, 1824, f.1.

109. Bradnock to WMMS, July 12, 1806, f.2.

110. Crofts to WMMS, October 16, 1824, f.3; Duncan to WMMS, September 3, 1824, f.1; and Duncan to WMMS, June 1, 1829, f.2.

111. Ratcliffe to WMMS, October 20, 1817, ff.2–3.

112. Cook to Johnston, August 15, 1809, ff.1–2.

113. Crofts to WMMS, October 16, 1824, f.3.

114. C. Hall, *Civilizing Subjects*.

115. Beilby Porteus, *Sermons on Several Subjects*, 5th ed. (London, 1787), p. 387. See also Bob Tennant, "Sentiment, Politics, and Empire: A Study of Beilby Porteus's Anti-Slavery Sermon," in *Discourses of Slavery and Abolition*, ed. Brycchan Carey, Markman Ellis and Sara Salih (London: Palgrave Macmillan, 2004), pp. 158–74.

116. "An Essay Towards a Plan for the More Effectual Civilization and Conversion of the Negro Slaves on the Trust Estate in Barbadoes," in *The Works of the Right Reverend Beilby Porteus, D.D. Late Bishop of London*, vol. 6, *Tracts* (London, 1823), p. 170.

117. "Essay Towards a Plan," p. 171.

118. "Essay Towards a Plan," p. 173.

119. "Essay Towards a Plan," pp. 177, 180, and 183–84. On the Moravians, see J. C. S. Mason, *The Moravian Church and the Missionary Awakening in England, 1760–1800* (Woodbridge, UK: Boydell Press, 2001).

120. Porteus, *Sermons on Several Subjects*, pp. 397 and 402; and "An Essay Towards a Plan," p. 192. On abolitionist versions of slavery, see Sasha Turner, *Contested Bodies: Pregnancy, Childrearing, and Slavery in Jamaica* (Philadelphia: Pennsylvania University Press, 2017); and Marisa J. Fuentes, *Dispossessed Lives: Enslaved Women, Violence, and the Archive* (Philadelphia: University of Pennsylvania Press, 2016), pp. 129–37.

121. Glasson, *Mastering Christianity*, p. 169.

122. Kenneth Morgan, "Slave Women and Reproduction in Jamaica, ca. 1776–1834," in *Women and Slavery*, vol. 2, *The Modern Atlantic*, ed. Gwyn Campbell, Suzanne Miers and Joseph C. Miller (Athens: Ohio University Press, 2008), pp. 27–53; and Sasha Turner, "Home Grown Slaves: Women, Reproduction, and the Abolition of the Slave Trade, Jamaica 1788–1807," *Journal of Women's History* 23, no. 3 (2011): 39–62.

123. Beilby Porteus, *A Letter to the Governors, Legislature, and Proprietors of Plantations, in the British West-India Islands* (London, 1808), pp. 22–23 and 26. It is not clear whether he differentiated between boys and girls.

124. M. Turner, *Slaves and Missionaries*.

125. WMMS FBN2, Letters Respecting the Wesleyan Methodist Missionaries in the Colonies, 1818, no. 24, "Letter from Dr Stewart West, Magistrate, Bath, Jamaica, 15th March 1818," p. 107, and "Letter from Thomas Thomson, Morant Bay, to George Cuthbert, 11th March 1818," pp. 215 and 219; Cook to Johnston, August 15, 1809, f.1; WMMS FBN1, Charges against the Methodist Missionaries in Jamaica, February 10, 1811, f.1; and C. Hall, *Civilizing Subjects*.

126. Misty G. Anderson, *Imagining Methodism in Eighteenth-Century Britain: Enthusiasm, Belief, and the Borders of the Self* (Baltimore: Johns Hopkins University Press, 2012).

127. Bradnock to WMMS, June 7, 1806, f.2.

128. Charges against the Methodist Missionaries in Jamaica, February 10, 1811, f.1.

129. Jasanoff, *Liberty's Exiles*, p. 270.

130. WMMS FBN1, Isaac Bradnock to WMMS, Bridgetown, Barbados, October 20, 1804; and Gilgrass to WMMS, May 25, 1810, f.2.

131. WMMS FBN1, James Whitworth, Bridgetown, to the Governor, February 15, 1813, f.2, and Richard Pattison, Barbados, to WMMS, March 21, 1805, f.5.

132. *Slave Law of Jamaica: With Proceedings and Documents Relative Thereunto* (London, 1828), p. 231; and M. Turner, *Slaves and Missionaries*, pp. 14–17.

133. WMMS FBN1, James Knowlan, Kingston, to WMMS, June 23, 1807, with included newspaper clipping from which quotation is taken.

134. Knowlan to WMMS, June 23, 1807, newspaper clipping.

135. WMMS FBN1, James Knowlan, Kingston, [June 1807], f.2; Shipman, "Plan of Instruction," pp. 118–22; and *Slave Law of Jamaica*, p. 231.

136. *Slave Law of Jamaica*, p. 258.

137. Knowlan to WMMS, June 23, 1807, f.1.

138. WMMS FBN1, William Fish, Kingston, February 8, 1805, and John Wiggins, Kingston, to WMMS, August 8, 1812, f.3.

139. WMMS FBN1, John Wiggins to WMMS, December 15, 1812, f.1; and Wiggins to WMMS, April 29, 1813, f.1.

140. For the intricate interdenominational politics of getting the Kingston Methodist chapel reopened in 1814, see WMMS FBN1, Stephen Ryan, Mount Olive, Jamaica, to Rev. Thomas Blanshard (WMMS), May 28, 1814.

141. WMMS FBN1, Address to the Mayor, Aldermen and Common Council of Kingston, June 11, 1807.

142. WMMS FBN1, John Wiggins, Morant Bay, to WMMS, January 1, 1817, ff.1–3.

143. WMMS FBN1, Mary Smith, Kingston, to WMMS, August 15, 1807.

144. Statement from Jamaica in WMMS FBN1, Thomas Coke, Sunderland, to WMMS, October 2, 1807, f.5.

145. M. Smith to WMMS, August 15, 1807, f.2; and "Minutes of the General Meeting of the Stewards, Leaders & several strangers in Kingston Chapel, 4th August 1807," in Coke to WMMS, October 2, 1807, f.2.

146. WMMS FBN1, Isaac Bradnock to WMMS, Kingston, Jamaica, October 30, 1807, f.1; WMMS FBN1, Isaac Bradnock to WMMS, Kingston, Jamaica, February 18, 1808, f.2; M. Smith to WMMS, August 15, 1807, f.1; Bradnock, Gilgrass, and Knowlan to WMMS, August 4, 1807, f.5; and WMMS FBN1, George Johnston to WMMS, Kingston, February 28, 1808, f.3.

147. The competing accounts are in M. Smith to WMMS, August 15, 1807, ff.2–3; and "Minutes of the General Meeting of . . . 4th August 1807."

148. Bradnock, Gilgrass, and Knowlan to WMMS, August 4, 1807, f.7.

149. Coke to WMMS, October 2, 1807, ff.1–3 and 5.

150. Shipman, "Thoughts on the Present State of Religion," p. 60.

151. WMMS FBN1, John Wiggins, Kingston, to WMMS, July 14, 1804, f.3; Wiggins to WMMS, March 10, 1814, f.4; and WMMS FBN1, John Wiggins, Kingston, to WMMS, April 3, 1814, f.2.

152. Duncan to WMMS, September 3, 1824, ff.2 and 3.

153. Evidence of Reverend William Knibb, pp. 765–66.

154. Charges against the Methodist Missionaries in Jamaica, February 10, 1811, f.1.

155. Evidence of Reverend William Knibb, p. 769.

156. Evidence of Reverend William Knibb, pp. 736 and 809.

157. "Letter from Dr Stewart West, 15th March 1818," pp. 107–8; and "Letter from Thomas Thomson, 11th March 1818," pp. 217–18.

158. "Letter from Dr Stewart West, 15th March 1818," p. 108; and "Letter from Stephen Drew, 12th March 1818," pp. 121–22.

159. Shipman, "Thoughts on the Present State of Religion," pp. i–ii.

160. Shipman, "Thoughts on the Present State of Religion," p. 41.

161. Shipman, "Thoughts on the Present State of Religion," pp. 42 and 44.

162. Shipman, "Thoughts on the Present State of Religion," pp. 47–48.

163. Shipman, "Thoughts on the Present State of Religion," pp. 47–49. On the McGhie Maroons, to whom there may be a connection, see Jean Besson, *Transformations of Freedom in the Land of the Maroons: Creolization in the Cockpits, Jamaica* (Kingston: Ian Randle, 2016), pp. 269–303.

164. Evidence of Reverend William Knibb, p. 744.

165. Evidence of Reverend William Knibb, p. 744.

166. Evidence of Reverend William Knibb, pp. 767–68 and 807–10.

CHAPTER FIVE

1. Matthew Gregory Lewis, *Journal of a West Indian Proprietor, Kept during a Residence in the Island of Jamaica* (London, 1834), pp. 227–28.

2. For example, *Slave Law of Jamaica: With Proceedings and Documents Relative Thereunto* (London, 1828).

3. By the 1820s, these wars of words were also being fought by Anglican clergymen. See "The Rev. W. M. Harte and the Parish of St. Lucy, Barbados," *Christian Remembrancer* 9 (1827): 766–81; and John T. Gilmore, "The Rev. William Harte and Attitudes to Slavery in Early Nineteenth-Century Barbados," *Journal of Ecclesiastical History* 30, no. 4 (1979): 461–74.

4. Manisha Sinha, *The Slave's Cause: A History of Abolition* (New Haven, CT: Yale University Press, 2016).

5. For a survey, see Seymour Drescher, *Abolition: A History of Slavery and Antislavery* (Cambridge: Cambridge University Press, 2009). On speech, see Brycchan Carey, "William Wilberforce's Sentimental Rhetoric: Parliamentary Reportage and the Abolition Speech of 1789," *Age of Johnson: A Scholarly Annual* 14 (2003): 281–305; and Bob Tennant, "Sentiment, Politics, and Empire: A Study of Beilby Porteus's Anti-Slavery Sermon," in *Discourses of Slavery and Abolition* ed. Brycchan Carey, Markman Ellis, and Sara Salih (London: Palgrave Macmillan, 2004), pp. 155–74. See also Brycchan Carey, *British Abolitionism and the Rhetoric of Sensibility: Writing, Sentiment and Slavery, 1760–1807* (Basingstoke, UK: Palgrave Macmilllan, 2005); and Srividhya Swaminathan, *Debating the Slave Trade: Rhetoric of British National Identity, 1759-1815* (Farnham, UK: Ashgate, 2009).

6. James Ramsay, *An Essay on the Treatment and Conversion of African Slaves in the British Sugar Colonies* (London, 1784), pp. 197 and 204.

7. W. Mason, *An Occasional Discourse, Preached in the Cathedral of St. Peter in York, January 27, 1788, on the subject of the African Slave Trade* (London, 1788), p. 16n.

8. Ramsay, *Essay*, pp. 206, 209, 238, 243–44, 245–46.

9. Ramsay, *Essay*, p. 282.

10. James Ramsay, *An Inquiry into the Effects of Putting a Stop to the African Slave Trade, and of Granting Liberty to the Slaves in the British Sugar Colonies* (London, 1784), pp. 7–8.

11. Ramsay, *Essay*, p. 276.

12. *An Inquiry into the Origin, Progress, and Present State of Slavery: With a Plan for the Gradual, Reasonable, & Secure Emancipation of Slaves* (London, 1789), pp. 19, 21, 28.

13. Ramsay, *Essay*, pp. 251–58.

14. Reprinted, probably in the early 1790s, as part of a single sheet entitled *A Subject for Conversation and Reflection at the Tea Table*, British Library (hereafter cited as BL) HS.74/1983(46).

15. See Elhanan Winchester, *The Reigning Abominations, especially the Slave Trade, considered as Causes of Lamentation* (London, 1788), p. 29; and, on tears, Carey, *British Abolitionism and the Rhetoric of Sensibility*.

16. *Remarks on the Slave Trade, and the Slavery of Negroes* (London, 1788), p. 2.

17. On Phillips, see Adam Hochschild, *Bury the Chains: The British Struggle to Abolish Slavery* (London: Macmillan, 2005).

18. Library of the Society of Friends, London (hereafter cited as LSF), Thompson-Clarkson Illustrations, II f.35.

19. *No Rum! No Sugar!* (London, 1792); and Julie L. Holcomb, "Blood-Stained Sugar: Gender, Commerce and the British Slave-Trade Debates," *Slavery and Abolition* 35, no. 4 (2014): 611–28.

20. *No Rum! No Sugar!* p. 12.

21. For visual culture, see Kay Dian Kriz, *Slavery, Sugar, and the Culture of Refinement* (New Haven, CT: Yale University Press, 2008), pp. 71–112.

22. John Carter Brown Library, Codex Eng 211, Theodore Barrell, *Book of Shorthand*, St. Christopher, West Indies, 1790, ff.2 and 3r.

23. Ramsay, *Essay*, p. 204.

24. Dorothy Couchman, "'Mungo Everywhere': How Anglophones Heard Chattel Slavery," *Slavery and Abolition* 36, no. 4 (2015): 704–20, at pp. 706–7 and 714.

25. [Samuel Disney], "Epitaph to The Padlock," *Gentleman's Magazine* 57, no. 2 (October 1787): 913–14. See Brycchan Carey, "To Force a Tear: British Abolitionism and the Eighteenth-Century London Stage," in *Affect and Abolition in the Anglo-Atlantic, 1770–1830*, ed. Stephen Ahern (Farnham, UK: Ashgate, 2013), pp. 109–28.

26. *No Rum! No Sugar!* p. 19.

27. Diana Paton, "'From His Own Lips': The Politics of Authenticity in A Narrative of Events since the First of August, 1834, by James Williams, an Apprenticed Labourer in Jamaica," in Carey, Ellis, and Salih, *Discourses of Slavery and Abolition*, pp. 108–22.

28. J. R. Oldfield, *Popular Politics and British Anti-Slavery: The Mobilisation of Popular Opinion Against the Slave Trade, 1787–1807* (Manchester, UK: Manchester University Press, 1995); Marcus Wood, *Blind Memory: Visual Representations of Slavery in England and America, 1760–1865* (London: Routledge, 2000); and Stephan Lenik and Christer Petley, "The Material Cultures of Slavery and Abolition in the British Caribbean," *Slavery and Abolition* 35, no. 3 (2014): 389–98.

29. Donna Andrew, ed., *London Debating Societies, 1776–1799*, vol 30 (London: London Record Society, 1994), British History Online, http://www.british-history.ac.uk/london-record-soc/vol30 (hereafter cited as *LDS*), item no. 408 (February 22, 1780).

30. *LDS*, item nos. 392 (February 2, 1780), 442 (March 16, 1780), and 538 (May 1, 1780).

31. *LDS*, item no. 1477 (March 6, 1789). Mary Thale, "Women in London Debating Societies," *Gender and History* 7, no. 1 (1995): 5–24; and Donna Andrew, "Popular Culture and Public Debate: London 1780," *Historical Journal* 39, no. 2 (1996): 405–23.

32. *LDS*, item nos. 1991 (August 29, 1796), 1993 (September 15, 1796), and 2004 (November 7, 1796). See also John Barrell, *Imagining the King's Death: Figurative Treason, Fantasies of Regicide, 1793–1796* (Oxford: Oxford University Press, 2000).

33. *LDS*, item nos. 341 (October 28, 1779) and 532 (April 24, 1780).

34. *LDS*, item no. 999 (March 22, 1785); see also item nos. 355 (November 15, 1779), 596 (June 6, 1780), 860 (May 22, 1782), 1116 (May 25, 1786), and 1144 (November 2, 1786).

35. *LDS*, item nos. 1309 (February 7, 1788) and 1312 (February 13, 1788); see also item nos. 1513–1516 (May 7–18, 1789).

36. *LDS*, item nos. 1327 (March 11, 1788), 1342 (April 8, 1788), 1379 (September 25, 1788), 1818 (November 24, 1791), 1820 (December 1, 1791), 1826 (January 5, 1792), 1832–1835 (February 6–March 18, 1792), and 1923 (March 23, 1795).

37. *LDS*, item nos. 394 (February 3, 1780), 427 (March 6, 1780), 972 (December 2, 1784), and 1513 (May 7, 1789).

38. *LDS*, item nos. 1309 (February 7, 1788) and 1316 (February 20, 1788).

39. *LDS*, item nos. 408 (February 22, 1780), 427 (March 6, 1780), 482 (April 6, 1780), and 559 (May 11, 1780).

40. *LDS*, item nos. 1318 (February 25, 1788), 1342 (April 8, 1788), and 1542 (October 26, 1789).

41. *LDS*, item nos. 1318 (February 25, 1788) and 1321 (February 28, 1788).

42. Paul E. Lovejoy, "Olaudah Equiano or Gustavus Vassa—What's in a Name?" *Atlantic Studies* 9, no. 2 (2012): 165–84.

43. *LDS*, item nos. 1312 (February 13, 1788) and 1514 (May 11, 1789); and Vincent Carretta, *Equiano the African: Biography of a Self-Made Man* (Athens: University of Georgia Press, 2005), p. 331.

44. Letters to Granville Sharp (December 15, 1787), William Dolben, William Pitt, and Charles James Fox (July 15, 1788,) and William Dickson (April 15, 1789) are reproduced in Olaudah Equiano, *The Interesting Narrative and Other Writings*, ed. Vincent Carretta (1789; repr., Harmondsworth, UK: Penguin, 1995), pp. 326–27 and 341–44, although they had probably been making collective interventions since at least 1785. See also Quobna Ottobah Cugoano, *Thoughts and Sentiments on the Evil of Slavery* (1787; repr., Harmondsworth, UK: Penguin Books, 1999).

45. *LDS*, item nos. 1513 (May 7, 1789) and 1818 (November 24, 1791).

46. Roger Anstey, *The Atlantic Slave Trade and British Abolition, 1760–1810* (Cambridge: Cambridge University Press, 1975); and Christopher J. Brown, *Moral Capital: Foundations of British Abolitionism* (Chapel Hill: University of North Carolina Press, 2006). In contrast, Cugoano, *Thoughts and Sentiments on the Evil of Slavery*, addresses slavery as much as the slave trade.

47. Prince Hoare, *Memoirs of Granville Sharp, Esq.* (London, 1820), p. 236 (quoting Sharp's diary); and *Morning Chronicle*, March 18, 1783.

48. James Walvin, *The Zong: A Massacre, the Law and the End of Slavery* (New Haven, CT: Yale University Press, 2011).

49. Carretta, *Equiano the African*.

50. York Minster Archives COLL 1896/1, Granville Sharp Letterbook, 1768–1773, 1793, "A Note sent with an Iron Gag-Muzzle in a green bag as a Brief against Slavery," London, February 5, 1772, pp. 34–35; Equiano, *Interesting Narrative*, p. 62 and *LDS*, p. 1826 (January 5, 1792).

51. Michelle Faubert, "Granville Sharp's Manuscript Letter to the Admiralty on the *Zong* Case: A New Discovery in the British Library," *Slavery and Abolition* 38, no. 1 (2017): 178–95, at p. 188; and Ian Baucom, *Spectres of the Atlantic: Finance Capital, Slavery, and the Philosophy of History* (Durham, NC: Duke University Press, 2005). See also Anita Rupprecht, "'A Very Uncommon Case': Representations of the Zong and the British Campaign to Abolish the Slave Trade," *Journal of Legal History* 28, no. 3 (2007): 329–46.

52. Walvin, *Zong*, p. 164; and Hoare, *Memoirs of Granville Sharp*, pp. 156–57.

53. The case was heard on May 21 and 22, 1783. See SFL Transcript of the Calendar of the Diaries of William Dillwyn, f.23; and Caretta, *Equiano the African*, p. 208. For Dillwyn, see Gwilym Games, "Dillwyn, William (1743–1824)," in *Encyclopedia of Emancipation and Abolition*, ed. Junius P. Rodriguez (London: Routledge, 2008), p. 176.

54. Brycchan Carey, *From Peace to Freedom: Quaker Rhetoric and the Birth of American Antislavery, 1657–1761* (New Haven, CT: Yale University Press, 2012).

55. Anstey, *Atlantic Slave Trade and British Abolition*; and C. Brown, *Moral Capital*.

56. John Pemberton to James Pemberton, Witney, Oxfordshire, May 14, 1783, Historical Society of Pennsylvania, Pemberton Papers, item 164, f.38.

57. Anthony Benezet to John Pemberton, Philadelphia, May 29, 1783, in George S. Brookes, *Friend Anthony Benezet* (Philadelphia: University of Pennsylvania Press,

1937), pp. 388–94. Brookes notes (p. 394) that the letter duplicates one to John Gough, also in Britain (pp. 375–77).

58. Thomas Clarkson, *The History of the Rise, Progress, and Accomplishment of the Abolition of the Slave-Trade by the British Parliament* (1808; repr., Teddington, UK: Echo Library, 2006), p. 61.

59. C. Brown, *Moral Capital.*

60. Richard Bauman, "Speaking in the Light: The Role of the Quaker Minister," in *Explorations in the Ethnography of Speaking*, 2nd ed., ed. Richard Bauman and Joel Sherzer (Cambridge: Cambridge University Press, 1989), pp. 144–60.

61. LSF MS Box 10/2(15), Journal of Yearly Meeting, 1783, ff.11v–12r.

62. LSF Thompson-Clarkson Collection, vol. 2, Minutes of the "Slave Association," ff.9–12; SFL MS Box E1/9(2) Minutes of "Committee of the Quarterly Meeting relative to distributing the case of the poor enslaved Africans, 1784–1786"; SFL MS Box F1/7, Minute Book of the Meeting for Sufferings Committee on the Slave Trade, 1783–1792, ff.3–5.

63. C. Brown, *Moral Capital.*

64. LSF YM/MFS/PHIL/1, Letters which passed betwixt the Meeting for Sufferings in London and the Meeting for Sufferings in Philadelphia, vol. 1, London Meeting to Philadelphia Meeting, November 3, 1786, f.262.

65. LSF MS Box X/2, John Kemp's Commonplace Book (1786), pp. 152–53; and LSF MS Box E1/9(2), pp. 9–17 (May 4, August 15, and December 21, 1785, and January 2, 1786).

66. LSF YM/MFS/PHIL/1, London Meeting to Philadelphia Meeting, November 3, 1786, f.262.

67. LSF YM/MFS/PHIL/1, London Meeting to Philadelphia Meeting, November 3, 1786, f.262–63, and see also December 2, 1785, ff.245–46; *The Case of Our Fellow Creatures, the Oppressed Africans* (London, 1784), p. 14; and Anthony Benezet, *A Caution to Great Britain and Her Colonies, in a Short Representation of the Calamitous State of the Enslaved Negroes in the British Dominions* (London, 1785), p. 33.

68. LSF YM/MFS/PHIL/1, Philadelphia Meeting to London Meeting, May 18, 1786, f.256.

69. Anstey, *Atlantic Slave Trade and British Abolition*; and Hochschild, *Bury the Chains.* See also BL Add. MS 21254, Fair Minute Book of the Committee of the Society for Effecting the Abolition of the Slave Trade, May 22, 1787, to February 26, 1788.

70. Oldfield, *Popular Politics and British Anti-Slavery.*

71. LSF Temp MSS 10/14, William Dickson: Diary of a Journey to Scotland, January to March 1792, pp. 24–26. *An Abstract of the Evidence Delivered before a Select Committee of the House of Commons in the Years 1790, and 1791; on the Part of the Petitioners for the Abolition of the Slave Trade*, 2nd ed. (London, 1792) was prepared by Clarkson for a general readership and printed by Phillips for wide circulation.

72. On the "Backlash of the 1790s," see Robin Blackburn, *The Overthrow of Colonial Slavery, 1776–1848* (London: Verso, 1988), pp. 131–60.

73. LSF Thompson-Clarkson Illustrations, vol. 2, William Dickson to James Phillips, Edinburgh, January 14, 1792, f.113. *Cur* and *quare* are "why" and "how" in Latin.

74. Clarkson, *History of the Slave Trade*, p. 165.

75. Dickson's Scottish diary contains several pages of advice from Clarkson based on his own experience; see LSF Temp MSS 10/14, pp. 4–5.

76. BL Add. MS 69038, Dropmore Papers (Series II), Special Correspondence of Lord

and Lady Grenville, William Wilberforce to William Grenville, November 23, 1788, f.141r.

77. Clare Midgley, *Women against Slavery: The British Campaigns, 1780–1870* (London: Routledge, 1992).

78. LSF Temp MSS 10/14, Tuesday January 17, 1792, p. 11.

79. Markman Ellis, *The Politics of Sensibility: Race, Gender and Commerce in the Sentimental Novel* (Cambridge: Cambridge University Press, 1996); Carey, *British Abolitionism and the Rhetoric of Sensibility*; and Clare Midgley, "Slave Sugar Boycotts, Female Activism and the Domestic Base of British Anti-Slavery Culture," *Slavery and Abolition* 17, no. 3 (1996): 137–62.

80. Robert Isaac Wilberforce and Samuel Wilberforce, *The Life of William Wilberforce*, vol. 1 (London: John Murray, 1838), pp. 143–46. On the Teston circle, see C. Brown, *Moral Capital*.

81. Latrobe's contribution was part of Wilberforce's sons' attempts to accuse Clarkson of overplaying his role. See J. R. Oldfield, *"Chords of Freedom": Commemoration, Ritual and British Transatlantic Slavery* (Manchester: Manchester University Press, 2007).

82. Hannah More to Lady Middleton, Cowslip Green, September 10, 1788, in *Memorials, Personal and Historical of Admiral Lord Gambier, G.C.B.* 2nd ed., vol. 1 (London, 1861), pp. 169–70; and Midgely, *Women against Slavery*, p. 32.

83. More to Lady Middleton, September 10, 1788, p. 169.

84. Jane Rendall, *The Origins of Modern Feminism: Women in Britain, France and the United States, 1780–1860* (Basingstoke, UK: Macmillan, 1985).

85. More to Lady Middleton, September 10, 1788, and Hannah More to Lady Middleton, June 14, 1786, pp. 169 and 155–58.

86. Kathryn Gleadle, "'Opinions Deliver'd in Conversation': Conversation, Politics, and Gender in the Late Eighteenth Century," in *Civil Society in British History: Ideas, Identities, Institutions*, ed. Jose Harris (Oxford: Oxford University Press, 2003), pp. 61–78, quotations at pp. 61 and 77; and BL HS.74/1983(46).

87. Gleadle, "'Opinions Delivered in Conversation,'" pp. 62, 68, and 72.

88. Gleadle, "'Opinions Delivered in Conversation,'" pp. 69, 73–74 and 77; Jon Mee, *Conversable Worlds: Literature, Contention, and Community 1762 to 1830* (Oxford: Oxford University Press, 2011), p. 118; and John Bugg, *Five Long Winters: The Trials of British Romanticism* (Stanford: Stanford University Press, 2014).

89. Judith Jennings, "A Trio of Talented Women: Abolition, Gender, and Political Participation, 1780–91," *Slavery and Abolition* 26, no. 1 (2005): 55–70, at p. 66.

90. Quoted in John Bugg, "The Other Interesting Narrative: Olaudah Equiano's Public Book Tour," *PMLA* 121, no. 5 (2006): 1424–42, pp. 1434–35.

91. *Chelmsford Chronicle*, May 5, 1786.

92. Carretta, *Equiano the African*, pp. 267 and 384n57.

93. LSF Thompson-Clarkson Illustrations, Volume III, James Ramsay to [Rev. John Baker?], Teston, September 6, 1788, f.157.

94. Bugg, "Other Interesting Narrative"; and Equiano, *Interesting Narrative*, pp. 41 and 329.

95. Carretta, *Equiano, the African*, pp. 280–92.

96. Acts 4:12 (AV); and Equiano, *Interesting Narrative*, pp. 190–91.

97. Acts 4:13, 18, and 20 (AV).

98. Carretta, *Equiano, the African*, pp. 330–67; Bugg, "Other Interesting Narrative"; and Nini Rogers, "Equiano in Belfast: A Study of the Anti-Slavery Ethos in a Northern Town," *Slavery and Abolition* 18, no. 2 (1997): 73–89.

99. *Northampton Mercury*, December 18, 1790; *Caledonian Mercury*, May 21, 1792; *Gloucester Journal*, July 15, 1793; *Bury and Norwich Post*, February 5, 1794; and *Norfolk Chronicle*, March 15, 1794.

100. *Sheffield Register*, August 20 and 27, 1790.

101. *Oracle*, June 15, 1790; and *General Evening Post*, August 10, 1790.

102. *Oracle*, June 15, 1790.

103. *Public Advertiser*, March 31, 1788; and *London Chronicle*, November 26–28, 1789, reviewing Samuel Stanhope Smith, *An Essay on the Causes of the Variety of Complexion and Figure in the Human Species* (London, 1789).

104. John Gardner Keymes, *Free and Candid Reflections Occasioned by the late Additional Duties on Sugar and on Rum* (London, 1783), and excerpted in *Public Advertiser*, October 25, 1784.

105. *Morning Chronicle and London Advertiser*, February 5, 1788; and John Adams, *Curious Thoughts on the History of Man* (London, 1789), title page.

106. Robert Boucher Nickolls, *Letter to the Treasurer of the Society Instituted for the Purpose of Effecting the Abolition of the Slave Trade* (London, 1788), p. 46.

107. *Morning Chronicle and London Advertiser*, September 6 and October 10 and 28, 1788. See also *St James's Chronicle or the British Evening Post*, April 28 and 30, 1789; *London Chronicle*, May 30–June 2, 1789; and *Diary or Woodfall's Register*, August 10, 1789.

108. For satires, see *Public Advertiser*, December 22, 1770, and July 4, 1775; *Morning Post and Daily Advertiser*, January 30, 1779; *St James's Chronicle*, August 19–21, 1779; *Lounger*, May 14, 1785; and *Morning Chronicle and London Advertiser*, September 12, 1787. For the Sheerness ape, see *Middlesex Journal and Evening Advertiser*, August 4–6, 1774.

109. *London Evening Post*, October 2–5, 1773; *London Chronicle*, August 25–27, 1772; and *Public Advertiser*, September 6, 1779.

110. On Madame Chimpanzee, exhibited in London in 1738, see Silvia Sebastiani, "Challenging Boundaries: Apes and Savages in Enlightenment," in *Simianization: Apes, Gender, Class and Race*, ed. Wulf D. Hund, Charles W. Mills, and Silvia Sebastiani (Zurich: Lit, 2015), pp. 105–38.

111. *London Chronicle and Morning Advertiser*, April 4, 1785, to *Gazeteer and New Daily Advertiser*, May 10, 1788, but also see *Morning Chronicle*, October 17, 1796. For the Monstrous Craws, see *Gazeteer and New Daily Advertiser*, August 13, 1787.

112. *Olla Podrida*, May 26 and June 18, 1787; *World and Fashionable Advertiser*, July 19, 1787; and *Gazetteer and New Daily Advertiser*, December 8, 1787.

113. *Morning Post and Daily Advertiser*, December 26, 1777; *Daily Advertiser*, June 4, 1778; and *Morning Chronicle*, December 14, 1791, and January 18, 1792.

114. Judith Jennings, "Trio of Talented Women," p. 65; and Alan Barnard, "*Orang Outang* and the Definition of Man: The Legacy of Lord Monboddo," in *Fieldwork and Footnotes: Studies in the History of European Anthropology*, ed. Hans F. Vermeulen and Arturo Alvarez Roldan (London: Routledge, 1995), pp. 95–112.

115. Katherine Paugh, "The Curious Case of Mary Hylas: Wives, Slaves and the Limits of British Abolitionism," *Slavery and Abolition* 35, no. 4 (2014): 629–51; Daniel Livesay, *Children of Uncertain Fortune: Mixed-Race Jamaicans in Britain and the Atlantic Family* (Chapel Hill: University of North Carolina Press, 2018); and Brooke N. Newman, *A Dark Inheritance: Blood, Race, and Sex in Colonial Jamaica* (New Haven, CT: Yale University Press, 2018).

116. *St James' Chronicle*, April 28–30, 1789.

117. Carretta, *Equiano, the African*, p. 217, argues that there is no record of discrimination

against Equiano or his family because of their marriage. However, "his *white* wife" is noted in the *Star*, May 30, 1792, and Equiano had favored "Intermarriages" in print in 1788 as a challenge to proslavery writers (p. 330).

118. *Sheffield Register*, August 27, 1790.

119. Bugg, "Other Interesting Narrative"; David Featherstone, "Contested Relationalities of Political Activism: The Democratic Spatial Practices of the London Corresponding Society," *Cultural Dynamics* 22, no. 2 (2010): 87–104; Peter Linebaugh and Marcus Rediker, *The Many Headed Hydra: Sailors, Slaves, Commoners, and the Hidden History of the Revolutionary Atlantic* (Boston: Beacon Press, 2000), pp. 334–41; and John Barrell, *The Spirit of Despotism: Invasions of Privacy in the 1790s* (Oxford: Oxford University Press, 2006).

120. Quoted in Midgely, *Women against Slavery*, p. 39. See also Jon Mee, *Print, Publicity and Popular Radicalism in the 1790s: The Laurel of Liberty* (Cambridge: Cambridge University Press, 2016), pp. 68–69.

121. *Telegraph*, November 7, 1796. On Tarleton, see Paula E. Dumas, *Proslavery Britain: Fighting for Slavery in an Era of Abolition* (Basingstoke, UK: Palgrave Macmillan, 2016).

122. *London Packet*, November 5, 1796; and *Morning Chronicle*, November 7, 1796. See also *Evening Mail*, November 18–21, 1796 (mentions "The African"); and *London Packet*, November 5, 1796 ("a black man took the chair"). Other reports did not mention Equiano: *Morning Post and Fashionable World*, November 7, 1796; *Sun*, November 22, 1796; and *Times*, November 8, 1796.

123. *Lloyds Evening Post*, November 7–9, 1796; *Times*, November 10 and 19, 1796; *True Briton*, November 10 and 22, 1796.

124. *Anti-Jacobin, or Weekly Examiner* 1, no. 4 (December 4, 1797): 126; Carretta, *Equiano, the African*, p. 362; and Stuart Andrews, *The British Periodical Press and the French Revolution, 1789–99* (Basingstoke, UK: Palgrave, 2000), pp. 74–75. For Henry Yorke, who was not mentioned as present at the meeting in *Telegraph*, November 7, 1796, see Amanda Goodrich, *Henry Redhead Yorke, Colonial Radical: Politics and Identity in the Atlantic World, 1772–1813* (London: Routledge, 2019).

125. Iain McCalman, ed., *The Horrors of Slavery and Other Writings by Robert Wedderburn* (Edinburgh: Edinburgh University Press, 1991), p. 33.

126. Dickson to Phillips, January 14, 1792, f.113.

127. Clarkson, *History of the Slave Trade*, p. 184.

128. BL Add. MS 27811, Original Letter Book of the London Corresponding Society, Thomas Hardy to Reverend Bogue, March 23, 1793, f.19r.

129. On Phillis Wheatley, see Paugh, "Curious Case of Mary Hylas"; and Sinha, *Slave's Cause*; and on Mary Prince, see Midgley, *Women against Slavery*.

130. Julius Sherrard Scott III, "The Common Wind: Currents of Afro-American Communication in the Era of the Haitian Revolution" (PhD diss., Duke University, 1986); and Ada Ferrer, *Freedom's Mirror: Cuba and Haiti in the Age of Revolution* (Cambridge: Cambridge University Press, 2014).

131. Michael Craton, *Testing the Chains: Resistance to Slavery in the British West Indies* (Ithaca, NY: Cornell University Press, 1982), p. 244.

132. Gelien Matthews, *Caribbean Slave Revolts and the British Abolitionist Movement* (Baton Rouge: Louisiana State University Press, 2006), p. 1. See also Richard Hart, *Slaves Who Abolished Slavery: Blacks in Rebellion* (Mona, Jamaica: University of the West Indies, 2006).

133. J. R. Oldfield, *Transatlantic Abolitionism in the Age of Revolution: An International His-*

tory of Anti-Slavery, c. 1787–1820 (Cambridge: Cambridge University Press, 2013); and Hilary McDonald Beckles, "The Wilberforce Song: How Enslaved Caribbean Blacks Heard British Abolitionists," *Parliamentary History* 26, no. S1 (2007): pp. 113–26.

134. Craton, *Testing the Chains*; Mary Turner, *Slaves and Missionaries: The Disintegration of Jamaican Slave Society, 1787–1834* (Kingston: University of West Indies Press, 1998); David Lambert, *White Creole Culture, Politics and Identity during the Age of Abolition* (Cambridge: Cambridge University Press, 2005), chap. 4; and Emilia Viotta da Costa, *Crowns of Glory, Tears of Blood: The Demerara Slave Rebellion of 1823* (New York: Oxford University Press, 1994).

135. William's evidence, December 19, 1823, in "Judicial Proceedings relative to the Trial and Punishment of Rebels, or alleged Rebels, in the Island of Jamaica, since the 1st of January 1823," in *Papers Relating to the Manumission, Government and Population of the Slaves in the West-Indies, 1822–1824*, House of Commons, *British Parliamentary Papers*, 1825 (66) xxv: 37–132 (hereafter cited as JPTPR), quotation at p. 40. See Paul Michael Brown, "Representations of Rebellion: Slavery in Jamaica, 1823–1831" (MA thesis, Clemson University, 2014).

136. Examination of William, December 16, 1823, and Ned's evidence, December 19, 1823, JPTPR, pp. 38 and 41.

137. JPTPR, pp. 41–42.

138. Mary's evidence, December 19, 1823, and Henry Cox to William Bullock, Industry, Jamaica, December 25, 1823, JPTPR, pp. 41–42 and 44.

139. Samuel Vaughn to William Bullock, St. James, October 9, 1823, and Examination of Samuel W. Sharpe, December 19, 1823, JPTPR, pp. 45–46.

140. Examination of Robert Bartibo, December 21, 1823, JPTPR, p. 47.

141. Examinations of William Stennett of Unity Hall and Pollydore, December 25, 1823; Daniel of Chatham Estate, December 29, 1823; William Glover and Amelia of Blue Hole, December 30, 1823; and Eleanor Brown of Bogue estate, Jane McDonald of Unity Hall, and Charles Sharpe, January 28, 1824, JPTPR, pp. 51–53, 57, 68, and 79–80.

142. Samuel Vaughn to William Bullock, Montego Bay, December 23, 1823; Report of the Magistrates' Committee, December 29, 1823; and Elizabeth Whittingham's evidence, January 3, 1824, JPTPR, pp. 48–49 and 59.

143. Examination of Robert Goldring, January 5, 1824; Samuel Vaughn to William Bullock, St James, February 2, 1824; and William Bullock to Samuel Vaughn, Spanish Town, February 9, 1824, JPTPR, pp. 59–60 and 82. On rumor, see James C. Scott, *Domination and the Arts of Resistance: Hidden Transcripts* (New Haven, CT: Yale University Press, 1992).

144. Examination of Charles Mack of Cambridge estate, January 7, 1824; and Jean Baptiste Corberand's evidence, February 2–3, 1824, JPTPR, pp. 85 and 93. Mack had been recaptured after escaping from enslavement, and Corberand turned king's evidence.

145. Examinations of Richard Montagnac and Jean Baptiste Corberand, December 26, 1823; and Jean Baptiste Corberand's evidence, January 19 and April 7, 1824, JPTPR, pp. 84, 88, and 103.

146. Examination of Jean Baptiste Corberand, January 8, 1824; and Corberand's evidence, January 19 and February 2, 1824, JPTPR, pp. 85–86, 89, and 96.

147. Jack's confession, April 8, 1824, and William A. Orgill to Bullock, Paradise, St. George's, January 10, 1824, JPTPR, pp. 109 and 82.

148. Luise White, *Speaking with Vampires: Rumor and History in Colonial Africa* (Berkeley: University of California Press, 2000), p. 62.

149. Aisha K. Finch, *Rethinking Slave Rebellion in Cuba: La Escalera and the Insurgencies of 1841-1844* (Chapel Hill: University of North Carolina Press, 2010).

150. Michael Craton, "Proto-Peasant Revolts? The Late Slave Rebellions in the British West Indies, 1816-1832," *Past and Present* 85, no. 1 (1979): 99-125, M. Turner, *Slaves and Missionaries*, pp. 148-78; and Lambert, *White Creole Culture*, pp. 105-39.

151. TNA CO28/85, Barbados, Original Correspondence, James Leith to Earl Bathurst, Barbados, April 30, 1816, ff.8v-9r.

152. TNA CO 28/85, f.21r. That others spoke transatlantically, even when professing not to be addressing other audiences, is clear from William Knibb's later denial that his public statement "as to the Impossibility of Christianity and Slavery being co-existent"—made in Birmingham and published in newspapers he knew would be read in the Caribbean—contradicted his own rule never to speak to the enslaved about Emancipation, on the grounds that "I consider myself, when in England, justified in using any Language which I consider consistent with Truth; [and] that I am not responsible for where my Language may go." Evidence of Reverend William Knibb to Select Committee of the House of Lords on the State of the West India Colonies, *British Parliamentary Papers*, 1831-32 (127) xxxvi, pp. 775-76.

153. TNA CO 28/85, f.33r.

154. "Circular Dispatch from Viscount Goderich to the Governors of the West India Colonies, London, 3rd June 1831," *West India Colonies: Slave Insurrection*, House of Commons, *British Parliamentary Papers*, 1831-32 (285) xlvii, p. 3.

155. Lewis, *Journal of a West Indian Proprietor*, p. 226. See also Ramesh Mallipeddi, *Spectacular Suffering: Witnessing Slavery in the Eighteenth-Century British Atlantic* (Charlottesville: University of Virginia Press, 2016).

156. J. L. Austin, *How to Do Things with Words*, 2nd ed. (Cambridge, MA: Harvard University Press, 1975).

157. Emily Zobel Marshall, "Anansi Tactics in Plantation Jamaica: Matthew Lewis's Record of Trickery," *Wadabagei* 12, no. 3 (2009): 126-50.

158. Confession of Robert, in *The Report from a Select Committee of the House of Assembly, Appointed to Inquire into the Origin, Causes, and Progress of the Late Insurrection* (Barbados and London, 1818) p. 29.

LAST WORDS

1. James M. Phillippo, *Jamaica: Its Past and Present State* (Philadelphia, 1843), pp. 70-71.

2. Phillippo, *Jamaica*, p. 71.

3. Knibb quoted in Catherine Hall, *Civilizing Subjects: Metropole and Colony in the English Imagination, 1830-1867* (Cambridge: Polity Press, 2002), pp. 117-18.

4. Phillippo, *Jamaica*, pp. 71 and 74. On protests in 1834 at the start of apprenticeship, see Gad Heuman, "Riots and Resistance in the Caribbean at the Moment of Freedom," in *After Slavery: Emancipation and its Discontents*, ed. Howard Temperley (London: Frank Cass, 2000), pp. 135-49.

5. Hall, *Civilizing Subjects*; and Thomas C. Holt, *The Problem of Freedom: Race, Labor, and Politics in Jamaica and Britain, 1832-1938* (Baltimore: Johns Hopkins University Press, 1992). See also Sidney W. Mintz, *Caribbean Transformations* (Chicago: Aldine Publishing Company, 1974); and Diana Paton, *No Bond but the Law: Punishment,*

Race, and Gender in Jamaican State Formation, 1780–1870 (Durham, NC: Duke University Press, 2004).

6. Bruno Latour, *An Inquiry into Modes of Existence: An Anthropology of the Moderns*, trans. Catherine Porter (Cambridge, MA: Harvard University Press, 2013).

7. Gian Domenico Iachini, "Pierre Eugene de Simitière and the First American National Museum," *RSA Journal* 23 (2012): 131–59.

8. LCP, du Simitière Papers, Series VI, Papers Relating to the West Indies [968], box 4, folder 20a–21, f.184.

9. Vincent Brown, *Tacky's Revolt: The Story of an Atlantic Slave War* (Cambridge MA: Harvard University Press, 2019).

10. LCP, du Simitière Papers, Series VI [968], box 4, folder 19h. I am grateful to Colin Jones for his translation from the French.

11. Somerset Record Office, Caleb Dickinson Letters, DD/DN 4/1/25(2), Francis Treble to Caleb Dickinson, Kingston, Jamaica, June 2, 1760, quotation is from a postscript dated June 12, 1760.

12. BL Sloane MS 2302, Thomas Walduck to James Petiver, Barbados, November 12, 1710, f.15v. See also Jenny Shaw, *Everyday Life in the Early English Caribbean: Irish, Africans, and the Construction of Difference* (Athens: University of Georgia Press, 2013).

13. Kay Dian Kriz, *Slavery, Sugar, and the Culture of Refinement* (New Haven, CT: Yale University Press, 2008); and John E. Crowley, "Sugar Machines: Picturing Industrialized Slavery," *American Historical Review* 212, no. 2 (2016): 403–36.

14. Hall, *Civilizing Subjects*; Jennifer L. Morgan, *Laboring Women: Reproduction and Gender in New World Slavery* (Philadelphia: University of Pennsylvania Press, 2004); and Katherine Paugh, *The Politics of Reproduction: Race, Medicine, and Fertility in the Age of Abolition* (Oxford: Oxford University Press, 2017).

15. Rachel G. Newman, "Conjuring Cane: The Art of William Berryman and Caribbean Sugar Plantations" (PhD thesis, Stanford University, 2016); and Tim Barringer, "Picturesque Prospects and the Labor of the Enslaved," in *Art and Emancipation: Isaac Mendes Belasario and His Worlds*, ed. Tim Barringer, Gillian Forrester, and Bárbaro Martínez-Ruiz (New Haven, CT: Yale University Press, 2007), pp. 41–63, and the catalog, pp. 326–31.

16. Trevor Burnard, *Mastery, Tyranny, and Desire: Thomas Thistlewood and His Slaves in the Anglo-Jamaican World* (Chapel Hill: University of North Carolina Press, 2004); and Randy M. Browne, *Surviving Slavery in the British Caribbean* (Philadelphia: University of Pennsylvania Press, 2017).

17. LCP, du Simitière Papers, Series VI [968], box 4, folder 20a–21, ff.179–80.

18. BL Add. MS 12405, f.365r.

19. BL Add. MS 12405, f.365r. See also Catherine Hall and Daniel Pick, "Thinking about Denial," *History Workshop Journal* 84 (2017): 1–23.

20. Treble to Dickinson, June 2, 1760 (postscript: June 12).

21. Michel-Rolph Trouillot, *Silencing the Past: Power and the Production of History* (Boston: Beacon Press, 1995); and Lauren Derby, "Beyond Fugitive Speech: Rumor and Affect in Caribbean History," *Small Axe* 44 (2014): 123–40.

22. LCP, du Simitière Papers, Series VI [968], box 4, folder 19h.

INDEX

Page numbers in italics refer to figures.